The Montana Mathematics Enthusiast

Monograph 5

Interdisciplinary Educational Research in Mathematics and Its Connections to the Arts and Sciences

INTERNATIONAL CONTRIBUTING EDITORS AND EDITORIAL ADVISORY BOARD

Miriam Amit, Ben-Gurion University of the Negev, Israel
Ziya Argun, Gazi University, Turkey
Ahmet Arikan, Gazi University, Turkey
Astrid Beckmann, University of Education, Schwäbisch Gmünd, Germany
John Berry, University of Plymouth, UK
Morten Blomhøj, Roskilde University, Denmark
Robert Carson, Montana State University- Bozeman, USA
Mohan Chinnappan, University of Wollongong, Australia
Constantinos Christou, University of Cyprus, Cyprus
Bettina Dahl Søndergaard, University of Aarhus, Denmark
Helen Doerr, Syracuse University, USA
Ted Eisenberg, Ben-Gurion University of the Negev, Israel
Lyn D. English, Queensland University of Technology, Australia
Paul Ernest, University of Exeter, UK
Viktor Freiman, Université de Moncton, Canada
Brian Greer, Portland State University, USA
Eric Gutstein, University of Illinois-Chicago, USA
Marja van den Heuvel-Panhuizen, University of Utrecht The Netherlands
Gabriele Kaiser, University of Hamburg, Germany
Libby Knott, The University of Montana, USA
Tinne Hoff Kjeldsen, Roskilde University, Denmark
Jean-Baptiste Lagrange, IUFM-Reims, France
Stephen Lerman, London South Bank University, UK
Frank Lester, Indiana University, USA
Richard Lesh, Indiana University, USA
Luis Moreno-Armella, University of Massachusetts-Dartmouth
Claus Michelsen, University of Southern Denmark, Denmark
Michael Mitchelmore, Macquarie University, Australia
Nicholas Mousoulides, University of Cyprus, Cyprus
Swapna Mukhopadhyay, Portland State University, USA
Norma Presmeg, Illinois State University, USA
Gudbjorg Palsdottir, Iceland University of Education, Iceland
Michael Pyryt, University of Calgary, Canada
Demetra Pitta Pantazi, University of Cyprus, Cyprus
Linda Sheffield, Northern Kentucky University, USA
Olof Bjorg Steinthorsdottir, University of North Carolina–Chapel Hill, USA
Günter Törner, University of Duisburg-Essen, Germany
Renuka Vithal, University of KwaZulu-Natal, South Africa
Dirk Wessels, Unisa, South Africa
Nurit Zehavi, The Weizmann Institute of Science, Rehovot, Israel

The Montana Mathematics Enthusiast

Monograph 5

Interdisciplinary Educational Research in Mathematics and Its Connections to the Arts and Sciences

Edited by

Bharath Sriraman
The University of Montana, USA

Claus Michelsen
University of Southern Denmark

Astrid Beckmann
University of Education- Schwäbisch Gmünd, Germany

Viktor Freiman
Université de Moncton, New Brunswick, Canada

INFORMATION AGE PUBLISHING, INC.
Charlotte, NC • www.infoagepub.com

Library of Congress Cataloging-in-Publication Data

International Symposium on Mathematics and Its Connections to the Arts and Sciences (2nd : 2007 : Odense, Denmark)
Proceedings of MACAS 2, Second International Symposium on Mathematics and Its Connections to the Arts and Sciences, Odense, Denmark / edited byBharath Sriraman ... [et al.].
p. cm. – (Montana mathematics enthusiast : monograph series in mathematics education)
Includes bibliographical references.
ISBN 978-1-59311-984-3 (hardcover) – ISBN 978-1-59311-983-6 (pbk.)
1. Mathematics–Study and teaching–Congresses. 2. Interdisciplinary approach in education–Congresses. 3. Science and the arts–Congresses. I. Sriraman, Bharath. II. Title. III. Title: Second International Symposium on Mathematics and Its Connections to the Arts and Sciences, Odense, Denmark.
QA11.A1I63 2008
510.71–dc22

2008030899

Copyright © 2008 Information Age Publishing Inc. & The Montana Council of Teachers of Mathematics

All rights reserved. No part of this publication may be reproduced, stored in a retrieval system, or transmitted, in any form or by any means, electronic, mechanical, photocopying, microfilming, recording or otherwise, without written permission from the publisher.

Permission to photocopy, microform, and distribute print or electronic copies may be obtained from:
Bharath Sriraman, Ph.D.
Editor, *The Montana Mathematics Enthusiast*
The University of Montana
Missoula, MT 59812
Email: sriramanb@mso.umt.edu
(406) 243-6714

Printed in the United States of America

PROCEEDINGS OF MACAS2

SECOND INTERNATIONAL SYMPOSIUM ON MATHEMATICS AND ITS CONNECTIONS TO THE ARTS AND SCIENCES, ODENSE, DENMARK

EDITED BY

BHARATH SRIRAMAN, The University of Montana
CLAUS MICHELSEN, University of Southern Denmark
ASTRID BECKMANN, University of Education- Schwäbisch Gmünd
VIKTOR FREIMAN, Université de Moncton

TABLE OF CONTENTS

PLENARY PAPERS

SYMPOSIUM PAPERS

GROUP PICTURE OF PARTICIPANTS

INTERDISCIPLINARITY IN MATHEMATICS, SCIENCE AND ARTS: STATE OF THE ART

Bharath Sriraman, The University of Montana
Claus Michelsen, University of Southern Denmark
Astrid Beckmann, University of Education- Schwäbisch Gmünd
Viktor Freiman, Université de Moncton

The Second International Symposium on Mathematics and its Connections to the Arts and Sciences (MACAS2) was held in Odense at the University of Southern Denmark, May 29-31, 2007, organized by Claus Michelsen (Univ. of Southern Denmark), Bharath Sriraman (Univ. of Montana) and Prof. Astrid Beckmann (Univ. of Education- Schwäbisch Gmünd). We took a leadership role in organizing this professional group at the conclusion of the 10th International Congress in Mathematics Education in Copenhagen in 2004 where we were part of a topics study group on this domain. MACAS2 attracted 45 participants from 15 different countries. These participants included math, science and art educators; physicists; historians and philosophers of mathematics and science, as well as some local classroom teachers. Four plenary lectures were delivered on the theme of the Symposium. In addition 21 papers were presented in an interactive atmosphere. These proceedings include the extended versions of all the papers presented at Odense. Several excursions were planned for the participants to expose them to Odense, the birthplace of Hans Christian Andersen, as well as the island of Fyn which is known as the garden of Denmark. Our long term goal is to build our interdisciplinary network by finding and sharing the best experiences; creating new collaborations; conducting new studies; reflecting on commonalities and differences – multidisciplinary, multicultural and divergent approaches to content, teaching and learning. Plans are already being made to hold the 3rd MACAS Symposium in Moncton, Canada in 2009 to be hosted by Viktor Freiman.

The papers in these proceedings are very diverse in nature. Some papers are theoretical musings on the relationship between mathematics, science and arts. However there are clear strands within which the papers could be clustered. Some focus on the role of mathematical modeling in fostering an interdisciplinary approach towards the learning of mathematics, others highlight the important role that art can play in helping students appreciate the interplay between mathematics and arts, as well as in the communication of unspeakable outrages occurring in the world today. Several papers address curricular initiatives aimed at integrating science and mathematics. Last but not least, the necessity to include cultural and ethnomathematical dimensions into the teaching of mathematics is also addressed in two papers.

Today's young learners have borne not only the fruits of enormous progress of science and technology that marked the 20th century but also the challenges surrounding complex and ill-defined real life problems that remain unsolved. Moreover, issues humanity faces such as climate change, health, environment, overpopulation, and so on are so complex that these problems can not be solved by a single person or even a single discipline. An interdisciplinary approach to teaching and learning is thus a key element for any successful educational enterprise which aims to prepare future generations to deal with the increasing complexity and

B.Sriraman, C.Michelsen, A. Beckmann & V. Freiman (Eds). (2008). *Proceedings of the Second International Symposium on Mathematics and its Connections to the Arts and Sciences* (MACAS2). Information Age Publishing, Charlotte: NC , pp.1-4

interconnectivity of our world. After two Symposia we have established a core group of researchers who share a vision of integrating the arts, mathematics, and the sciences in the school and university curricula by engaging in joint interdisciplinary projects. The four authors are consolidating efforts at collecting research evidence based on similar classroom experiments involving interdisciplinary problem solving in the school curriculum at different sites (Canada, USA, Denmark and Germany).

In the USA, several models of connected and integrated mathematics curriculum have been implemented and evaluated by researches. Standard concepts of mathematics curriculum have been applied to the solving of real-life problems that deal with finding of the most efficient use of resources for a manufacturing company. The context provided necessary "transfer" opportunities to other real-world situations involving chemistry, scheduling of services, financial business, and ecology. However, in the dominant school practice in the USA interdisciplinarity is seen more as an exception, which is often challenged by the so-called 'traditionalists' who regard interdisciplinary learning as a loss of integrity and thematic curriculum as 'squishy' (Ferrero, 2006). In Germany one can observe that the special position of interdisciplinarity is disintegrating since the beginning of this century. Impulses for this re-initiating this tradition have come from the international working application-oriented research groups. Since that time there are great efforts to demand interdisciplinarity also in German schools. A central problem in school implementation of the last years was the often missing cooperation between teachers and educational researchers. Although there are plenty of single interdisciplinary ideas, there is no theoretical help for implementation (Beckmann, 2007a, b, Michelsen, 2007).

In Denmark several research projects describe mathematics, not solely as various areas of content knowledge but also discuss eight mathematical competencies that comprise intuition and creativity across educational levels and topic areas throughout the education system. In a similar research project (Andersen et. al., 2003) four science competences are identified and described. But although the two projects only to a modest degree deal with the relations between mathematics and science two competences are present in both projects – the modelling and the representational competence. This indicates that the modelling and representational competences are links between mathematics and science, and according to Michelsen (2005) these interdisciplinary competences could be the generic methodology that acts as a common denominator for disciplines, such as mathematics and the subjects of natural sciences. This approach to interdisciplinary activities is applied in Danish in-service teacher training programmes (Michelsen 2007).

In Canada, the transition in the school reform movement between the 20th and 21st centuries was marked with more explicit emphasis on more integrative school curriculum. New Brunswick French schools go even beyond it putting a common K-12 theoretical framework for all school subjects which prioritizes the development of a new learning culture of 'learning to learn', making sense of learning, getting equilibrium between individual engagement and collaborative work in an interdisciplinary learning environment. Interesting results regarding interdisciplinary connections have been obtained in numerous studies conducted by the authors such as the successful integration of literature, art and mathematics in the high school curricula ; physics and mathematics; paradoxes and mathematics; and an analysis of polymathic traits of students in interdisciplinary problem situations (see Beckmann & Sriraman, 2007; Sriraman & Adrian, 2004,a,b; Sriraman, 2003, 2004a,b 2005, 2007a,b).

Data from New Brunswick Laptop Initiative show how technology can be an agent of change in the classroom learning and teaching culture when interdisciplinary problem based scenarios have been used to track the process of use of science, mathematics and language art to solve a real-world problem (Freiman, et al., 2007). Another experience also related to technology is a creation of virtual interdisciplinary interactive collaborative learning community of problem solvers in math, science, and French called CASMI that after one year of existence has attracted more than 5000 schoolchildren, university students, and teachers (Freiman, Lirette-Pitre, and Manuel, 2007).

From psychological and educational points of view there are many arguments for interdisciplinarity: Interdisciplinarity enables more connections to existing knowledge and thus leads to more complex and integrated learning. Interdiscplinarity allows more students' centred lessons and increases motivation. It can also nourish reflection on specific methods of the own subject and to understand its importance. At higher educational levels, research argues for the need of so called interdisciplinary conversations that may provide a fruitful new area of exploration involving broader patterns invisible from a strictly disciplinary view (Dalke, Grobstein, & McCormack, 2006). Dalke et al., also suggest that the most generative exchange between individuals occurs as a reciprocal loop between the metaphoric relations of one individual and the metonymic structures of another, such interplay shifts perpetually generating new questions and new understandings.

From its historical and cultural development, the heritage of mathematics reveals itself as a highly connected field of study. Recognition and use of connectivity among mathematical ideas, understanding how to build interconnected mathematical ideas into a whole, as well as capacity of their applications in deferent contexts outside of mathematics are now seen as core competence of every mathematically educated individual. In this vein, many educational systems in Canada and worldwide are implementing a more integrated and real- life connected mathematics curricula. Beyond these curricular statements, we see the utopian goal to create a humanistic approach of education, one that unifies various strands of the curricula as opposed to dividing it (Beckmann, Michelsen, Sriraman, 2005; Sriraman & Dahl, in press). How can schooling create well-rounded individuals akin to the great thinkers of the Renaissance (Italian, Islamic)? That is, individuals who are able to pursue multiple fields of research and appreciate both the aesthetic and structural/ scientific connections between the arts and the sciences. The history of model building in science conveys epistemological awareness of domain limitations. Arts imagine possibilities, science attempts to generate models to test possibilities, mathematics serves as the tool (Sriraman, 2005). The implications for education today is to move away from the post Renaissance snobbery rampant within individual disciplines at the school and university levels. By building bridges today between disciplines, the greatest benefactors are today's gifted children, the potential innovators of tomorrow (Sriraman & Dahl, in press). We are hopeful that the MACAS group will evolve into a network of finding and sharing the best experiences; creating new collaborations; conducting new studies; reflecting on commonalities and differences – multidisciplinary, multicultural and divergent approached to mathematics and its teaching and learning.

References

Andersen, N.O., Busch, H., Troelsen, R. & Horst, S. (2003). Fremtidens naturfaglige uddannelser. (in Danish) Copenhagen: Undervisningsministeriet

Beckmann, A. (2007a). Interdisciplinary lessons – background, arguments and possible forms of cooperation Minisymposium. In: *Beiträge zum Mathematikunterricht.* Hildesheim, Berlin (Franzbecker Verlag)

Beckmann, A. (2007b). ScienceMath – an interdisciplinary European project. ScienceMath - ein fächerübergreifendes europäisches Projekt. In: *Beiträge zum Mathematikunterricht* Hildesheim, Berlin (Franzbecker Verlag).

Beckmann, C. Michelsen & B.Sriraman (Eds.) (2007) , Proceedings of the 1st International Symposium on Mathematics and its Connections to the Arts and Sciences. University of Schwaebisch Gmuend: Germany. Franzbecker Verlag.

Dalke, A., Grobstein, P., & McCormack, E. (2006). Exploring Interdisciplinairty: The Significance of Metaphoric and Metonymic Exchange. *Journal of Research Practice*, Vol.2, no. 2.

Ferrero (2006) Having it all: Two Chicago-area high schools demonstrate that educators don't have to choose between innovation and traditionalism. *Educational Leadership*, May, pp. 8-14.

Freiman, V., Manuel, D et Lirette-Pitre, N. (2007). CASMI Virtual Learning Collaborative Environment for Mathematical Enrichment. *Understanding our Gifted*, Summer 2007, pp. 20-23.

Freiman, V., Lirette-Pitre, N. et Manuel, D. (2007). Building virtual learning community of problem solvers: example of CASMI Community. Communication présentée au colloque Second International Symposium on Mathematics and its Connections to the Arts and Sciences, The University of Southern Denmark, Odense, Denmark, (29 au 31 mai).

Michelsen, C. (2005). Expanding the domain: Variables and functions in an interdisciplinary context between mathematics and physics. In A. Beckmann, C. Michelsen & B. Sriraman (eds.) *Proceedings of The First International Symposium of Mathematics and its Connection to the Arts and Sciences.* Verlag Franzbecker, Hildesheim/ Berlin, pp. 201-214

Michelsen, C. (2007). Modellbildungsprozesse und Integration von Mathematik, Physik und Biologie. *Gemeinsame Jahrestagung DMV und GDM 2007.*

Sriraman, B. (2003a). Mathematics and Literature: Synonyms, Antonyms or the Perfect Amalgam. *The Australian Mathematics Teacher*, 59 (4), 26-31.

Sriraman, B (2003b) Can mathematical discovery fill the existential void? The use of Conjecture, Proof and Refutation in a high school classroom (feature article). *Mathematics in School*, 32 (2), 2-6.

Sriraman, B. (2004a). Mathematics and Literature (the sequel): Imagination as a pathway to advanced mathematical ideas and philosophy. *The Australian Mathematics Teacher*. 60 (1), 17-23.

Sriraman,B. (2004b). Re-creating the Renaissance. In M. Anaya, C. Michelsen (Eds), *Proceedings of the Topic Study Group 21: Relations between mathematics and others subjects of art and science*: The 10th International Congress of Mathematics Education, Copenhagen, Denmark, pp.14-19.

Sriraman, B. (2005). Philosophy as a bridge between mathematics arts and the sciences. In A. Beckmann, C. Michelsen & B.Sriraman (Eds.), Proceedings of the 1st International Symposium on Mathematics and its Connections to the Arts and Sciences. University of Schwaebisch Gmuend: Germany. Franzbecker Verlag, 7-31.

Sriraman, B., & Adrian, H. (2004a). The Pedagogical Value and the Interdisciplinary Nature of Inductive Processes in Forming Generalizations. *Interchange: A Quarterly Review of Education*, 35(4), 407-422.

Sriraman, B., & Adrian, H. (2004b). The use of fiction as a didactic tool to examine existential problems. *The Journal of Secondary Gifted Education.* 15(3), 96-106.

Sriraman, B., & Beckmann, A. (2007). Mathematics and Literature: Perspectives for interdisciplinary classroom pedagogy. *Proceedings of the 9th International History, Philosophy and Science Teaching Group (IHPST)* 2007, Calgary, June 22-26, 2007.

Sriraman, B., & Dahl, B. (in press). On bringing interdisciplinary Ideas to Gifted Education. To appear in L.V. Shavinina (Ed). *The International Handbook of Giftedness.* Springer Science

MATHEMATICAL MODELLING: LINKING MATHEMATICS, SCIENCE, AND THE ARTS IN THE PRIMARY CURRICULUM

Lyn D. English

Queensland University of Technology

This paper presents one approach to incorporating interdisciplinary experiences in the primary school mathematics curriculum, namely, the creation of realistic mathematical modelling problems that draw on other disciplines for their contexts and data. The paper first considers the nature of modelling with complex systems and how such experiences differ from existing problem-solving activities in the primary mathematics curriculum. Principles for designing interdisciplinary modelling problems are then presented, with reference to two mathematical modelling problems, one based in the scientific domain and the other in the literary domain. Examples of the models children have created in solving these problems follow. Finally, a reflection on the differences in the diversity and sophistication of these models raises issues regarding the design of interdisciplinary modelling problems.

> Modelling and theory building lie at the intersection of art-science-mathematics. The history of model building in science conveys epistemological awareness of domain limitations. Arts imagine possibilities, science attempts to generate models to test possibilities, mathematics serves as the tool (Sriraman & Dahl, in press).

INTRODUCTION

Our world is increasingly governed by complex systems. Financial corporations, education and health systems, the World Wide Web, the human body, and our own families are all examples of complex systems. Complexity—the study of systems of interconnected components whose behaviour cannot be explained solely by the properties of their parts but from the behavior that arises from their interconnectedness—is a field that has led to significant scientific methodological advances (Sabelli, 2006).

In the 21st century, such systems are becoming increasingly important in the everyday lives of both children and adults. For all citizens, an appreciation and understanding of the world as interlocked complex systems is critical for making effective decisions about one's life as both an individual and as a community member (Bar-Yam, 2004; Davis & Sumara, 2006; Jacobson & Wilensky, 2006; Lesh, 2006).

Educational leaders from different walks of life are emphasizing the need to develop students' abilities to deal with complex systems for success beyond school. These abilities include: constructing, describing, explaining, manipulating, and predicting complex systems (such as sophisticated buying, leasing, and loan plans); working on multi-phase and multi-

B.Sriraman, C.Michelsen, A. Beckmann & V. Freiman (Eds). (2008). *Proceedings of the Second International Symposium on Mathematics and its Connections to the Arts and Sciences* (MACAS2). Information Age Publishing, Charlotte: NC, pp.5-36

component projects in which planning, monitoring, and communicating are critical for success; and adapting rapidly to ever-evolving conceptual tools (or complex artifacts) and resources (English, 2002; Gainsburg, 2006; Lesh & Doerr, 2003). One approach to developing such abilities is through mathematical modelling, which is central to the study of complexity and to modern science. Meaningful inquiry, which involves cycles of model construction, evaluation, and revision, is fundamental to mathematical and scientific understanding and to the professional practice of mathematicians and scientists (Lesh & Zawojewski, 2007; Romberg, Carpenter, & Kwako, 2005).

Modelling is not just confined to mathematics and science, however. Other disciplines including economics, information systems, finance, medicine, and the arts have also contributed in large part to the powerful mathematical models we have in place for dealing with a range of complex systems (Steen, 2001; Lesh & Sriraman, 2005; Sriraman & Dahl, in press). Unfortunately, our mathematics curricula do not capitalize on the contributions of these external disciplines. A more interdisciplinary and unifying model-based approach to students' mathematics learning could go some way towards alleviating the well-known "one inch deep and one mile wide" problem in many of our curricula (Sabelli, 2006, p. 7; Sriraman & Dahl, in press; Sriraman & Steinthorsdottir, in press). We have limited research, however, on ways in which we might incorporate other disciplines within the primary mathematics curriculum. I offer one such approach here, namely, the creation of realistic mathematical modelling problems that draw on other disciplines for their contexts and data.

This paper first considers the nature of modelling with complex systems and how such experiences differ from existing problem-solving activities in the primary mathematics curriculum. Principles for designing interdisciplinary modelling problems are then presented, with reference to two mathematical modelling problems, one based in the scientific domain and the other in the literary domain. Examples of the models children have created in solving these problems follow. Finally, a reflection on the differences in the diversity and sophistication of these models raises issues regarding the design of interdisciplinary modelling problems.

THE POWER OF MODELLING

Modelling is increasingly recognized as providing students with a "sense of agency" in appreciating the potential of mathematics as a critical tool for analyzing important issues in their lives, their communities, and in society in general (Greer, Verschaffel, & Mukhopadhyay, in press). Indeed, new research is showing that modelling promotes students' understanding of a wide range of key mathematical and scientific concepts and "should be fostered at every age and grade...as a powerful way to accomplish learning with understanding in mathematics and science classrooms" (Romberg et al., 2005, p. 10). Students' development of potent models should be regarded as among the most significant goals of mathematics and science education (Lesh & Sriraman, 2005).

As Greer et al. point out, mathematical modelling has traditionally been reserved for the secondary and tertiary levels, with the assumption that primary school children are incapable of developing their own models and sense-making systems for dealing with complex situations. However, recent research (e.g., English, 2006; English & Watters, 2005) is showing that younger children can and should deal with situations that involve more than just simple counts and measures, and that entertain core ideas from other disciplines.

Modelling for the Primary Classroom

The terms, models and modelling, have been used variously in the literature, including in reference to solving word problems, conducting mathematical simulations, creating representations of problem situations (including constructing explanations of natural phenomena), and creating internal, psychological representations while solving a particular problem (e.g., Doerr & Tripp, 1999; English & Halford, 1995; Gravemeijer, 1999; Greer, 1997; Lesh & Doerr, 2003; Romberg et al., 2005; Van den Heuvel-Panhuizen, 2003).

The perspective adopted here is that models are "systems of elements, operations, relationships, and rules that can be used to describe, explain, or predict the behavior of some other familiar system" (Doerr & English, 2003, p.112). From this perspective, modelling problems are realistically complex situations where the problem solver engages in mathematical thinking beyond the traditional school experience and where the products to be generated often include complex artifacts or conceptual tools that are needed for some purpose, or to accomplish some goal (Lesh & Zawojewski, 2007).

A focus on modelling differs from existing approaches to the teaching of mathematics in the primary classroom. First, *the quantities and operations that are needed to mathematize realistic situations often go beyond what is taught traditionally in school mathematics.* The types of quantities needed in realistic situations include accumulations, probabilities, frequencies, ranks, and vectors, while the operations needed include sorting, organizing, selecting, quantifying, weighting, and transforming large data sets (Doerr & English, 2001; English, 2006; Lesh, Zawojewski, & Carmona, 2003). As indicated next, modelling problems provide children with opportunities to generate these important constructs for themselves.

Second, *modelling problems offer richer learning experiences than the standard classroom word problems ("concept-then-word problem,"* Hamilton, in press). In solving such word problems, children generally engage in a one- or two-step process of mapping problem information onto arithmetic quantities and operations. In most cases, the problem information has already been carefully mathematized for the children. Their goal is to unmask the mathematics by mapping the problem information in such a way as to produce an answer using familiar quantities and basic operations. These word problems constrain problem-solving contexts to those that often artificially house and highlight the relevant concept (Hamilton, in press). They thus preclude children from creating their own mathematical constructs out of necessity. Indeed, as Hamilton (in press) notes, there is little evidence to suggest that solving standard textbook problems leads to improved competencies in using mathematics to solve problems beyond the classroom.

In contrast, modelling provides opportunities for children to elicit their own mathematics as they work the problem. That is, the problems require children to make sense of the situation so that they can mathematize it themselves in ways that are meaningful to them. This involves a cyclic process of interpreting the problem information, selecting relevant quantities, identifying operations that may lead to new quantities, and creating meaningful representations (Lesh & Doerr, 2003).

Third, *mathematical modelling explicitly uses real-world contexts that elicit the creation of useful systems or models and draw upon several topic areas not only from mathematics but also from other disciplines.* For example, in the "*Creek Watch*" problem (Appendix A [1-4], children developed models to determine the water quality of their local creek. In doing so, they engaged with core ideas from the "Life and Living," "Science and Society," and "Systems, Resources, and Power" strands of their primary science and social sciences curricula. In the problem in Appendix B ("*Summer Reading*") children develop a model to determine a fair way to assign points to readers who enrol in a summer reading program. Children need to consider the number and variety of books read, their difficulty levels, the length of the books, and the quality of the readers' reports. "Experientially real" contexts such as children's local creek or community/school library provide a platform for the growth of their mathematization skills, thus enabling children to use mathematics as a "generative resource" in life beyond the classroom (Freudenthal, 1973; Stevens, 2000; Streefland, 1993).

Fourth, *modelling problems encourage the development of generalizable models.* In my research with elementary and middle school children, I have implemented sequences of modelling problems that encourage the creation of models that are applicable to a range of related situations (e.g., Doerr & English, 2003; Doerr & English, 2006; English & Watters, 2005). Children are initially presented with a problem that confronts them with the need to develop a model to describe, explain, or predict the behavior of a given system (a model-eliciting problem). Given that re-using and generalizing models are central activities in a modelling approach to learning mathematics and science, the children then work related problems that enable them to extend, explore, and refine those constructs developed in the initial problem (model-exploration and model-application problems). Because the children's final products embody the factors, relationships, and operations that they considered important, powerful insights can be gained into the children's mathematical and scientific thinking as they work the problem sequence.

Fifth, *modelling problems are designed for small-group work where members of the group act as a "local community of practice" solving a complex situation* (Lesh & Zawojewski, 2007). Numerous questions, issues, conflicts, revisions, and resolutions arise as the children develop, assess, and prepare to communicate their products to their peers. Because the products are to be shared with and used by others, they must hold up under the scrutiny of the team and other class members.

DESIGNING INTERDISCIPLINARY MODELLING PROBLEMS FOR THE PRIMARY SCHOOL

My research has involved working with teachers to design mathematical modelling problems that align themselves with the learning themes being implemented in the classroom. Such themes have included, among others, natural disasters, the local environment, classroom gardens, early colonisation, the gold-rush days, the Olympic and Commonwealth Games, book reading clubs, and class excursions to fun parks. The instructional design principles of Lesh and his colleagues (e.g., Lesh, Cramer, Doerr, Post, & Zawojewski, 2003, p. 43) are applied in designing these problems. These principles include the following:

1. The Personal Meaningfulness Principle

It is important that children can relate to and make sense of the complex system being presented in the problem. That is, the system should be one that reflects a real-life situation and that builds on children's existing knowledge and experiences. By designing problems that are integrated within the classroom's particular learning theme, the activities are less likely to be treated as "add-ons" in an already crowded curriculum. The modelling problems serve to not only enrich the problem-solving component of the mathematics curriculum but to also help children link their learning meaningfully across disciplines. For example, the Creek Watch problem incorporates core scientific and societal understandings such as *environments are dynamic and have living and non-living components that interact, living organisms depend on others and the environment for survival*; and *the activities of people can change the balance of nature.*

2. The Model Construction Principle

A modelling problem should require children to develop an explicit mathematical construction, description, explanation, or prediction of a meaningful complex system. The models children create should be mathematically significant. That is, they should focus on the underlying structural characteristics (key ideas and their relationships), rather than the surface features, of the system being addressed. For example, in the Summer Reading problem, as indicated later, children developed models that involved significant mathematisation processes such as assigning value points, using interval quantities, weighting selected factors, aggregating quantities, and applying informal measures of rate.

3. The Model Documentation Principle

Modelling problems should encourage children to externalise their thinking and reasoning as much as possible and in a variety of ways. The need to create representations such as lists, tables, graphs, diagrams, and drawings should be a feature of the problem. Furthermore, the models children construct need to involve more than a brief answer: descriptions and explanations of the steps children took in constructing their models should be included.

4. The Self-Assessment Principle

Modelling activities should provide children with sufficient criteria for determining whether their final model is an effective one and adequately meets the client's needs in dealing with the given complex system (and related systems). Such criteria also enable children to progressively assess and revise their creations as they work the problem. For example, in the Summer Reading problem, children are informed that the students who enrol in a reading program often read between ten and twenty books over the summer and that they can select books from any grade level. In developing their model for assigning points to each reader, children need to take into account the number and variety of books read, the difficulty level, the book length, and the quality of the written reports.

5. The Model Generalisation Principle

The models that children generate should be applicable to other related problem situations. As previously noted, by implementing a sequence of related modelling problems, children are provided with opportunities to extend, explore, and refine the models they developed in an initial problem (e.g., Doerr & English, 2003).

CHILDREN'S MODEL CONSTRUCTIONS FOR THE CREEK WATCH AND SUMMER READING PROBLEMS

This section addresses two longitudinal studies in which the Creek Watch and Summer Reading problems were implemented respectively. The models that the children created in solving each problem are presented.

In analysing the children's model developmental as they work these problems, I considered several aspects of their responses. These include the ways in which the children interpreted and re-interpreted the problem, the nature of the problem factors children chose to work with, the shifts in their thinking that led them to more sophisticated conceptual understanding, the mathematisation processes they applied to the given data including ways in which they quantified qualitative data, the types of data transformations they made through these processes, the representations they used in documenting and supporting their thinking, and the diversity in the final models they created.

Creek Watch Problem

Participants, Design, and Procedures

Four classes of third-grade children (7-8 year-olds) and their teachers from a suburban Brisbane state school participated in this longitudinal study, which spanned the children's third-, fourth-, and fifth-grade levels. The Creek Watch problem was implemented at the beginning of the fifth grade.

A teaching experiment involving multilevel collaboration (English, 2003; Lesh & Kelly, 2000) was employed. At the first level, children engage in mathematical modelling, as discussed previously. At the second and third levels, the classroom teacher works collaboratively with the researchers in designing and implementing the modelling problems. These problems also serve as challenging and thought-provoking experiences for the teachers as they explore the mathematical ideas being developed, consider appropriate implementation strategies, and promote learning communities within their classrooms. To facilitate the teachers' development, a number of workshops, meetings, and debriefing sessions were held throughout each year of the study.

At the beginning of each of the first two years of the study the children completed a number of preparatory activities prior to completing 2-3 comprehensive modelling problems during the remainder of the year. These preparatory activities included interpreting mathematical and scientific information presented in text and diagrammatic form; reading simple tables of data; collecting, analysing, and representing data; preparing written reports from data analysis; working collaboratively in group situations; and sharing end products with class peers by means of verbal and written reports.

For the Creek Watch problem, implemented at the beginning of the study's third year, the class teachers were given background reading on modelling and on the core scientific concepts embedded in the problem. Prior to implementing the problem, the teachers conducted a preparatory lesson addressing the main scientific ideas. During the first problem session, the children were presented with the following information:

(a) The *Creek Watch Activity* (see Appendix A[1]), which included background information on river ecology, a set of "readiness" questions, and the problem itself;

(b) A *Creek Watch "Fact Sheet,"* which provided information on water quality as determined by: (i) the presence of macro-invertebrates (diversity and sensitivity); (ii) the type and quantity of fish (native or exotic); (iii) the presence of algae and other plants including weeds; and (iv) the concentration or amounts of various chemicals in the creek. For example, the higher the level of nitrogen the poorer the quality of water; dissolved oxygen is most critical to water quality; high levels of turbidity are detrimental to the water quality; and most organisms are able to survive in water with a pH of between pH6.5 and pH9.0 (normal pH is 7.0).

(c) An explanation of the *Pollution Index* (see Appendix A[2]);

(d) A *Water Bugs Identification Chart* (see appendix A[3]); and

(e) *Moggill Creek Data Table* (see Appendix A[4]).

The children completed the problem activities in four sessions, each of 40-60 minutes duration over a period of 2.5 weeks. The children worked the problem in small groups (2-4 children per group), without direct instruction from the teacher or researchers. During the last session the children presented group reports to the class on the models they had created. In each classroom, 2-3 groups were videotaped and 2-4 groups were audiotaped as they worked the problem; all whole class reports were also videotaped. Each of the tapes was transcribed for analysis and all of the children's written artifacts were also examined.

Following their completion of this activity, the classes applied their models in exploring their local creek behind their school.

Children's Models for the Creek Watch Problem

A selection of the children's models follows. These models ranged from little or no application of mathematisation processes through to transformation of data using ranking and assigning of scores. Some models comprise inappropriate operations on the data, as can be seen in models 3 and 4.

Model 1: Selecting the healthiest site for each factor

The groups who developed this model listed all the problem factors one under the other (i.e., each of the macro-invertebrates, fish, weeds, and chemicals) and identified the healthiest site for each factor. For example, site A was selected as the best site for dragonfly nymphs, which are highly sensitive creatures, while site E was chosen as the cleanest with respect to each weed. The number of times each site was chosen was tallied and the site/s with the highest total was identified as the cleanest. This model did not include the calculation of a pollution index for each site, nor did model 2.

Model 2: Classifying each factor as "good" or "bad" for each site

A variation of model 1 entailed classifying each factor as "good" (√) or "bad" (x) for each site. The macro-invertebrates were first classified as "sensitive," "tolerant" and "very tolerant," with "sensitive" and "tolerant" considered good and the "very tolerant" as bad. For example, for dragonfly nymphs, sites A and C were rated as good while the remaining sites were rated as bad. A table was drawn for each set of factors across the five sites (i.e., a table for macro-invertebrates for sites A-E, a table for fish for sites A-E etc.) and each site awarded a tick or cross accordingly. The cleanest site/s had the greatest number of ticks across all factors.

Model 3: Determining the PI for the macro-invertebrates and a "chemical index" for each site

Although the majority of groups determined the pollution index for the macro-invertebrates for each site, there were variations in how they dealt with the remaining factors. Two groups calculated a "chemical index" for each site by adding the units of all the chemicals, ignoring the differences in the nature of the chemicals and their units. The two indexes were then totalled to determine the water quality of the sites. As one group explained in their final letter:

> Dear Jack and your group,
>
> We are writing to you to help you work out how to test and find out how polluted the water is. First you have to collect the data from the sites (animals, fish, chemicals etc) that you and your group have been to. Secondly you will have to write down the animals, chemicals etc. and you add the sensitivity number for the animals and for the chemicals. You add the numbers for different chemicals to find out how polluted the water is. Once you have added these up (animals) you will end up with the pollution index total and you will end up with a pollution index also for the chemicals. If the answer (animals) is quite high, then the water will be quite clean. If the answer (chemicals) is quite low then the water is quite dirty. When we worked it out we found that Site A was quite clean, site B was sort of dirty, and sort of clean. For site C it was also quite dirty and quite clean. Site D was very clean and site E was very clean.

The group did not consider the fish and weeds.

Model 4: Determining the PI for each site and quantifying the fish and weed data

A variation of model 3 involved a consideration of the PI for each site plus a consideration of the fish and weed data only. One group simply totalled the fish and weed data, ignoring the fact that some fish were not desirable. The fish and weed totals for each site were combined with the site's PI to give an overall score as an indication of the water quality (the site with the highest score was deemed the cleanest).

Model 5: Determining the PI for each site and describing impact of remaining factors

In this model only the PI was calculated, with the remaining factors considered individually for their respective impact on the water quality at each site. This impact was not quantified, however. For example, one group presented the following report, which stops short of indicating which site is the cleanest when the remaining factors are taken into account:

> Dear Jack Simpson,
>
> We heard about your problem and we would like to help solve your problem. First find the pollution index of the Moggill Creek by adding the points of the macro-invertebrates. The points we got were: site A- 51, site B- 35, site C- 17, site D- 30, site E- 49. After finding the pollution index of the area look at the other data. If an area has less bad fish (platy, guppy, swordtail) and more good fish (eel fish, black mangrove, carp, purple-spotted gudgeon) the area is good. With the macro-invertebrates and the fish there are weeds. The weeds are alligator weeds, Chinese elm, and camphor laurel. The less weeds an area has the less it is polluted. Also there are chemicals, good chemicals like dissolved oxygen, OK chemicals salinity (salt) and bad chemicals like turbidity, phosphorous, nitrogen, and pH (a good pH level is 7.0-7.05).

Model 6: Transforming the data through ranking

This model represented a more sophisticated version of model 2 and involved a ranking of the factors across the sites. This group of children constructed tables for each set of factors and ranked each site from 1 to 5 according to the desirability/non-desirability of the factor. For the macro-invertebrates, the children only considered selected examples and first recorded their sensitivity score. The children then identified the best site for each of these macro-invertebrates and ranked that site as 1. Each of the remaining sites was then ranked accordingly from 2 through 5. The site with the greatest number of 1s was chosen as the best site for macro-invertebrates.

A similar process was followed for the chemical analysis factors (see Figure 1). For dissolved oxygen, the sites were ranked from 1 (highest amount) to 5 (lowest amount). For turbidity and salinity, the sites were ranked from 1 (lowest amount) to 5 (highest amount). The remaining chemicals were ranked in the same way.

For the fish, only the exotic were considered, with each site ranked from 1 (least amount) to 5 (highest amount). Finally, the number of weeds at each site was aggregated, with site E clearly the best site with respect to lack of weeds (the children omitted their rankings from their weed table). The site with the greatest number of 1s was declared the cleanest site. The children explained their system in the following report, although their description of the ascending and descending ranking is not entirely clear.

> Dear Jack Simpson
>
> We found that Site E is the healthiest site of the creek. We did this by drawing graphs of macro-invertebrates, fish, weeds, and chemical analysis. Using a fact sheet we found out whether they are good or bad for the environment. Depending on whether they were good or bad, we ranked them. If they are good we ranked them in descending order but if they were bad we'd rank them in ascending order.

For the dissolved oxygen we put a number for the one that has the most, which is site A and then for the turhity, salinity, phosphorous, nitoergen, and other notorgen we put the one that was the least because they are all bad. Then we counted up the one that has the most ones. It was site E. We did that to the other ones (factors) as well.

Then we put a tally beside the number that was ranked first. The site with the most tallies is the most healthiest. Site E. We hope our guidelines help you.

INSERT FIGURE 1 ABOUT HERE

Model 7: Transforming the data by assigning scores

This model is a variation of models 6 and 1. In developing their model, the children suggested rating each site out of five for each of the factors (*You know how there's five sites. We'll give them a rating out of five for each of them, like, one, two, three, four....we'll look at our data and if that is bad or not and we'll tally at the end, once we've gone through all of it, would be the most healthiest*). However, the children did not continue with this suggestion, rather, for each desired factor, they decided to award one point to the site that had the "best" of that factor. If the factor was not desirable, the site that had the least of that factor was given one point (*If it was bad we looked for the least and if it was good we looked for the most...we didn't really give any points for bad*). All points were then tallied and the site with greatest number of points was chosen as the cleanest. The children completed a bar graph showing the overall number of points awarded to each site.

The children's approaches to solving the Creek Watch problem are revisited in the discussion section. Consideration is now given to the study in which the Summer Reading problem was implemented and the models created by seventh-grade children in solving the problem.

Summer Reading Problem

Participants, Design, and Procedures

In this longitudinal study I worked with one class of children and their teachers from the fifth grade (9-10 years-old) through to the end of the seventh grade. The children attended a suburban, private co-educational P-12 college. As in the previous study, a teaching experiment involving multilevel collaboration was used.

In the first year of the study a number of preparatory modelling activities were implemented, followed by a model-eliciting problem and a model-exploration problem. In each of the second and third years an initial model-eliciting problem, a model-exploration problem, and two model-application problems were implemented. The problems involved interpreting and dealing with multiple tables of data; creating, using, modifying, and transforming quantities; exploring relationships and trends; and representing findings in visual and text forms. Also inherent in the problems were the key mathematical ideas of rate and proportion, and ranks and weighted ranks. The children had had no formal instruction on these core mathematical ideas and processes prior to commencing the problem activities.

The Summer Reading Problem (see Appendix B) was the final activity that the children completed in their seventh grade; the problem was a model-application activity where the children extended, explored, and refined constructs they had developed in the previous modelling problems. The problem required the children to develop a fair rating system to award points to students participating in a summer reading program. The children worked the problem in small groups during two 50-minute sessions. They then presented their models to the class in a third session, where they explained and justified the models they had developed and subsequently invited feedback from their peers. This group reporting was followed by a whole class discussion that compared the features of the mathematical models produced by the various groups.

As for the previous study, data sources included audiotapes and videotapes of the children's group work and classroom presentations. Field notes, children's work sheets, and final reports detailing their models and how they developed them were also important data sources. The tapes were transcribed and analysed for evidence of the mathematical understandings and mathematisation processes that the children used in building their models. To assist in the analysis, a modified version of Carmona's (2004) assessment tool for describing students' mathematical knowledge was used. The analysis addressed the nature of the problem factors that the children chose to consider, the operations they applied, the types of transformations they made through these operations, and the representations they used in documenting their final model.

Children's Model Creations

Table 1 displays the problem factors, mathematical operations, and representational formats used by each of five student groups in developing their models. The representational formats included the use of tables, text, lists, and formulae. In constructing their models, children chose to work with some or all of the problem factors, namely, number and variety of books read, reading level and length of the books, a student's grade level, and the quality of written reports. Children's operations on these factors included assigning valuc points, using interval quantities, using weighting, aggregating quantities, and using informal notions of rate. One group of children also imposed constraints on the use of their operations.

INSERT TABLE 1 ABOUT HERE

As can be seen in Table 1, the variety of books read was considered by only one group in creating a fair rating system while four out of the five groups took into account a student's grade level and a book's reading level. It is perhaps not surprising that the variety of books read presented a challenge—the notion of variety can be interpreted in different ways and its measure involves a consideration of several factors. As one group explained, "*With different variety of books—just say like if you read three different variety of books in subject or level or pages—just variety, so varieties include level, pages, and subject.*" It is also interesting to note that all but one group created formulae as part of their model, while only one group used any form of a table (which they used in developing their model but chose not to display in their report.)

The children displayed various data transformation processes in developing their models, as indicated in Table 1. Group 1 created new, interval quantities (e.g., "Year 9 reads 4 books," "Year 9 reads 5 books etc.") and transformed these into other quantities by assigning value points (e.g., "= 1 point"), as can be seen in their documented account below.

Group 1's Model

Dear Margaret Scott
We have found a solution to the problem you have given us. Our information is below.

Year 9 reads 4 books = 1 point	8=4=2
Year 9 reads 5 books = 2 points	8=5=3
Year 9 reads 6 books = 3 points	8=6=4
Year 9 reads 7 books = 4 points	8=7=5
9=8=5	8=8=6
9=9=6	8=9=7
9=10=7	8=10=8

(The group continued the pattern for the grade 7 and grade 6 reading books.)

Group 2's Model

Group 2 quantified selected problem factors and transformed quantities into other quantities. In doing so, the children made use of weighting (multiplying a book's grade by 10), used interval quantities, assigned value points, and aggregated quantities.

Dear Mrs Scott
We have found a solution to your problem.
We think that the grade of the book times by 10 and then add the amount of pages to that. After doing this you should look at the book report grade and if it is between an A+ to an A- they get another 50 points, B+ to a B- 40 points, C+ to a C- 30 points, D+ to a D- 10 points and no points for an F. The three scores should be added together and that will be the score of the book. We chose this way because the person involved will easily get a high score. There for lifting their high esteem and they will be encouraged to read more which is the whole point of this activity. P.S the total score of all the books read should be added up in the end and that will be the total score of all of the books.

- The grade of the book x 10
- Plus the amount of pages in it
- Add the grade of the book report

A+ to A− = 50 points
B+ to B− = 40 points
C+ to C− = 30 points
D+ to D− = 20 points
E+ to E− = 10 points
F = 0 points

Group 3's Model

Group 3 quantified only two factors, namely, a student's grade level and a book's reading level. The group also transformed quantities into other quantities by considering the difference between the student's grade level and the book's reading level, and then assigning points according to the extent of difference.

We have resolved the problem about the points issue. We have come up with a fair point system. We surround it by the level you read at. 2 points if you read at your level, 3 points above your level 2 levels higher. You get four points if it is 4 levels above your level. You get one point if it is one below your level. You get zero points if it is two levels below you. We think this is the easiest way to reward the children.

2 = year level
3 = above your level twice
4 = above your level 4 times
1 = below your level one times
0 = below your level two times

Group 4's Model

Group 4 created the most comprehensive model, taking into account all the problem factors. The group commenced the problem with a child commenting, "*Well, let's go over the possibilities like the number of books they read... Well, the number of books they read is up to them and what grades they do. So there should be like one point of order for each book they read – like, one bonus point. And depending on the grade, if they read a grade 4 book they get 8 points – like, you double it.*" Another group member responded, "*But it depends. If they're in grade 10 and they read a grade 4 book, then they wouldn't really get any points for that*." In subsequently constructing their model the children weighted each book read (one bonus point awarded), assigned value points to the variety of books read (variety as determined by books from several grade levels having been read), and assigned points to the difficulty level of a book as it related to a student's grade level (informally toying with the idea of rate). It is interesting to note the "rules" and constraints on rules that the children included in their model.

Dear Miss Margaret Scott,
We have found a solution to your problem. But first we will tell you our strategy. For every book a student reads they would get 1 bonus point per book. Next if they read three different grades of books or more they would get 5 points. Eg. If a grade 4 girl/boy reads a grade 4, 5 and 6 book then he/she would get 5 points. We gave points for the difficulty of books by if you were in grade 6 and you read a grade 6 then you would get 3 points because you half the grade. Although there is a rule of you can only read two levels below your grade to receive points and as many levels above that you are capable of. If you were in grade 5 and you read a grade 4 book you would still get 4 (2?) points because you are still halving it. But if you were in grade 9 and you read a grade 5 books you would not get any points. Because the book is too easy. We went by a code for the lengths of books: 50-70 = 3 points, 71-100 = 4 points, 101-170 = 5 points, 171-220 = 6 points and 221 and up = 7 points. We also made a code for the written reports as well which is: F=0, D=1, C- =2, C=3, C+ = 4, B- = 5, B=6, B+ = 7, A- = 8, A=9 and A+ = 10. we hope this helps you decide on how to give out points and awards during the summer.

Group 5's Model

Group 5 considered the factors of grade level, book reading level, and report quality. Value points were awarded according to the level of a book read and the difference between a student's grade level and the book level. An informal notion of rate was evident in their assigning of points here.

Dear Margaret Scott
We have formulated a point system to determine results for your reading marathon. The system works on a point basis. If you read a book that is based for your grade level you receive 10 points although if you read a book higher than your grade level you receive 2 points for every grade level you read up and you 2 points go down for every grade level lower. If you get an F for the report you don't get bonus points but from every grade that goes up from F you receive one point.

DISCUSSION

Children's Models and Implications for Problem Design

This paper has addressed one way in which mathematics, science, and the arts can be brought together in meaningful learning experiences for primary school children. A model-based approach to students' mathematical learning, which takes children beyond their usual problem-solving experiences, provides rich opportunities for interdisciplinary experiences. Such an approach also engages children in future-oriented learning, both with respect to the more sophisticated mathematics they encounter and the collaborative team work they develop in dealing with complex situations. In contrast to children's regular mathematical activities, modelling places children at the centre of their own learning. That is, children are placed in authentic problem situations where they have to construct their own mathematics to solve the problem.

Modelling in both mathematics and science should not be confined to the secondary school years and beyond. As this paper and others have shown, primary school children are capable of engaging successfully with modelling problems that involve complex data systems. In working such problems children cycle through interpreting and re-interpreting the problem and the data sets, identify key problem factors, determine and apply quantification processes to transform the data, and document and support their actions in various representational formats. For example, in the Summer Reading problem the children debated which factors to include in their models and how to quantify them. In quantifying their data they assigned value points, used interval quantities, weighted some factors, aggregated quantities, and applied informal measures of rate. In so doing, the children created new quantities, transformed qualitative factors into quantities, transformed quantities into other quantities, and created formulae and lists to support their descriptions of their model construction.

Although it is difficult to make comparisons between the two studies, given the different student populations and prior modelling experiences, it is nevertheless a useful exercise to consider differences in the design of the two problems. These differences could account in part for the variation in diversity and sophistication of the models the children created on the two problems. The Creek Watch problem presented the children with a substantial amount of data (a Creek Watch Fact Sheet, an explanation of Pollution Index, a water bugs identification chart, a Moggill Creek data table, and the problem goal and background information). In dealing with the Moggill Creek data table (see Appendix A [4]), children need to consider the diversity and sensitivity of the macro-invertebrates at each site, the number and nature of fish present at each site (whether each fish is exotic or native), and the quantity of weeds found at each site and their impact on the water quality. The presence of chemicals at each site and their effect on the water quality also must be taken into account. In working with these data, children need to understand the pollution index and how (and whether) to apply it. Some groups ignored the PI while others calculated it for each site. The issue of how to deal with the remaining factors and how these might impact on the PI of the creek was a challenge for many children.

Although the children were able to create a variety of models to solve the problem, a few issues remain for attention. Why did some children simply aggregate all the fish and all the chemicals, ignoring their different features? Why didn't more children demonstrate an understanding that a balance of certain levels of chemicals in the water leads to optimum numbers and types of macro-invertebrates? Why didn't more models incorporate mathematisation processes and relationships between data? One possible explanation for these issues is that some children lacked an adequate understanding of the core scientific ideas, even though the teachers were asked to undertake preliminary work on these ideas. Another explanation lies in the nature and presentation of the data in the Moggill Creek data table. All of the data are presented in a quantitative format, in contrast to the Summer Reading problem where children have to quantify some qualitative factors (e.g., variety of books read). Perhaps the amount of additional scientific information given to the children in the Creek Watch problem was too excessive and required too much additional interpretation, drawing children's attention away from the important mathematical ideas inherent in the problem. The design of the problem itself could also be a contributing factor here. That is, it might be that the self-assessment principle was not applied adequately in the design of the problem—there might have been insufficient criteria for the children to assess their progress and determine whether their final model met the client's needs. The children were simply told to utilise "all the data collected over the year in five locations along the creek." A redesign of this problem could address this issue and also include some qualitative data to be quantified, as well as a restructuring of the data (e.g., fewer macro-invertebrates and a greater variety of chemicals and plants, some detrimental and others beneficial).

The Summer Reading problem, on the other hand, elicited more diverse and more sophisticated models than the Creek Watch problem. Although the difference in student populations is an obvious factor here, there are other reasons why this might be the case. The children were not presented with additional data to interpret and apply to problem solution as in the CreekWatch problem. Furthermore, the nature of the problem factors to be considered included both qualitative and quantitative data (e.g., variety of books versus number of books). The children thus had to come up with some way of determining how to quantify all of the factors, suggesting that this problem met the model construction principle better than the Creek Watch problem. It could also be that the Summer Reading problem more effectively met the self-assessment principle, that is, children were told explicitly the factors to consider in creating their model, not all of which were defined for them. Finally, although the problem was couched within a literary discipline, the problem did not draw upon as much additional disciplinary content as did the Creek Watch problem.

Interdisciplinary Projects with Modelling

As previously noted modelling problems can serve as unifying vehicles for the primary curriculum, bringing together key ideas from several disciplines. For example, in the Creek Watch problem, core concepts are drawn not only from mathematics, but also from science and from studies of society and the environment. Such concepts include the dynamic nature of environments and how living and non-living components interact, the ways in which living organisms depend on others and the environment for survival, and how the activities of people can change the balance of nature. In providing a literary context, the Summer Reading problem draws children's attention to the features of an array of books and encourages them to consider how book length, difficulty level, variety of books read, and quality of book reports could determine points awarded to readers.

Because of their interdisciplinary nature, modelling problems provide a rich platform for student research projects. There are numerous opportunities for generating such projects within the regular curriculum. One such opportunity, which I utilised recently in three fifth-grade classes, engaged children in an investigation of their country's settlement. The children were commencing a study of the arrival of the First Fleet on the east coast of Australia in 1787. A modelling problem was created to tie in with their study of this settlement. As part of the problem, the children explored the reasons behind the first settlement, the composition of the First Fleet (comprising 11 ships), the difficulties faced by the Fleet en route to their destination, the types of supplies on board the ships, and the various conditions required for the settlement of a new colony. The problem text explained that, on his return from Australia to the UK in 1770, Captain James Cook reported that Botany Bay had lush pastures and well-watered and fertile ground suitable for crops and for the grazing of cattle. But when Captain Phillip (commander of the First Fleet) arrived in Botany Bay in January 1788, he thought it was unsuitable for the new settlement. Captain Phillip headed north in search of a better place for settlement. The children's task was as follows:

Where to locate the first settlement was a difficult decision to make for Captain Phillip as there were so many factors to consider. If you could turn a time machine back to 1788, how would you advise Captain Phillip? Was Botany Bay a poor choice or not? Early settlements occurred in Sydney Cove Port Jackson, at Rose Hill along the Parramatta River, on Norfolk Island, Port Hacking, and in Botany Bay. Which of these five sites would have been Captain Phillip's best choice? Your job is to create a system or model that could be used to help decide where it was best to anchor their boats and settle. Use the data given in the table and the list of provisions on board to determine which location was best for settlement. Whilst Captain Phillip was the first commander to settle in Australia many more ships were planning to make the journey and settle on the shores of Australia. Your system or model should be able to assist future settlers make informed decisions about where to locate their townships.

As for the Creek Watch problem, the First Fleet problem elicits similar core ideas from science and from environmental and societal studies; these ideas serve as springboards for new investigations of other early settlements in Australia. The problem can also lead nicely into a more in-depth study of the interrelationship between ecological systems and economies, and a consideration of ways to promote and attain ecologically sustainable development.

In another study, fourth- and fifth-grade classes worked two modelling problems exploring Australia's cyclones as part of their study of natural disasters (English, Fox, & Watters, 2005). In the first problem the children explored data pertaining to cyclone categories and how these are determined, the wind speeds of cyclones, their impact on the environment (e.g., landslides, flooding, destruction of buildings, loss of power etc), and also the locations and severity of cyclones in Queensland (the children's home state) in the last 12 years. The children were to take on the role of assisting a Project Resort Development Committee to assess possible locations for building new holiday destinations in Queensland. Using the data supplied, the children were to develop a model that would determine which two locations the resort development company should avoid. The follow-up problem required children to do their own research on cyclone activity in another Australian state, namely, Western Australia. The children researched and identified data that they considered important in determining suitable sites for new resorts. The models the children developed for the first problem were then applied to this second example, with some groups of children making improvements to their original models.

CONCLUDING POINTS

Designing appropriate interdisciplinary modelling experiences is not an easy task for teachers or researchers. Numerous cycles of creating, testing, re-building, and refining a problem are usually required. The instructional design principles cited previously provide an effective guide in creating these modelling problems. Consideration needs to be given not only to the mathematical ideas to be embedded in the problem but also to the other discipline content. Determining an appropriate balance of interdisciplinary ideas is an important aspect of the problem design and will be governed in part by the nature of the disciplines and the age of the children involved. Also of importance is the need to capitalize on the myriad opportunities across the curriculum for children to undertake interdisciplinary research projects involving modelling problems. Such projects both develop and extend multi-disciplinary learnings and enable children to review, adapt, and apply their model creations.

References

Bar-Yam, Y. (2004). *Making things work: Solving complex problems in a complex world.* NECSI: Knowledge Press.

Davis, B., & Sumara, D. (2006). *Complexity and education: Inquiries into learning, teaching, and research.* Mahwah, NJ: Lawrence Erlbaum Associates.

Doerr, H. M., & English, L. D. (2003). A modeling perspective on students' mathematical reasoning about data. *Journal for Research in Mathematics Education, 34*(2), 110-137.

Doerr, H. M, & English, L. D. (2001). A modelling perspective on students' learning through data analysis. In M. van den Heuvel-Panhuizen (Ed.), *Proceedings of the 25th Annual Conference of the International Group for the Psychology of Mathematics Education* (pp. 361-368). Utrecht University.

Doerr, H. M., & Tripp, J. S. (1999). Understanding how students develop mathematical models. *Mathematical Thinking and Learning, 1*(3), 231-254.

English, L. D. (2002). Promoting learning access to powerful mathematics for a knowledge-based era. In D. Edge & Y. Ban Har (Eds.), *Mathematics Education for a Knowledge-based Era* (vol. 1, pp100-107). National Institute of Education, Singapore: Association of Mathematics Educators.

English, L. D. (2003). Reconciling theory, research, and practice: A models and modeling perspective. *Educational Studies in Mathematics, 54*, 2 & 3, 225-248.

English, L. D. (2006). Mathematical modeling in the primary school: Children's construction of a consumer guide. *Educational Studies in Mathematics, 62* (3), 303-323.

English, L. D., & Halford, G. S. (1995). *Mathematics education: Models and processes.* Mahwah, New Jersey: Lawrence Erlbaum Associates.

English, L. D., & Watters, J. J. (2005). Mathematical modeling in the early school years. *Mathematics Education Research Journal, 16* (3), 58 – 79.

English, L. D., Fox, J. L., & Watters, J. J. (2005). Problem posing and solving with mathematical modelling. *Teaching Children Mathematics, 12*(3), 156-163.

Freudenthal, H. (1973). *Didactical phenomenology of mathematical structures*, Kluwer, Boston.

Gainsburg, J. (2006). The mathematical modeling of structural engineers. *Mathematical Thinking and Learning, 8*(1), 3-36.

Gravemeijer, K. (1999). How emergent models may foster the construction of formal mathematics. *Mathematical Thinking and Learning, 1*, 155-177.

Greer, B. (1997.) Modeling Reality in Mathematics Classroom: The Case of Word Problems. *Learning and Instruction 7*, 293-307.

Greer, B., Verschaffel, L., & Mukhopadhyay, S. (in press). Modelling for life: Mathematics and children's experience. In W. Blum, W. Henne, & M. Niss (Eds.), *Applications and modelling in mathematics education* (ICMI Study 14). Dordrecht: Kluwer.

Hamilton, E. (In press). What changes are needed in the kind of problem solving situations where mathematical thinking is needed beyond school? In R. Lesh, E. Hamilton, & J. Kaput (Eds.), *Foundations for the future in mathematics education.* Mahwah, NJ: Lawrence Erlbaum.

Jacobson, M., & Wilensky, U. (2006). Complex systems in education: Scientific and educational importance and implications for the learning sciences. *The Journal of the Learning Sciences, 15* (1), 11-34.

Lesh, R. (2006). Modeling students modeling abilities: The teaching and learning of complex systems in education. *The Journal of the Learning Sciences, 15* (1), 45-52.

Lesh, R., & Doerr, H. M. (Eds.). (2003). *Beyond constructivism: Models and modeling perspectives on mathematic problem solving, learning and teaching.* Mahwah, NJ: Lawrence Erlbaum.

Lesh, R. A., & Kelly, A. E. (2000). Multi-tiered teaching experiments. In R. A. Lesh & A. Kelly (Eds.), *Handbook of research design in mathematics and science education* (pp. 197-230). Mahwah, NJ: Lawrence Erlbaum.

Lesh, R. A., & Sriraman, B. (2005). John Dewey revisited—pragmatisim and the models-modeling perspective on mathematical learning. In A. Beckmann, C. Michelsen, & B. Sriraman (Eds.). *Proceedings of the 1st international symposium of mathematics and its connections to the arts and sciences* (pp. 7-31). The University of Education, Schwöbisch Gmund, Germany.

Lesh, R., & Zawojewski, J. S. (2007). Problem solving and modeling. In F. Lester (Ed.), *Second Handbook of research on mathematics teaching and learning*. Greenwich, CT: Information Age Publishing.

Lesh, R., Zawojewski, J. S., & Carmona, G. (2003). What mathematical abilities are needed for success beyond school in a technology-based age of information? In R. Lesh & H. Doerr (Eds.). Beyond constructivism: Models and modeling perspectives on mathematic problem solving, learning and teaching (pp.205-222). Mahwah, NJ: Lawrence Erlbaum.

Lesh, R., Cramer, K., Doerr, H. M., Post, T., & Zawojewski, J. S. (2003). Model development sequences. In R. Lesh & H. M. Doerr, (Eds.). (2003). *Beyond constructivism: Models and modeling perspectives on mathematic problem solving, learning and teaching* (pp.35-58). Mahwah, NJ: Lawrence Erlbaum.

Romberg, T. A., Carpenter, T. P., & Kwako, J. (2005). Standards-based reform and teaching for understanding. In T. A. Romberg, T. P. Carpenter, & F. Dremock (Eds.), *Understanding mathematics and science matters*. Mahwah, NJ: Lawrence Erlbaum Associates.

Sabelli, N. H. (2006). Complexity, technology, science, and education. *The Journal of the Learning Sciences, 15*(1), 5-9.

Sriraman, B., & Dahl, B. (In press). On Bringing Interdisciplinary Ideas to Gifted Education. In L.V. Shavinina (Ed). *The International Handbook of Giftedness*. Springer Science.

Sriraman, B., & Steinthorsdottir (2007, in press). Implications of research on mathematics gifted education for the secondary curriculum. To appear in C. Callahan & J. Plucker (Eds.), *What research says: Encyclopedia on research in gifted education*. Prufrock Press.

Steen, L. A. (Ed.). (2001). *Mathematics and democracy: The case for quantitative literacy*. National Council on Education and the Disciplines. USA

Stevens, R. (2000). Who counts what as mathematics: Emergent and assigned mathematics problems in a project-based classroom. In J. Boaler (Ed.), *Multiple perspectives on mathematics teaching and learning* (pp. 105-144). Westport, CT: Ablex Publishing.

Van den Heuvel-Panhuzen, M. (2003). The didactical use of models in realistic mathematics education: An example from a longitudinal trajectory on percentage. *Educational Studies in Mathematics, 54,* 9-35.

APPENDICES

Appendix A (1) The Creek Watch Problem[1]

INDOOROOPILLY TIMES

State department and local school kids work together to keep watch over Moggill Creek

Increased urban development in the Western Suburbs of Brisbane is threatening the quality of the local environment. In particular, runoff is impacting on water quality and visible signs such as rising salinity and blue-green algal blooms are becoming more prevalent.

Since 1995, scientists from the Queensland Environmental Protection agency (EPA) and Brisbane City Council (BCC) have monitored the water quality in many creeks in South East Queensland.

In 2004 Year 5 students from Indooroopilly State School (ISS) assisted these authorities to collect important data about Moggill Creek. The Creek flows into the Brisbane River in the Western Suburbs. The members of the class collected water samples from the creek to test for chemicals. Jack Simpson, a student from ISS said, "The class also took samples of different species of fish and macro-invertebrates. I loved looking at the macro-invertebrates the best, they are great water bugs."

Mrs. Jones from the EPA said the students' assistance had been invaluable and the information they provided helpful. "By monitoring the waterways we can gain a picture of catchment health. Monitoring over time can provide information on the state of the catchment which can assist with the maintenance and rehabilitation of our waterways."

When deciding how healthy Moggill Creek is, many factors have to be investigated and recorded over a period of time. A healthy river has high amounts of dissolved oxygen and low amounts of phosphorous and nitrogen. It also has relatively low salinity. "We have to combine all of these different chemical and biological measurements to come up with an indication" said Mrs. Jones.

Below: A stretch of Moggill Creek where students from ISS collected valuable data for the study.

Apart from the water quality components, other factors help to determine how healthy the creek is. "Putting the chemical data together with the information that we gather about the fish and macro-invertebrates tells us how healthy the river is at that particular moment" said Mrs. Jones.

[1] *The activities displayed in each appendix were developed by Lyn English and James Watters with assistance from Jo Macri*

Mrs Jones said "We know that certain macro-invertebrates are highly sensitive to pollution whereas others have low sensitivity and survive in polluted water. By counting the different types of macro-invertebrates and knowing their sensitivity we can work out the pollution index of a creek."
The river is considered clean and healthy when the river has lots of different species of water bugs, particularly highly pollution sensitive macro-invertebrates.
Obtaining samples is a time-consuming job. Having students gather data from different sites along Moggill Creek helps the department identify which areas are healthy and which sites need the most help in getting cleaned up.
While the children are supporting the local community through their investigations, they are also learning about river ecology and why it is important to keep pollution down in waterways.

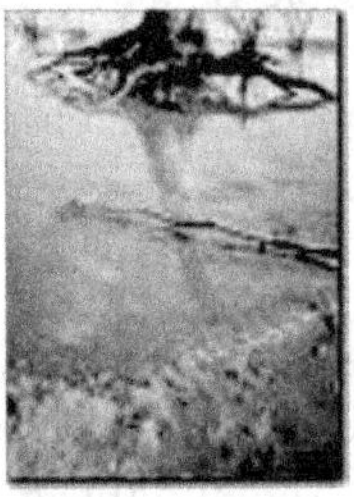

Left: A creek highly polluted showing an algal bloom

READINESS QUESTIONS

1. What are macro-invertebrates? (draw a picture of an example)
2. Why does the EPA want students to help collect data for them about the conditions of Moggill Creek?
3. When scientists want information on "dissolved oxygen, phosphorous and nitrogen" what are they looking for?
4. What levels of dissolved oxygen, phosphorus, nitrogen, and total salinity does a healthy river have?
5. Why is it important to know both the number of organisms and the amount of each species that a creek site has?
6. Why is the students' involvement in the creek monitoring task a good idea?
7. What is the pollution index of a creek and how do you calculate it?

MODELING ACTIVITY

Jack Simpson's class is presenting their information at a community meeting where other interested groups are presenting their conclusions. The meeting organisers are offering a prize for the group that develops the best system that describes the most important criteria in establishing the total water quality of a creek.

Jack's group needs your help to construct a model or set of guidelines that indicates the health of the creek. Your system should make use of all the data collected over the year in five locations along the creek. These data are shown in the Table attached. Jack's class started collecting near the source of the Creek (Site E) and took samples all the way to the mouth of the Creek where it entered the Brisbane River (Site A)

Write a letter to Jack's group that describes how you developed your system so that it can be used by others in determining the health of any creek.

Appendix A(2) Explanation of Pollution Index

Pollution index[2]

There are a number of different things that can pollute water and consequently affect the distribution of macro-invertebrates. Pollutants include domestic waste and animal wastes (e.g. from paddocks, dairies, horse stables and yards.) These wastes can contribute to the development of toxins, bacteria, and viruses. They enter water courses through run off, or seep in through ground water. The quality of the water can be determined by calculating a pollution index.

Pollution index

Macro-invertebrates can be divided into three groups according to how sensitive they are to pollution and assigned a number related to their group:

Sensitive	5-10
Tolerant	3-4
Very Tolerant	1-2

Each animal has a number or score next to it in the water bug table.

When you have completed the collection and identification, add the numbers assigned to each animal. For the index, only count each type of animal once. Clean water will have a high total score because it can support a lot of pollution sensitive bugs.

High abundance of only a few species might indicate poorer conditions.

[2] Water and Rivers Commission of Western Australia:
http://www.wrc.wa.gov.au/public/waterfacts/2_macro/water_condition.html

Appendix A(2) Fact Sheet for the Water Watch problem

Fact Sheet

Water quality is determined by measuring the presence of macro-invertebrates in the water, the presence of fish, algae and weeds, and the concentration or amounts of various chemicals.

Macroinvertebrates are animals without backbones that are big enough to see with the naked eye. Examples include most aquatic insects, snails and crustaceans such as yabbies. Some macro-invertebrates tolerate polluted water well while others are very sensitive to pollution. Thus the diversity of macro-invertebrates and types of species are good indicators of the quality of water in a creek. By counting the different species of macro-invertebrates Scientists have devised a pollution index to describe the state of pollution of the creek.

Fish: There are many hundreds of species of native Australian freshwater fish. Some species have been around for over 60 million years and have become adapted to the unique Australian conditions. However many species are now threatened by the introduction of Exotic or Alien fish. ***Exotic species*** are those species brought into Australia from a foreign country mostly as aquarium fish. These have been released into the wild from home fish tanks. Common exotic species are Guppies, Swordtails and Platys. Although these species are small fish, they compete for available food resources to the disadvantage of some native fish. The Mosquito Fish (*Gambusia holbrookii*) was introduced in the past in the hope of providing insect pest control services but it has proven itself to be a voracious controller of tadpoles, too. The health of our creeks is indicated by the abundance and diversity of native fish.

Plants: Various plants grow in and around Queensland creeks and billabongs. Plants are important as they produce oxygen which dissolves in the water and helps fish and animals survive. Bulrushes and reeds are common indigenous plants which have root systems whereas water lilies are floating plants. Unfortunately there are many introduced exotic species. Exotic species of plants have caused havoc in some areas. Waterlettuce, alligator weed, fanwort (Camomba), water hyacinths, and salvinia are particular unwelcome weeds because they grow rapidly and choke out native plants and fishes. Chinese elm and Camphor Laurel are also considered as weeds. Others such as cape blue water lily are useful as they contribute to the health of the water way by stimulating oxygen production.

Algae: Blue green algae are an important part of a healthy creek. Although quite rare, some species have the potential to produce toxins in low quantities which kill fish and can harm animals and swimmers. In normal conditions these plants and algae exist in low numbers in the waterways at no detriment to the environment or to human health. However under favourable conditions native plant species can be out competed. Weeds and algal colonies can over grow natural habitats. In large numbers this harms waterways attacking native aquatic plants and animals.

Chemicals: The presence of various chemicals and matter is an important indicator of water quality. Pure water is H_2O. Chemicals from the soil, such as sodium chloride, contribute to salinity. Thus sea water is described as salty because it contains lots of common salt or sodium chloride. Two important chemicals are phosphorus and nitrogen. Compounds containing these chemicals often contribute to the growth of phytoplankton which are microscopic animals that form algae blooms. After storms, saltwater swimming pools either overflow or their owners drain them to lower water level. If this coincides with reduced stream flow the salinity can rise markedly.

Phosphates are common polluting chemicals found in fertilizers. So run off from farms and gardens add excess amounts of phosphorus to the water. However, phosphates also come from the natural weathering of rocks and the decomposition of organic material. Sewerage is rich in phosphorus. The higher the level of phosphate the poorer the quality of water.

Nitrogen containing substances also occur naturally in water however human activity contributes excess nitrogen to our waterways. Nitrogen is usually measured as nitrate and occurs in nitrates, nitrites and ammonia. Nitrates occur in fertilizer and hence run off containing nitrates from farms and gardens leads to a proliferation of plant life which competes with other life. The higher the level of nitrogen the poorer the quality of water.

Oxygen: Oxygen is essential for all animals. Oxygen is what we breathe from the air. Dissolved oxygen is probably the most critical water quality variable in freshwater creeks. Fish need oxygen. Oxygen levels in creek systems depend on water temperature, salinity, and the amount of aquatic vegetation and number of aquatic animals in the creek. It also depends on the flow rate of the creek and the extent to which the water gets churned up in passing over rocks. When there is too much growth of microscopic algae (phytoplankton) and other oxygen consuming micro-organisms larger species of animals die off. High levels of oxygen are important for good quality water.

T**urbidity**: The clarity of water is also a useful indicator of quality although being cloudy does not necessarily mean the water is polluted. Muddiness or *turbidity* could be caused by silt or dirt but it could also be caused by bacteria and algae. Turbidity prevents light from getting to plants to help them grow. High levels of turbidity are detrimental.

A**cid:** pH is the measure of the amount of acid in water. pH describes the acidity of water and low pH means highly acidic. Vinegar and lemon juice are acidic and have low pH whereas soap and washing powder are described as alkaline – thc opposite of acidic – and have high pH. The strange thing is that the more acid the lower the pH. Normal pH is about 7.0 and most organisms can survive between pH6.5 and pH9.0. Acid gets into creeks through increased pollution from burning fossil fuels. Alkaline materials are often leached from rocks that the creek passes over.

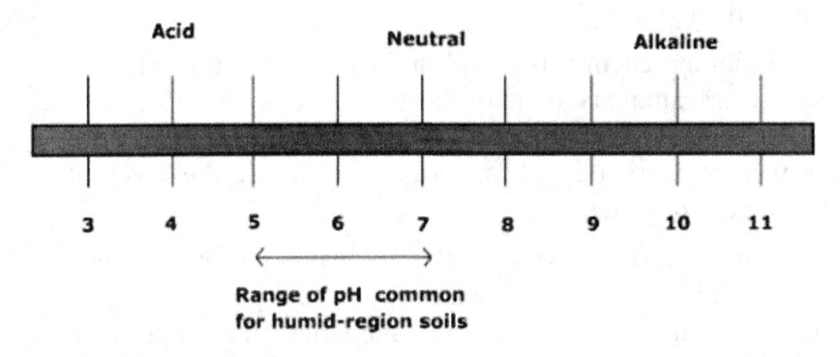

Weeds: Invasive weed infestations crowd out and smother native plants and hence have an adverse impact on the health of the ecosystem. Native butterflies, birds and other insects rely on native plants. These exotic invasive weed are distributed by wind, birds, or as the result of dumping of garden rubbish. Poor quality water ways have dense growths of weeds.

Appendix A(3) Water Bugs Identification Chart

Water Bugs Identification Chart

The health of a stream is given in terms of the diversity of species and the relative tolerance level of various species.

Species	Size	Features	Drawing	Score
High Sensitivity				
Caddisfly larvae	Up to 20mm	They live in a wide range of environments from fast flowing streams to freshwater ponds. Their soft bodies are usually covered in a protective silky case		6
Dragonfly nymph	18-50mm	Dragonfly Nymphs are short and chunky with wing pads and internal gills. Their six legs are all located near the head		6
Mayfly Nymph	10-20 mm	These are only found in very clean water containing lots of oxygen. They absorb oxygen from the water through their gills.		7
Stonefly nymph	Up to 50 mm	They have three segmented tarsi and long antennae. Require lots of oxygen		8
Water mite	5mm	Adults are free-swimming or crawling. Most common in heavily vegetated wetlands – often parasitic on other insects		5
Medium Sensivity				
Fairy shrimp	10-30 mm	Related to brine shrimp, copepods and Daphnia		3
Freshwater mussel		Mussels are soft bodied animals enclosed in two hinged shells		3
Leech	3-15mm	Leeches are segmented worms with a sucker on one or both ends. They are found free swimming in water as well as on plants or on the bottom.		3

Planarian	2-5mm	These are free living flat worms which possess a remarkable ability to regenerate their lost body parts		3
Pond snail	10-20mm	Aquatic snails are similar in form to land snails but smaller		3
Water boatman		Water boatmen and water striders are bugs. These tend to be shield shaped when viewed from above. Their soft front wings are folded and overlap to leave a small triangle on their back.		3
Water strider		Water boatmen and water striders are bugs. These tend to be shield shaped when viewed from above. Their soft front wings are folded and overlap to leave a small triangle on their back.	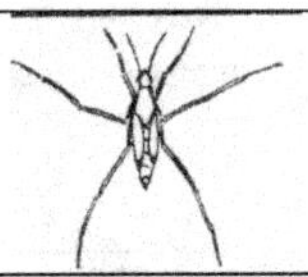	3
Water tiger beetle		Beetle larvae are segmented, have three distinct pairs of legs. They are usually active with large mouth parts.		3
Waterflea	1 mm	Also known as daphnia these are related to crabs and prawns.		3
Whirligig beetle	3-35mm	They congregate in large numbers and scurry about the water surface in a random pattern. Shiny to dull black, often with a bronzy sheen.	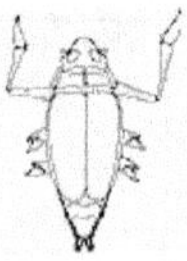	3
Shrimp		Shrimp are small crustations that look similar to prawns.	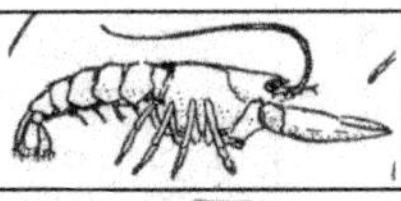	5
Yabby		Freshwater crayfish that are commonly found in ponds and streams.	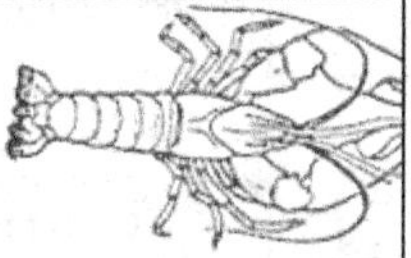	5
Low sensitivity (Tolerant)				
Water Scorpion	30mm	Noted for the first pair of legs which are modified into prehensile organs for grasping prey. They are carnivorous and feed on smaller insects. The prey is held securely between their first pair of legs while the water scorpion sucks up its body fluids. Tends to be found on the muddy bottom of creeks.	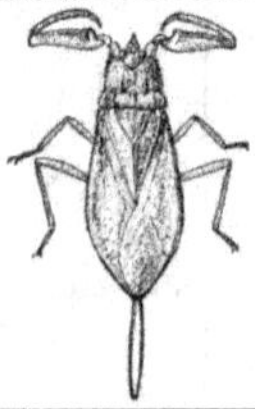	2

Midge		These are small pesky biting insects as adults but are slender worm-like creatures, sometimes red, with no legs.	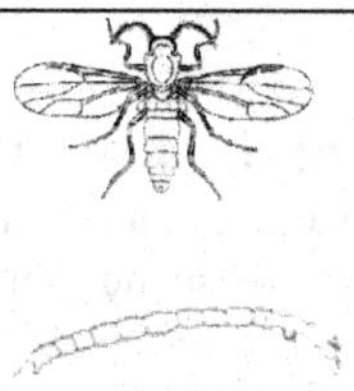	2
Mosquito lavae		These animals twist and squirm just below the surface of the water. The larvae look like hairy maggots with siphons.	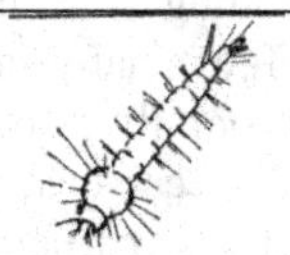	2

APPENDIX B(1)

SUMMER READING PROBLEM

Brisbane – While a long hot summer may be ahead of us, the Brisbane City Council Library (BCCL) is offering a chance for patrons to stay cool this year.

The annual "Reading is Cool" summer reading program will officially start at noon, June 1, in the Indooroopilly Room. Students from St. Peters will receive a free library card that will let them participate in the program.

Students can choose from an approved collection of books that the library has placed on reserve. The books have been classified by grade level (according to difficulty of the book), to help the students choose which books to read. However, students may read any of the books, regardless of their current grade level.

St. Peters students who participate will have the chance to not only earn prizes from the library, but also prizes from their school. The St. Peters School and the BCCL have teamed up to provide prizes for overall winners and classroom winners.

Some prizes that the students can win, based on a point system, include bookmarks, books, T-shirts, hats, meals from local restaurants, and compact discs. Classroom winners will also be eligible for a chance to win a $300 savings account.

To register, simply stop by the Brisbane City Council Library, 318 Moggill Rd, between 9 a.m. and 9 p.m. The contest ends Aug. 12, with the final day to collect prizes Aug. 25.

The library is accepting suggestions for this year's reading contest. To give your input, please send a letter to Lynn, the Indooroopilly reading coordinator. All suggestions must be received by May 1.

Ready to Go: The books are all shelved at the Brisbane City Council Library in Indooroopilly. Participating students can choose from over 250 books for extra bonus points in this year's summer reading program.

Summer Reading Program Readiness Questions

Answer the following questions using the journal article and the tables given below.

1. Drew read The Tell-Tale Heart and Roll of Thunder, Hear My Cry. Should he receive the same number of points for each book? Why or why not?

2. If a sixth grader and a ninth grader both read A Tale of Two Cities, should they both earn the same number of points? Why or why not?

3. If Shelly reads Jurassic Park and Much Ado About Nothing, should she get the same number of points for each?

EXAMPLES OF APPROVED BOOKS

TITLE	AUTHOR	READING LEVEL (BY GRADE)	PAGES
Sarah, Plain and Tall	Patricia MacLachlan	4	58
Are You There God? It's Me Margaret.	Judy Blume	4	149
Awesome Athletes	Multiple Authors	5	288
Encyclopedia Brown and the Case of Pablo's Nose	Donald J. Sobol	5	80
Get Real (Sweet Valley Jr. High, No.1)	Francine Pascal, Jamie Suzanne	6	144
Roll of Thunder, Hear My Cry	Mildred Taylor	6	276
The Tell-Tale Heart	Edgar Allan Poe	6	64
Little Women	Louisa Mae Alcott	7	388
The Scarlet Letter	Nathaniel Hawthorne	7	202
Aftershock (Sweet Valley High)	Kate Williams, Francine Pascal	8	208
Jurassic Park	Michael Crichton	8	400
A Tale of Two Cities	Charles Dickens	9	384
Lord of the Flies	William Golding	9	184
Much Ado About Nothing	William Shakespeare	10	75

TITLE	*BRIEF DESCRIPTION OF BOOK*
Sarah, Plain and Tall	When their father invites a mail-order bride to come and live with them in their prairie home, Caleb and Anna are captivated by their new mother and hope that she will stay.
Are You There God? It's Me Margaret	*Faced with the difficulties of growing up and choosing a religion, a twelve-year-old girl talks over her problems with her own private God.*
Awesome Athletes	*Sports Illustrated for Kids*
Encyclopedia Brown and The Case of Pablo's Nose	*America's Sherlock Holmes in sneakers continues his war on crime in ten more cases.*
Get Real (Sweet Valley Jr. High, No. 1)	*Describes the trials and tribulations of twins that moved to a new junior high school.*
Roll of Thunder, Hear My Cry	*A black family living in the South during the 1930s is faced with prejudice and discrimination that its children do not understand.*
The Tell-Tale Heart	*The murder of an old man is revealed by the continuous beating of his heart.*
Little Women	*A story of family, of hope, of dreams, and of growing up as four devoted sisters search for romance and find maturity in Civil-War era 19th century New England.*
The Scarlet Letter	*Hawthorne's masterpiece about Hester Prynne, hapless victim of sin, guilt and hypocrisy in Puritan New England.*
Aftershock	*Twins deal with the pain and shock of an earthquake.*
Jurassic Park	*A modern-day scientist brings to life a horde of prehistoric animals and dinosaurs.*
A Tale of Two Cities	*A highly charged examination of human suffering and human sacrifice, private experience and public history, during the French Revolution.*
Lord of the Flies	*The classic tale of a group of English school-boys who are left stranded on an unpopulated island.*
Much Ado About Nothing	*Shakespeare comedy.*

Summer Reading Problem

Information: The Brisbane City Council Library and St. Peters School are sponsoring a summer reading program. Students in grades 6-9 will read books and prepare written reports about each book to collect points and win prizes. The winner in each class will be the student who has earned the most reading points. The overall winner will be the student that earns the most points. A collection of approved books has already been selected and put on reserve. See the previous page for a sample of this collection.

Students who enroll in the program often read between ten and twenty books over the summer. The contest committee is trying to figure out a fair way to assign points to each student. Margaret Scott, the program director, said, "Whatever procedure is used, we want to take into account: (a) the number of books, (b) the variety of the books, (c) the difficulty of the books, (d) the lengths of the books, and (e) the quality of the written reports.

Note: The students are given grades of A+, A, A-, B+, B, B-, C+, C, C-, D, or F for the quality of their written reports.

Your Mission . . .

Write a letter to Margaret Scott explaining how to assign points to each student for all of the books that the student reads and writes about during the summer reading program.

Table 1 Representation Formats, Problem Factors, and Mathematical Operations used by each of Five Student Groups in Model Development

Group	Representation Format					Problem Factors						Mathematical Operations				
	Table	Text	List	Computation	Formulae	No. of books	Variety	Length	Student's grade level	Book's reading level	Report's quality	Assigning value points	Using interval quantities	Using weighting	Aggregating quantities	Informal rate
1		✔	✔		✔	✔			✔			✔	✔			
2		✔	✔	✔	✔			✔		✔	✔	✔	✔	✔	✔	
3		✔	✔		✔				✔	✔		✔				
4		✔			✔	✔	✔	✔	✔	✔	✔	✔	✔	✔		✔
5	✔	✔							✔	✔	✔	✔	✔			✔

LEARNING MATHEMATICS THROUGH APPLICATIONS BY EMERGENT MODELING: THE CASE OF SLOPE AND VELOCITY

L.M. Doorman & *K.P.E. Gravemeijer*

Freudenthal Institute for science and mathematics education
Utrecht University, The Netherlands

In this paper we focus on an instructional sequence to support a reinvention process by the students. The conjectured process of teaching and learning is supposed to ensure that the basic principles of calculus and kinematics are rooted in the students' understanding of everyday-life phenomena. Students' inventions are supported by carefully planned activities and tools that fit their reasoning. The central design heuristic of the instructional sequence is emergent modeling. With this sequence we create an educational setting to investigate tenth grade students' learning. Classroom events and computer activities are video-taped, group work is audio-taped and student materials are collected. The data is analyzed with a qualitative grounded theory method. We conclude that with the emergent modeling approach students can reinvent mathematics in connection with physics supported by consecutive manifestations of graphical models of motion.

1. INTRODUCTION

The content aimed at in this report is calculus and kinematics. These two topics originated from observing and organizing motion phenomena. Knowledge of how we might symbolize motion is used to understand how the related concepts and inscriptions were invented, and to obtain clues for a local instruction theory. Especially for mathematics education, we need to apply such knowledge to prevent students from acquiring an instrumental use of mathematical symbols without understanding the represented concepts. In addition, domain specific instruction theories and research on the use of computer tools in mathematics education were used to develop a scenario for the teaching and learning of calculus and kinematics.

Traditionally graphs play an important role in the teaching of calculus and kinematics. Often, distance-time graphs are used to give meaning to the difference quotient as a measure of velocity. This presupposes that students understand the relation between velocity and distance traveled. However, this relationship is taught in physics education with the use of graphs and the difference quotient. We will argue that an overestimation of the explanatory power of graphs may be the cause of many problems with the learning of these topics for students. We will suggest an approach of modeling motion where students are involved in a development from informal inscriptions to formal representations. This approach and the role of models and modeling in the learning process will be addressed in this paper. More specifically, the instructional design heuristic, 'emergent modeling' that is part of the domain-specific

B.Sriraman, C.Michelsen, A. Beckmann & V. Freiman (Eds). (2008). *Proceedings of the Second International Symposium on Mathematics and its Connections to the Arts and Sciences* (MACAS2). Information Age Publishing, Charlotte: NC , pp.37-56

instruction theory for realistic mathematics education, RME, will be the central topic of discussion.

We designed an instructional sequence for the teaching and learning of calculus and kinematics based upon emergent modeling. We created an educational setting with which we could understand and improve the learning processes and the means to support them. A research approach that consists of planning and creating innovative educational settings for analyzing teaching and learning processes is design research. The instructional sequence was investigated according to a design research approach in two tenth-grade classes. The interpretative framework for the teaching experiments was primarily an instructional design perspective inspired by the realistic mathematics education. Interpretations of classroom events guided the instructional design decisions and aimed at understanding and improving the teaching and learning processes. We analyzed the role of the teacher and of the computer tools in the teaching and learning processes.

We especially focused if and how tool-use supported students in making the desired steps on the learning route. This analysis offers insight in the possibilities of an approach to calculus and kinematics that prevents the described problems with graphs. It appears that students develop language and reasoning during the modeling activities, and that this development can concur with a development in understanding mathematical and physical representations and concepts.

2. THE DIDACTICAL PROBLEM

For many years, research has shown that students who followed calculus and physics classes have problems with interpreting graphical situations like

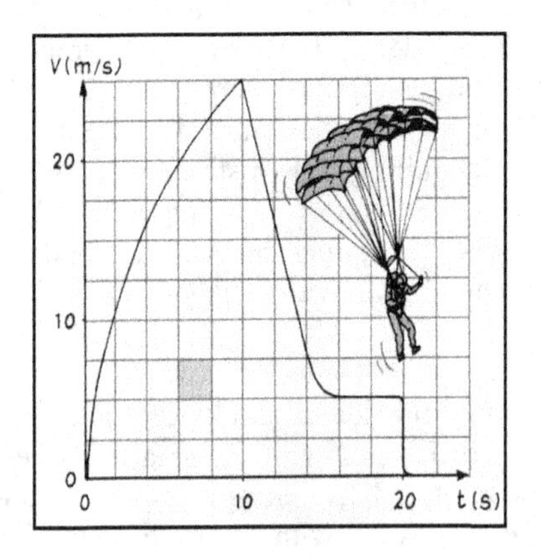

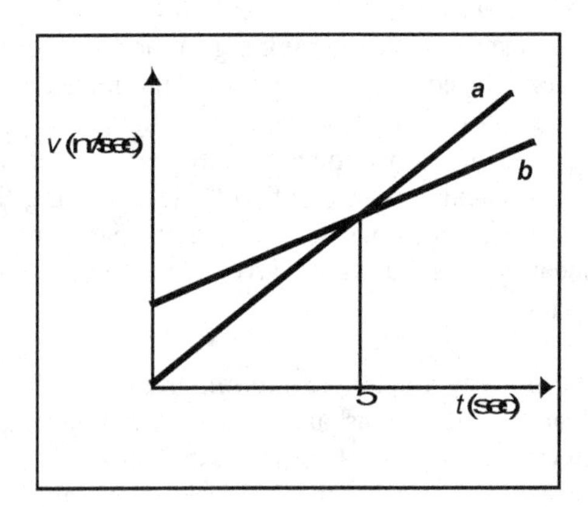

Figure 1: Two examples from Dutch textbooks

Student reactions on the left picture are:

"After 10 seconds he is on his highest point . . ."

"The falling distance is 50 squares, which is 250 cm2."

A reaction on the right picture is:

"After 5 seconds ***a*** catches up with ***b***"

In calculus and kinematics courses, a lot of attention is paid to how to do the calculations with, and manipulations of formulas, instead of attention to why they work (Clement 1985, Dall'Alba e.a. 1993). This may be one of the causes of the above mentioned problems of students. A consequence is that education focuses on seemingly straightforward manipulations of formulas, while the concepts are graphical and not trivial.

We first argue that model building is essential for understanding and does not receive enough attention in education. Although the importance of modeling activities often is stressed, students are hardly involved in the building of the models, their purposes, conventions, representations and their meanings in terms of situations being represented and problems being solved.

Secondly, for investigating how students can participate in a modeling process where those models are built, a learning trajectory is presented. The idea is that students get the possibilities to invent representations and meanings and they are guided by the teacher and by carefully introduced models. This should result in the development of concepts which are rooted in the students' common sense reasoning, and should prevent the before mentioned wrong interpretations.

3. THEORETICAL FRAMEWORK

Continuous velocity-time and distance-time graphs are – often theoretical – descriptions of a situation for math¬ematical reasoning. Their appearance and conventions (e.g. time-axis horizontal) are the result of a long period of scientific research on the calculus of change. During this period – prior to the continuous graphs – other models were developed and used for different purposes. After a period of almost 2000 years the graphs that we use nowadays were created (e.g. Clagett 1959). These continuous time graphs are models to be used for reasoning about motion with graphical properties like area and slope. Our claim is that many problems of students with calculus and kinematics are due to the fact that they don't really understand why they can use these graphs for their reasoning.

We investigated an alternative approach that aims at a process in which the mathematics stays related to their understanding of the physical properties of motion, and emerges from the modeling activities of the students. This is also an objective of realistic mathematics education, where instructional design is aimed at creating optimal opportunities for the emergence of formal mathematical knowledge. During this process students can preserve the connection between the mathematical concepts and what is described by these concepts. The students' final understanding of the formal mathematics should stay connected with, or as Freudenthal would say, should be "rooted in", their understanding of these experientially real, everyday-life phenomena (Freudenthal, 1991). The core mathematical activity is "mathematizing", which stands for organizing from a mathematical perspective. Freudenthal sees this activity of the students as a way to reinvent mathematics. However, the students are not expected to reinvent everything by themselves. In relation to this, Freudenthal (1991) speaks of guided reinvention; the emphasis is on the character of the learning process rather than on the invention as such. The idea is to allow learners to come to regard the knowledge they acquire as their own private knowledge, knowledge for which they themselves are responsible.

In a reinvention approach, the problem situations for the students play a key role. Well-chosen context problems offer opportunities for the students to develop informal, highly context-specific models and solving strategies. These informal solving procedures then may function as foothold inventions for formalization and generalization, in other words: for progressive mathematization. The instructional designer tries to construe a set of problems that can lead to a series of processes that together result in the reinvention of the intended mathematics. Basically, the guiding question for the designers is: How could I have invented this?

Research on the design of primary-school RME sequences has shown, that the concept of emergent models can function as a powerful design heuristic (Gravemeijer, 1998). First, context problems that offer the students the opportunity to develop situation-specific methods are selected. Second, these methods are modeled during classroom discussions and subsequent activities. In this sense, the models emerge from the activity of the students. Even if the models are not actually invented by the students, great care is taken to approximate student invention as closely as possible by choosing models that link up with the learning history of the students. Another criterion is in the potential of the models to support mathematization. The idea is to look for models that can be generalized and formalized to develop into entities of their own, which as such can become models for mathematical reasoning (Gravemeijer & Doorman, 1999). The shift from informal *models of* realistic situations to *models for* mathematical reasoning concurs with a shift in the way the student thinks about the model, from models that derive their meaning from the modeled context situation, to thinking about mathematical relations. In this context the term 'model' must be understood in a broad sense. It is not just the physical representation, but everything that comes with it (e.g. activity and purpose) that constitutes the model (Cobb 1999). As a consequence, during the activities of the students the model and the situation being modeled co-evolve. Modeling in this view is a process of reorganizing both activities and the situation. The situation becomes to be structured in terms of mathematical concepts and relationships.
The theoretical foundation of this research concerns also the use of computer minitools in mathematics education. In general, tools influence the process of students' mathematical sense making. Cobb (1999) illustrated this by describing an interplay between the students' ways of symbolizing and the development of mathematical meaning. In relation to a series of computer tools in teaching materials for statistics, he mentioned the notion of affordances. These didactical minitools afforded the students' reasoning on statistical problems. However, this appeared not to be a trivial process. The students have to see and understand the tool as an affordance for a specific problem. To achieve this, instructional designers should take into account how students might use and reason with the tools in the software as they participate in a sequence of mathematical practices. Moreover, teachers need to maintain a delicate balance between constructive activities at the computer and reflections upon these activities (Hoyles and Noss, 2003).

We adopted these ideas for integrating two computer minitools in the instructional materials for the teaching and learning of calculus and kinematics. The tools were supposed to create opportunities for the students to trace subsequent positions and to picture displacements and total distances traveled in two-dimensional graphs. We conjectured that the tools supported students in organizing motion with graphs, in finding patterns, and in developing physical and mathematical concepts related to motion and change.

This use of tools differs from an approach in which the primary goal of the activities is that students come to understand the external representations in the tool as a result of dynamic linkages in the software between these representations and known phenomena like realistic animations (Doerr 1997, Doorman 2002). The difference can be referred to as an emphasis on expressive modeling in our case versus explorative modeling in the other approach. This difference has many similarities with the dichotomy between providing models and designing models in co-construction (Van Dijk et al., 2003).

Finally, research into the use of hand-held calculators and computer tools also points at the importance of the teacher's role. The process of tool appropriation and learning mathematics has both an individual and a collective aspect, and needs guidance by the teacher (e.g. Doerr & Zangor, 2000). As a consequence, we added to the emergent modeling heuristic the design heuristic of problem posing (Klaassen, 1995). The problem posing heuristic aims at suggestions for the teacher to evoke content-related motives for the way to proceed. These content-specific motives are supported by an overarching global problem and successive local motives related to the activities. The teacher should understand the relation between the global and the local motives to be able to deal with classroom situations, to filter and rephrase questions and suggestions and to frame the students' intentions.

We investigated the possibilities of a modeling process from informal and intuitive notions to the basic principles of calculus and kinematics: an approach that progressively builds on students' symbolizations. We conjectured that the design heuristics of emergent modeling and of problem posing have the potential for realizing the intended teaching and learning processes with computer tools.

4. METHODOLOGY

The aim of this research is to investigate (*i*) the process of students' learning in the domain of calculus and kinematics, and (*ii*) the means of supporting and organizing that learning process. To be able to answer the research question, we had to create an educational setting with which we could investigate to what extent and how this dialectic process of symbolizing and meaning making could be fostered. We designed an instructional sequence for creating this educational setting. Consequently, the aim is primarily the understanding of the learning processes in connection with this instructional sequence. The approach is characterized as design research and consists of planning and creating innovative educational settings for investigating teaching and learning processes.

This interpretation of design research has proved itself suitable for developing empirically grounded, local instruction theories in the areas of science and mathematics education (Gravemeijer, 1994; Klaassen, 1995; Lijnse, 1995). This approach aims at generating empirically grounded theories. The main result is not a design that works, but the rationales how, why and to what extent it works (Cobb et al., 2003; Edelson, 2002; Gravemeijer, 1994; Gravemeijer, 2004).

The first teaching experiment took place in two 10th grade classes in two comprehensive schools in provincial towns. For the first experiment we wanted to be able to triangulate data within one class, and to compare the two different schools. The comparison enabled us to observe teacher- and school-specific norms and procedures. The student activities and the guidelines for the teacher, together with our intentions, were discussed beforehand with the teachers of both schools in two meetings. During the experiments we made notes and audio-taped the lessons. During the computer lessons, one of the pairs was video-taped, and the classroom discussions were also video-taped. The pairs were selected with help from the teachers using the criteria of clear speech and capabilities varying from low to high achievers. We used the video-tapes to be able to analyze gestures and reasoning with graphs on the computer screen and on the blackboard. After the teaching experiments we collected the students' written materials. An analysis of these results was needed to investigate to what extent we had reached content-specific goals with all the students.

First, the analysis of these data provided information on how to optimize the activities with respect to formulating student texts, contexts used, and information provided. Second, conjectures that paralleled the instructional sequence could be verified as far as the students were taught as intended. For example, some students appeared to have difficulties with interpreting and using the graphs provided produced by the tools. This led to adjustments to the sequence and to a new research question for the second experiment. We conjectured that class discussions about the use of specific graphs would prepare all students for the computer lessons. As a result of this, we looked for ways to support such class discussions for reaching whole class consensus about a way to proceed with the computer tools. A consensus based upon the students' activities. The adjustments and the new hypotheses were objects of study in the second teaching experiment.

The second teaching experiment was confined to eight lessons in one 10th grade class. During these lessons we focused our data collection on adjusting the instructional sequence and the tools and computer tools used. As a result of this focus, we expected one classroom experiment would provide us with enough data for analyzing the teaching and learning processes in the specified situations. This experiment took place in a comprehensive city school with a teacher who was experienced in discussing mathematical problems with students without presenting them with the intended approach or answer. We expected him to understand the teacher role we aimed at in our approach, but still took more time preparing the experiment with this teacher. We discussed the planning of the activities and especially the scope of the classroom discussions.

We wanted to analyze the development in reasoning shown by both weak students and high achievers with the computer tools. We audio-taped a weak pair, an average pair, and a high achieving pair during the computer lessons, and video-taped the average pair. Classroom discussions were video-taped and field notes were made as in the first teaching experiment. In these experiments we expected to collect enough information to reconstruct and to understand the classroom process of the teaching and learning of calculus and kinematics.

The data were organized into case studies of class discussions and of students' work during the computer lessons. We interpreted these case studies in terms of what preceded the lessons, the student activities, the teaching, and the tools provided. Interpretations were compared with other available data, such as students' written materials and data from another experiment in our research project.

The interpretative framework for the teaching experiments was primarily an instructional design perspective. The theory of realistic mathematics education and theories about how humans symbolize guided interpretations of classroom events and aimed at understanding and improving the teaching and learning processes. In addition, we tried to incorporate both a social and an individual perspective. We tried to assign such meanings to students' expressions that they came out as consistent with their history. The behavior and the students' contributions were related to classroom norms and students' beliefs about what was expected from them (Cobb, Yackel & Wood, 1992). For a more detailed discussion of the methodology and its validity see (Doorman, 2005).

5. THE INSTRUCTIONAL DESIGN

We designed an instructional sequence in which discrete graphs should come to the fore as *models of* changing location of a hurricane, and develop into *models for* reasoning about the relation between displacements in time intervals and total distance traveled. A situation in which it makes sense to describe modeling motion seems the weather forecast, especially the change of position of hurricanes: when will it reach land? This problem is posed as a leading question throughout the unit as a context for the need of grasping change (see Figure 2).

After being introduced to time series, students work with situations that are described with stroboscopic photographs. The idea is that students come up with measurements of displacements as basic structure elements of motion and that it makes sense to display them graphically (based upon Boyd et al. 1996). The key issue that should arise in the discussion is how to describe change (of position) in order to view patterns and to be able to do predictions.

The introductory part of the instructional sequence comprised one lesson. Predicting and describing change were introduced in the context of weather. The central question for the students was: how can you describe change so you can make predictions?

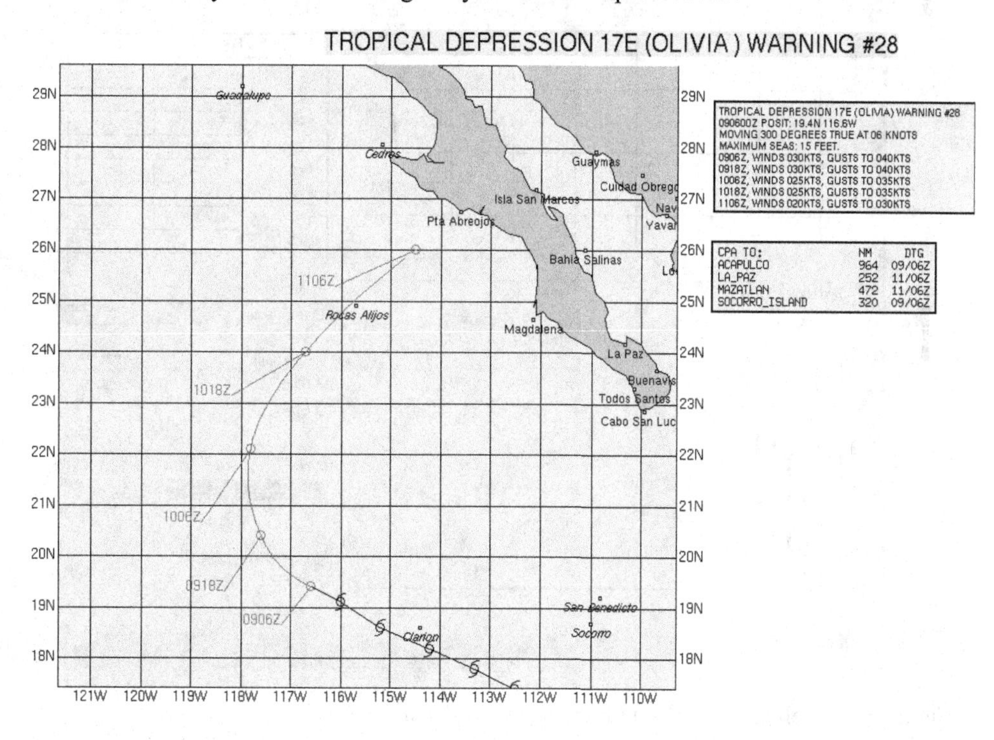

Figure 2: A tropical storm approaching the coast

A representation of a time series is a trace graph of points and connecting lines. This representation in the weather context is the starting point for understanding the importance of gaining insight into patterns of trace graphs for doing predictions. The class discussion based on students' reasoning should result in a consensus on proceeding with two-dimensional displacement graphs and total distance-traveled graphs. To achieve the transition to reasoning with 2-dimensional graphs of motion we designed an open-ended activity about a falling ball before the computer lesson.

In the computer activities we gave students the opportunity to investigate various situations with the computer program Flash. We wanted them to construct the relation between patterns in trace graphs and graphical characteristics during their investigations. More specifically, the use of the computer tool should enable the students to invent properties like the relation between average displacement and total distance traveled, and to find the relation between the linearity of a distance traveled graph for a motion with constant displacements.

For instance, the students could click on successive positions of an object in a stroboscopic picture. Simultaneously, the program shows the distances between these positions in a table, and displays them in a displacement graph or in a graph of total distances. The values are displayed as vertical bars instead of dots to preserve the link with the displayed measurements.

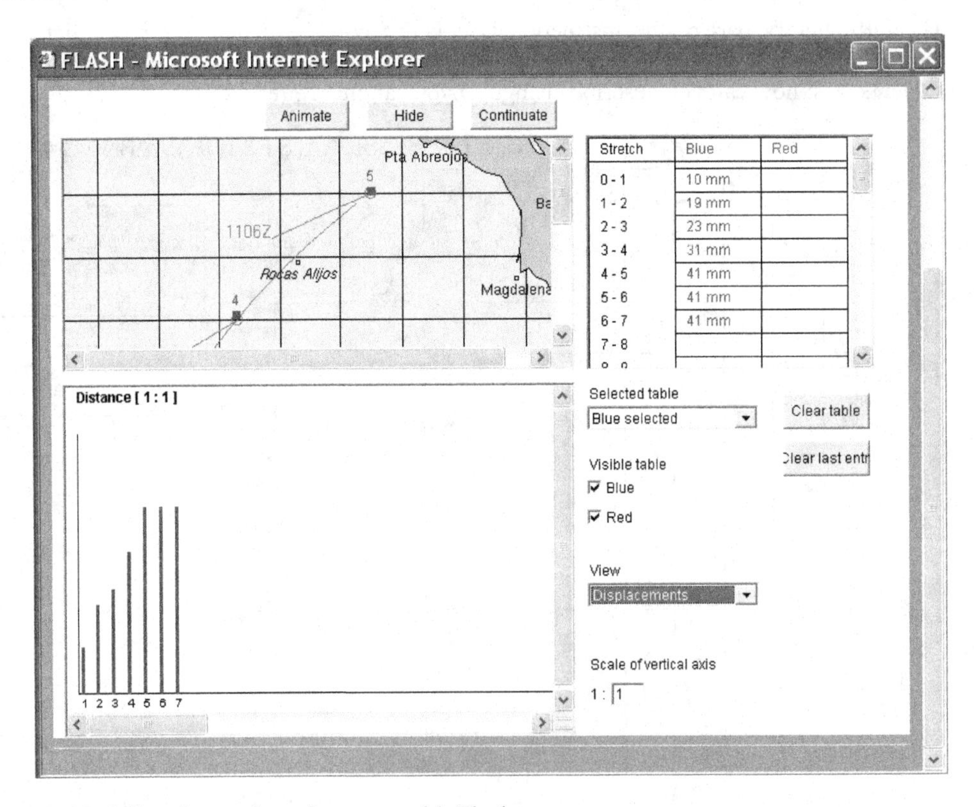

Figure 3: Modeling the motion of a storm with Flash

The clicking of successive positions signified measuring displacements in successive time-intervals. Activities for the students consisted of various situations in which they could click on successive positions and could reason with the table and the two graphs for solving the problem stated (e.g. "when will the hurricane hit the coast?").

The conjectured change in the students' reasoning from situation specific to reasoning about graphs should parallel a shift in the way the students think about the graphical model. A shift from a model deriving its meaning from the modeled context situation, to thinking about mathematical relations. First, these graphs are used to describe the situations and are related with measurements in the photograph. Second, the use of the graphs is dominated by thinking about graphical and conceptual relations between displacements and distance traveled (e.g. "a horizontal line of summit in the displacement graph signifies a constant velocity"). What used to be a record of measurements is now used as a tool for reasoning about graphical patterns. Note that a key element of the notion of reinvention is that the models first come to the fore as models of situations that are experientially real for the students. It is in line with this notion that discrete graphs are not introduced as an arbitrary symbol system, but as consecutive models of discrete approximations of a motion that link up with prior activities and students' experiences (see Figure 4).

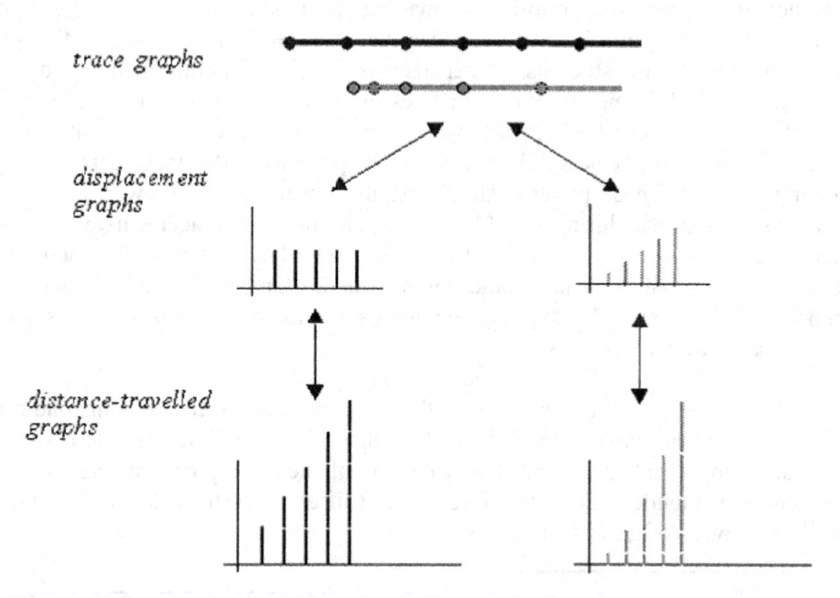

Figure 4: Consecutive manifestations of a model of motion

The starting point for the transition of displacements to modeling velocity, is in the medieval notion of instantaneous speed, which is introduced in the context of a narrative about Galileo's work. Instantaneous velocity will be defined in accordance with this medieval notion, in terms of the distance covered if the moving object maintains its instantaneous velocity for a given period of time. In this context, the problem is posed of how to verify Galileo's hypothesis on free fall: velocity increases constantly, and is proportional to time. Figuring this out demands of the students that they come to grips with the relationship between the motion, the representation, and the approximation. During this process, the way of modeling motion, and the conceptualization of the motion that is being modeled, co-evolve.

A shift is made from problems cast in terms of everyday-life contexts to a focus on the mathematical and physical concepts and relations. In order to make such a shift possible for the students, they have to develop a mathematical framework of reference that enables them to look at these types of problems mathematically (see also Simon 1995). It is exactly the emergence of such a framework that this sequence tries to foster. It is this framework that enables the students to trace the origin of the mathematical models and to anticipate on what is to come.

6. TEACHING EXPERIMENTS

The conjectured local instruction theory was elaborated in an instructional sequence for ten lessons including two computer lessons. This sequence was tested in two teaching experiments.

As a result of the character of the introductory class discussion we were of the opinion that the importance of a graphical description for making predictions had started to become clear to the students. Students contributed to the idea that time series play a role in this process. With the contributions of the students, consensus was established about the model of the time series: the trace graph. In their reasoning, changes in velocity signified changes in lengths of displacements in subsequent time intervals. During the class discussion about the introductory activities the teachers didn't focus on the patterns in displacements as a problem for predicting weather. These patterns should result in a motive to draw two-dimensional graphs. We concluded that during this lesson the patterns in displacements and the use of two-dimensional graphs were not successfully addressed. The aim of the second lesson, a computer lesson, was that students should begin to understand the need to display patterns and to understand the relation between these patterns and the characteristics of displacement graphs and distance-traveled graphs.

As a result of the open-ended activity before the computer lesson, the students had a means available for organizing and thinking about the subsequent problem situations. During the activity about seeing a pattern in the displacements of the falling ball on the stroboscopic photo, we saw the intended diversity of solutions (different vertical and horizontal axes), including the two types of graphs from Flash.

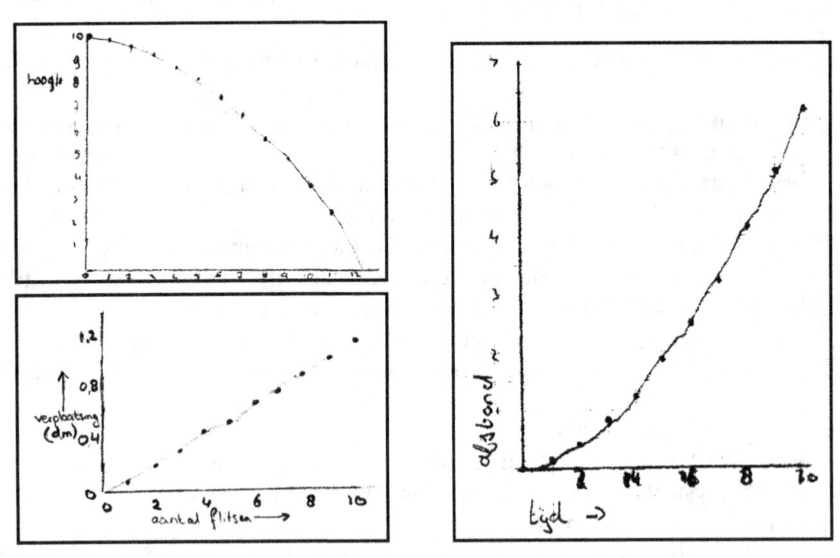

Figure 5: Graphs by students describing free fall

The teacher asked the students to show their solutions and discussed the differences between the students' solutions. Many students were involved and seemed to reach a consensus with the teacher about the usefulness of the two discrete graphs as tools for describing and predicting motion phenomena. This was a productive preparation for the computer lesson with the computer program Flash.

Measuring by clicking and seeing the graph appear did focus them on change in displacements and reasoning with the Flash-graphs. Due to this building upon stroboscopic pictures and trace graphs, the process of learning did not remain to trial and error. Most students deduced that the lengths of the vertical bars signified displacements on the trace graph.

At the beginning of the computer activities, the students' language primarily referred to successive positions in the stroboscopic photograph. As the lesson progressed, their language increasingly involved characteristics of the graphs of displacements and of distance traveled. The students appeared to develop insight into the relation between a graph of constant displacements and constant velocity and the accompanying linear graph of distance traveled. However, it was unclear for us how explicitly this took place. The only indications we could find in the records concerned the participation of an observer who requested clarification about what the students thought. After the computer lesson the teacher discussed the activities of the students.

Two students work on an activity about a thrown and rotating stick with Flash. The problem is how the distances traveled of the endpoint and the midpoint of the stick vary (see Figure 6).

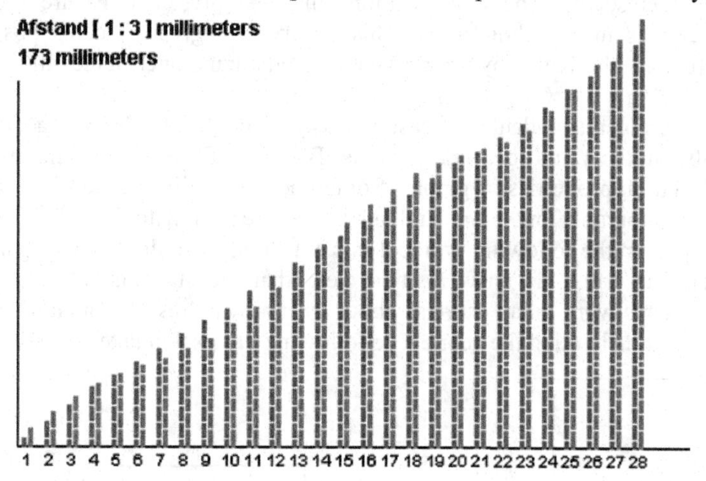

Figure 6: Graphs of midpoint and of endpoint of a rotating stick

The observer (O) questions the two students (J&M).

J: That's right, then, hey. What does it mean then, that the red is wavy?
O: That's the distance between the dots.
M: That sometimes the ends go faster and sometimes slower.
O: Yes. (turns to J) Do you see that too?
J: Yes.
O: Yes, that's what you really see in the photo, that - at least can you see that in the photo that the end sometimes goes faster? (-) How can you see that in the photo?
M: Then there's less space between two sticks. I mean, between two photos.
O: If it goes faster?
M: If it goes slower. (J confirms this)

O: (...) can you tell, from this table, whether the center moves further than the end in total?
M: No, the red does. [the end points]
O: What makes you think that?
M: Why do you think that? Well, go down a time - [J goes to the graph with displacements under the photo with the mouse]. The red trajectory that's more if you add it all up than the blue one. [She changes the graph to the graph of the total distance traveled]
J: It doesn't get any higher.... [he points to the blue one]
M: The red one has traveled further. [O and J confirm this]
O: So your estimate was correct. Do you understand what's happened now?
J: Yes, yes. It's added all these things together.
O: What you just saw was that the red waved around the blue and that the blue was fairly constant. How can you see that the blue, for example, is fairly constant here?
J: Because if you drew a line it would be straight. (M makes a straight line on the screen)

This last remark of J is precisely what we hoped for. In the continuous situation it is not easy to explain why a constant velocity results in a linear distance traveled graph. Traditionally, when the basic consideration is how to do the calculations, this property is proven by giving the graph and some calculations to confirm the 'fact'. In the discrete case students can find the property by themselves. The interpretations of the representations are rooted in the marked displacements in the photograph. This observation gives students possibilities to explain the continuous alternative by progressive mathematizing later in the unit.

During following problems students increasingly use the change of the displacements in the graph to describe motions and to do predictions. By using Flash their attention during the activities is focused on properties of graphs. Sometimes they still refer to distances between the dots in the photograph, but more and more they reason with the global shapes and properties of graphs. In the following protocol this reasoning with the graphs, again, is forced by the observer. But from the answers it is clear that the students interpret the graphs correctly. The students work on an exercise about a zebra which is running at constant speed and a cheetah that starts hunting the zebra. The following discussion takes place.

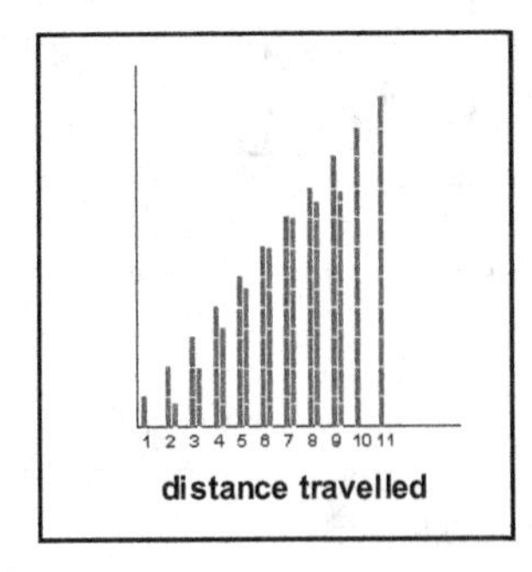

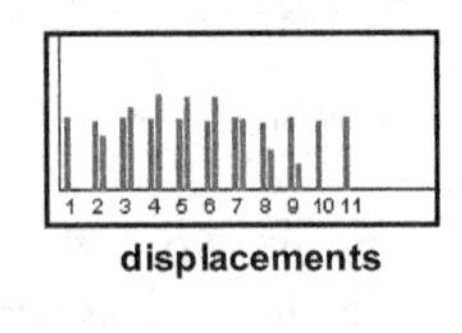

Figure 7: Graphs of a cheetah chasing a zebra

O: Oh yes, that was the question about whether if he started later, whether he would still -
M: Yes, the cheetah would still catch up.
O: Really? How did you conclude that?

M: Uhm, then you put two here together and then you can see here that they're both equal.
O: Yes. And which of the two graphs is that?
J+M: That's the total distance covered [they point at the left graph below].
O: Oh yes. So why did you choose the one for the total distance?
J: Because it's the total distance that they cover and then you can-
M: Then you can see if they catch up with each other.
O: And can't you see that in the other? There you can also see that the red catches up with blue?
J: Yes, but -
M: Yes, but that's at one moment. That only means that it's going faster at that moment but not that it'll catch up with the zebra.

An important difference between the displacements graph and the distance traveled graph is the difference between the interpretation of the horizontal (time) axis. A value in the distance traveled graph represents a distance from the start until the corresponding time, while a value in the displacements graph represents a distance in the corresponding time interval. The observation of the student is an important step in the process of building the model of a velocity-time graph (and everything that comes with it). This should prevent a wrong interpretation of velocity graphs as shown in the beginning of this paper, where the intersection of two velocity graphs must be interpreted in terms of velocities at one moment. These two protocols illustrate progress by two students while working with Flash.

During the computer lesson, the students increasingly reasoned about the characteristics of, and relation between, the two graphs and their meaning for the specific situations. A few students took a little longer to work this out. This group continued for some time to relate the characteristics of the graphs to the displacements in the stroboscopic situation. After the computer lesson, the teacher first discussed their experiences. He used overhead sheets with screen shots for this part of the discussion. Many students participated actively in the discussion.

The teacher was especially capable of guiding whole class discussions within which students contributed a diversity of approaches. This diversity in reasoning and approaches was created by problems for which the students didn't have standard solution procedures yet (e.g. the activity about the falling ball), and by explorative computer lessons during which the students were confronted with many problem-situations and the tools in the program that allowed for a variety in reasoning. In the latter cases, students and teacher referred to these tools in reflective discussions afterwards.
An important aspect of the class discussions was that the teacher did not answer the students' questions, but limited his contribution to making the students' arguments more explicit, and posing questions about these arguments to the class.

7. THE EMERGING LOCAL INSTRUCTION THEORY

The qualitative analyses from the teaching experiments show that many students used a strategy based on their learning history and imagery of the tools provided in the software. We conjectured that activities with computer tools need careful preparation beforehand, and need reflection afterwards. The model-eliciting activity before the Flash lesson supported the students' inventions of, and reasoning with graphs. The variety in solutions enabled the teacher to discuss graphical models of motion and to reach consensus about the way to proceed. Their understanding of the graphs provided by the tool, was an important condition for their meaningful and flexible reasoning with the tool. The tools afforded students' expressions with the graphs about the presented situations, and of relationships between the graphs.

An intended model shift – associated with emergent modeling – did take place at a different pace and to a different extent during the computer activities. Consequently, the small group activity afterwards, together with the students' presentations, class discussion and teacher's guidance promoted the whole-class understanding of the acquired notions. The relation between the inscriptions, the intended activities and conceptual development of the students can be summarized as a chain of signification in a table. In previous design research such a table is used for describing a learning trajectory on measurement and flexible arithmetic (Gravemeijer, 2004).

The first column is labeled 'tool' to emphasize the tool-character of inscriptions. The 'imagery' column refers to the history that frames students' perception. By providing this column we make it plausible how the tool derives its meaning in relation to previous activities. The activities and discussions that address specific concepts should result in motives to proceed. In our approach we focused especially on evoking these motives, and we have therefore added them to the table.

The table gives an overview of (*i*) how students are expected to act and reason with the tools, (*ii*) how an activity relates to preceding activities, and (*iii*) the conceptual development aimed at by that activity. The first half of the table which underpins the instructional sequence of this study is displayed in Table 1 (next page).

8. DISCUSSION

In the classroom learning processes, we noticed that students reasoned and wrote about mathematical and kinematical notions with a tentative language and inscriptions, which were not as precise as the notions aimed at. This seems comparable with Goldin's description of three main stages in the development of representational systems: (*i*) an inventive and semiotic stage, (*ii*) structural development and establishment of relationships, and (*iii*) an autonomous stage (Goldin, 2003). In the autonomous stage the system can function flexibly in new contexts. During the inventive stage, students tentatively used inscriptions and language to communicate their developing ideas. We observed that the teacher played an important role in these stages.

In RME approaches, the final model is the result of modeling activities in which students' constructions play a central role. The related inscriptions and language are progressively developed through these activities, and tool-use supports students' inventions and the construction of a chain of signification. It is difficult – maybe even impossible – to design learning processes for classroom situations in which all students experience their learning process as invention. However, a learning trajectory that supports invention and which makes it possible to trace meaning through a series of inscriptions that built progressively on each other provides the teacher with possibilities for guiding the students' reasoning. This can be realized when the consecutive models as well as the tools in the software are part of an emergent modeling process and are – as much as possible – compatible with the students' current reasoning. To further clarify the idea of emergent modeling, we will continue this discussion by discerning three different kinds of modeling: mathematical modeling, didactical modeling and emergent modeling.

With the label *mathematical modeling* we refer to the role of models in the context of mathematics as a discipline. In mathematics, models function as systems of quantitative or spatial relations that can be transformed in prescribed ways. The primary use of those mathematical models is that they enable one to explore the consequences of manipulating the situation being modeled without the necessity to actually manipulate that situation. In this type of modeling the mathematical model and the situation being modeled are treated as separate entities.

Table 1: Overview of the instructional sequence

Tool	Imagery	Activity	Concepts
time series (e.g. satellite photos & stroboscopic pictures)	real world representations signify real world situations	predicting motion (e.g. in the context of weather predictions)	displacements in equal time intervals as an aid for describing and predicting change
		should result in a feeling that the ability to predict motion with discrete data is an important issue	
trace graphs of successive locations	signifies a series of successive displacements in equal time intervals	compare, look for patterns in displacements and make predictions by extrapolating these patterns	displacements as a measure of speed, of changing positions, but difficult to extrapolate
		resulting in a willingness to find other ways to display displacements for viewing and extrapolating patterns in them	
discrete 2-dim graphs	signifies patterns in displacements of trace graphs (and cumulative)	compare patterns and use graphs for reasoning and making predictions about motion (also at certain moments: interpolate graphs) refine your measurements for a better prediction: displacements decrease	displacements depicting patterns in motion; linear line of summit in graph of displacements or graph of distances traveled; problems with predictions of instantaneous velocity
		should result in the need to know more about the relation between sums and differences, and in the need to know how to determine and to depict velocity	

In education, other kinds of modeling play an important role as well. One concerns what we may call *didactical modeling*, an activity of mathematics educators and instructional designers, who design manipulative materials and visual models to make the abstract mathematics to be taught more concrete and accessible for the students. Gradually, however, these types of use of ready-made models have come under fire. Apart from problems with student understanding and proficiency, doubts have emerged about the tenability of the underlying theory. The most fundamental critique is that external representations do not come with intrinsic meaning, but that the meaning of external representations is dependent on the knowledge and understanding of the interpreter. This creates a problem for the use of didactical models in instruction, since this would imply that the students should already have at their disposal, the knowledge and understanding that is to be conveyed by the concrete models. In response to this dilemma, one has started to look for approaches in which the development of meaning is conceived as a dynamic process. The static models are to be replaced by the dynamics of symbolizing and modeling. Here the point of departure is in a reflexive relation between symbolizing and the development of meaning (Meira, 1995).

The design heuristic, called *emergent modeling*, fits this directive (Gravemeijer, 1999). The emergent-modeling heuristic assigns a role to models that differ from the aforementioned didactical role of models in mathematics education: instead of trying to concretize abstract mathematical knowledge, the objective is to try to help students model their own informal mathematical activity. Here we have to add that the model we are referring to is more an overarching concept, than one specific model. In practice, "the model" in the emergent modeling heuristic is actually shaped as a series of consecutive local models that can be described as a chain of signification. From a more global perspective, these local models can be seen as various manifestations of the same model. So when we speak of a shift in the role of the model, we are talking about "the model" on a more general level. On a more detailed level, this transition may encompass various local models that gradually take on different roles.

The label 'emergent' refers both to the character of the process by which models emerge within realistic mathematics education, and to the process by which these models support the emergence of formal mathematical ways of knowing. From this perspective, the process of constructing models is not one of abstracting relationships from situations, but one of progressively reorganizing situations. The model and the situation being modeled co-evolve and are mutually constituted in the course of modeling activity. Instead of thinking in terms of "cutting bonds with (everyday-life) reality", the construction of mathematical knowledge in connection with contextual knowledge gets more emphasis. In this conception, informal, situated knowledge is the basis upon which more formal, abstract mathematical knowledge is build and connections with original contexts are preserved.

Although we contrast emergent modeling with mathematical modeling, it should be clear, that the issue at hand is not that of deciding which typology is the better. Instead, the distinction is made to support our thinking about the design of mathematics education. In this regard, we note that these two kinds of modeling are located in different phases of the learning process. The emergent modeling heuristic is part of an instructional approach in which students are expected to make sense of unfamiliar situations by mathematizing them. In such an approach, modeling serves not only as an instructional goal but also as a means of supporting the reinvention of mathematics.

The integration of physics and mathematics in the teaching experiments was confined to the integration of kinematical notions in an instructional sequence for mathematics lessons. Ideally, this instructional sequence is acted out in both physics and mathematics lessons.
Our research indicated that students' conceptual problems in applying mathematical notions in other topics can be prevented by integrating the learning and teaching in an application. In our approach, the learning of calculus and kinematics was rooted in organizing motion. The activities for making predictions and describing motion graphically helped the students to develop a notion of velocity as a compound quantity, which supported the notion of proportionality that underlies the understanding of the difference quotient as a measure for change.

We do not believe that all mathematical topics can be developed through an integrated approach. Some topics are essentially the result of a process of organization within or between mathematical systems or structures. In fact, the trajectory in this research should also be followed both by a series of lessons where the mathematics of change is developed as a generalizing principle for many applications, and by elaborating the topic within a mathematical context.

Other possibilities for an integrated approach lie, for instance, in the field of discrete and continuous dynamic modeling. Investigating the dynamics of various situations (population growth, cooling down, consumer-production dynamics) could be the source for developing both mathematical models and knowledge of their applications (e.g. Michelsen, 1998).
The integrated approach contributes to a trend within education to develop skills (general and topic-specific) through problem-oriented case studies. The experiences we gained in our research raised two points of concern. First, the teacher needs didactical knowledge of both disciplines for dealing with class discussions and guiding students' tentative ideas. Second, the period of integrated lessons should be alternated with topic-oriented lessons to develop relations with the topic's systematics: relations that (*i*) support its understanding, (*ii*) create possibilities to trace meaning, and (*iii*) provide opportunities for new inventions.

The integration of mathematics and science topics is an important issue for education. In current scientific research many breakthroughs happen on the border of different topics. However, integration of mathematics and science in secondary education should take into account that didactical problems within the various topics have more consequences than a superficial consideration of the respective curricula might suggest. Integration has more implications than just the tuning of standards, and needs schools, teachers and researchers to invest in understanding each other's didactical problems and cultures.

This study has provided some insight into the constraints and possibilities for the integration of physics and mathematics. We recommend further research on this integration for the understanding of the teaching and learning of science and mathematics as closely related disciplines, and for implementing real changes in the way these topics are covered in schools.

Acknowledgements
This research was funded by NWO, the Dutch National Science Foundation, under grant no. 575-36-03C.

References

Bakker, A. (2004). *Design research in statistics education; On symbolizing and computer minitools*. Utrecht: CD- Press.

Boyd, A. & Rubin, A. (1996). Interactive video: A bridge between motion and math. *International Journal of Computers for Mathematical Learning*, 1, 57-93.

Clagett, M. (1959). *Science of mechanics in the middle ages*. Madison: The University of Wisconsin Press.

Clement, J. (1985). Misconceptions in graphing. In: L. Streefland (Ed.), *Proceedings of the Ninth International Conference for the Psychology of Mathematics Education* (pp. 369-375). Utrecht: Utrecht University.

Cobb, P., Yackel, E. & Wood, T. (1992). A Constructivist Alternative to the Representational View of Mind in Mathematics Education. *Journal for Research in Mathematics Education*, 23, 2-33.

Cobb, P. (1999). Individual and Collective Mathematical Development: The Case of Statistical Data Analysis. *Mathematical Thinking and Learning*, 1(1), 5-43.

Cobb, P., Confrey, J., diSessa, A.A., Lehrer, R. & Schauble, L. (2003). Design experiments in educational research. *Educational Researcher*, 32, 9-13.

Dall'Alba, Gloria (et al.) (1993). Textbook Treatment of Students' Understanding acceleration. *Journal of Research in Science Teaching*, 30(7), 621-635.

Doerr, H.M. (1997). Experiment, simulation and analysis: an integrated instructional approach to the concept of force. *International Journal of Science Education*, 19(3), 265-282.

Doerr, H.M. & Zangor, R. (2000). Creating meaning for and with the graphing calculator. *Educational Studies in Mathematics*, 41, 143-163.

Doorman, L.M. (2002). How to Guide Students? A Reinvention Course on Modeling Motion. In: Fou-Lai Lin (Eds.), *Common sense in Mathematics education*, Taipei, Taiwan: National Taiwan Normal University, pp. 97-114.

Doorman, L.M. (2005). *Modeling motion: from trace graphs to instantaneous change.* Utrecht, the Netherlands: CD- Press.

Edelson, D.C. (2002). Design Research: What We Learn When We Engage in Design. The *Journal of the Learning Sciences*, 11(1), 105-121.

Freudenthal, H. (1991). *Revisiting Mathematics Education – China Lectures*. Dordrecht: Kluwer Academic Publishers.

Glaser, B. G. and Strauss, A. L. (1967). *The Discovery of Grounded Theory: Strategies for Qualitative Research.* Chicago: Aldine Publishing Company.

Goldin, G.A. (2003). Representation in school mathematics: a unifying research perspective. In: J. Kilpatrick, W.G. Martin & D. Schifter (Eds.) *A research companion to Principles and Standards for School Mathematics*. Reston: NCTM.

Gravemeijer, K.P.E. (1994). Developing realistic mathematics education. Utrecht: CD-b Press.

Gravemeijer, K. (1998). 'Developmental Research as a Research Method', In: J. Kilpa¬trick and A. Sierpinska (Eds) *What is research in mathematics education and what are its results?* (ICMI Study Publication). Kluwer Academic Publishers, Dordrecht, 277-296.

Gravemeijer, K.P.E. & Doorman, L.M. (1999). Context problems in realistic mathematics education: A calculus course as an example. *Educational Studies in Mathematics*, 39(1-3), pp. 111-129.

Gravemeijer, K. (1999). How emergent models may foster the constitution of formal mathematics. *Mathematical Thinking and Learning*.1 (2), 155-177.

Gravemeijer, K.P.E., Lehrer, R., Van Oers, B. & Verschaffel, L. (Eds.). (2002). *Symbolizing, modeling and tool use in mathematics education.* Dordrecht: Kluwer Academic Publishers.

Gravemeijer, K. (2004). Local Instruction Theories as Means of Support for Teachers in Reform Mathematics Education. *Mathematical Thinking and Learning*, 6, 105-128.

Hoyles, C., & Noss, R. (2003). What can digital technologies take from and bring to research in mathematics education? In: Bishop, A.J., Clements, K., Keitel, C, Kilpatrick, J., & Leung, F.K.S. (Eds.) *Second International Handbook of Mathematics Education* (323-349). Dordrecht, the Netherlands: Kluwer Academic Publishers

Klaassen, C.W.J.M. (1995). *A Problem-Posing Approach to Teaching the Topic of Radioactivity*. Utrecht: CD- Press.

Lijnse, P.L. (1995). 'Developmental research' as a way to an empirically based 'didactic structure' of science. *Science Education*, 79, 189-199.

Michelsen, C. (1998). Expanding Context and Domain: A Cross-curricular Activity in Mathematics and Physics. *ZDM*, 30(4).

Meira, L. (1995). The microevolution of mathematical representations in children's activities. *Cognition and Instruction*, 13(2), 269-313.

Resnick, L.B. and S.F. Omanson (1987). Learning to Understand Arithmetic. In: Glaser, R. *Advances in Instructional Psychology*, Vol. 3. London: Lawrence Erlbaum Ass.

Roth, W. and Tobin, K. (1997). Cascades of inscriptions and the re-presentation of nature: how numbers, tables, graphs, and money come to re-present a rolling ball. *International Journal of Science Education*, vol 19 no 9, p. 1075-1091.

Sfard, A. (1991). On the dual nature of mathematical conceptions: Reflections on processes and objects as different sides of the same coin. *Educational Studies in Mathematics*, 22, 1-36.

Simon, M.A. (1995). Reconstructing mathematics pedagogy from a constructivist perspective, *Journal for Research in Mathematics Education*, 26(2), 114-145.

Van Dijk, I. M. A. W., Van Oers, B., & Terwel, J. (2003). Providing or designing? Constructing models in primary maths education. *Learning and Instruction*, 13, 53-72.

THE DECORATIVE IMPULSE: ETHNOMATHEMATICS AND TLINGIT BASKETRY[1]

Swapna Mukhopadhyay
Portland State University
Portland, Oregon.

Do you doodle?
Have you been trained in "doodling"?
Have you thought about where your designs come from?
Have you looked at other people's doodles?
Is doodling the same as designing and adorning?

These questions may sound flippant, but they have a point. As humans, we are drawn to, and have a desire to create, objects that are aesthetically pleasing to us. In short, we share a decorative impulse. Depending on where we live and what we do, we visit museums to appreciate and admire the work of people from different parts of the world, from different eras of history. We travel to different places – even to the local markets – to acquire artifacts that catch our fancy. Sometimes these are everyday utilitarian objects and sometimes they are adornment. We are attracted to the forms, the shapes, the colors, and the designs. But when we look at the designs on artifacts made by people who are not "formally" trained, or, as Jean Lave (1988) would characterize as "just plain folks", do we wonder where the designs come from? How do they create an inherent sense of beauty that pleases any viewer? What motivates one to "design"?

Figure 1: A basket with two bands of the "leaves of fireweed" pattern and a dropper of the "double around the cross" pattern. (ID: 1-612).

[1] The research for this paper was supported by Portland State University's Diversity Action Council Faculty Mini-grant. 2003.

B.Sriraman, C.Michelsen, A. Beckmann & V. Freiman (Eds). (2008). *Proceedings of the Second International Symposium on Mathematics and its Connections to the Arts and Sciences* (MACAS2). Information Age Publishing, Charlotte: NC pp.57- 74

Take a look at the above basket. It is a Tlingit basket, made up of twined spruce root. It is 32 cm high; the top diameter is 31 cm, and the bottom diameter is 29 cm. It has two small leather straps on the sides. The basket has a very fine weave and is flared at the top. There are three bands of embroidery, with the top and the bottom bands wider than the middle band. There is also a "dropper" – an individual design dropping down the base. All the patterns are woven in a technique called "false embroidery", using bear grass. The pattern on the bands is called the "the leaves of fireweed" (Emmons, 1903/1993; Paul, 1944). It looks like a set of five skinny parallelograms in dark and light colors, stacked one on top of the other to form a larger parallelogram, which then tessellates with another parallelogram of the same dimensions, composed of five skinny parallelograms, but stacked as light and dark in an opposite order. The entire band consists of repeating parallelograms. The dropper is a "half-cross pattern", a design with horizontal symmetry. The band within the two "fireweed" is a set of five skinny rectangles, stacked up like the wider band in alternate dark and light colors. According to the records[2] the basket was made in 1925. The basket, with its patterns, is a prime example of artistic excellence. Along with the aesthetic element, one doesn't miss the complex geometry of the pattern. It is as precise as an illustration in a schoolbook on geometry. How did the maker of the basket come up with the design? How was she taught the geometry of pattern making? Were all designers well versed in Euclidean geometry? Clearly not, in any conventional sense. Who, then, taught them? What comprised their teaching material? What was the curriculum like? How did the teacher know that the pupil had mastered the concepts? How were the other designs of Tlingit baskets, which are rich and varied, generated? A deeper issue is whether the above questions, which come naturally to those of us steeped in a particular conception of what education – more particularly, schooling – have meaning in the context of the makers of these baskets.

As a part of colonization by the European settlers, "formal education" and Christianity as a religion was introduced to the Indian tribes to "civilize" the natives (Lomawaima, 1999). Did the Tlingit basket makers have access to formal western mathematics? How did the indigenous knowledge of making designs with geometric complexities interact with their school learning? These are just a few questions that I have as I look at the baskets.

In this paper I share my investigations of embedded mathematics of the Tlingit and interrogate some of the questions above. My title is intended to underscore the dynamic element of designed artifacts and its connection to mathematical thinking of "just plain folks".

In describing mathematical artifacts, generally speaking, one conjures up images of symbols and formulas, diagrams from geometry texts. The graphs of daily temperature fluctuations from the local newspapers, or from the business sections, often are categorized as mathematical artifacts that surround us. Cultural artifacts, especially baskets and blankets, pottery and sculpture rarely evoke mathematical curiosity in most people. These artifacts are considered "pretty" and "artistic", whereas in public perception mathematics doesn't hold the same descriptors. In any casual conversations with otherwise highly functioning adults, it is not rare to hear comments on mathematics and its connection to culture as "hmmm... never thought about it", "you mean math phobia?", and so on. The connection to the arts may be expressed more enthusiastically by these people, "oh yeah – art history – golden rectangle!"

[2] http://www.washington.edu/burkemuseum/collections/ethnology/collections/display.php?ID=39686

Given that throughout the known history of humankind people have always created music and dance, designed household objects, adorned their bodies with paints and designs, it is not unusual for anthropologists and others to examine where patterns come from. Franz Boas (1927/1955, p. 9) wrote that:

> No people… however hard their lives may be, spend all their time, all their energies in the acquisition of food and shelter …
> Even the poorest tribes have produced work that gives them esthetic pleasure …[they] devote much of their energy to the creation of works of beauty"
> …No matter how diverse the ideals may be, the general character of the enjoyment of beauty is of the same order everywhere.

This realization points to the link between indigenous craftwork and formal mathematics, namely an instinctive fascination with patterns. After all, one definition of mathematics is as "the classification and study of all possible patterns" (Sawyer, 1955, p. 12).

The perspective of Ethnomathematics

My work in ethnomathematics has been inspired most by Ubiratan D'Ambrosio. D'Ambrosio, a Brazilian scholar, coined the term and was the first to lead a series of conversations on ethnomathematics, describing it (D'Ambrosio, 1985, p. 45) as:

> …the mathematics practiced among identifiable cultural groups, such as national-tribal societies, labor groups, children of certain age bracket, professional classes, and so on. Its identity depends largely on focuses of interest, on motivation, and on certain codes and jargons which do not belong to the realm of academic mathematics.

The knowledge of an indigenous "craft" – basket weaving, canoe making, for example – is complex and situated in cultural values and everyday living. For example, the construction of canoes for a non-Western group using indigenous material evolves over a substantial period of time incorporating cultural traditions of design, understanding of engineering, and adaptation to the environment. To an individual who does not belong to the community, the canoes, although functional and efficient, may look crude and primitive. If the individual happens to be trained in, and used to, only technologically-sophisticated building tools, the process of construction might seem to him/her as simplistic and rudimentary. As a consequence, both the process and the product of creating an artifact is not recognized as a cognitively complex process.

Ethnomathematics, as a discipline, studies the intersection of cultural anthropology and mathematics. Starting from a broad and dynamic interpretation of the terms *ethno* and *mathematics*, one of the main aims of ethnomathematics is to continually interrogate the assumptions of the paradigm that denies or disregards the intellectual contributions of non-Western people particularly in relation to the hegemonic control of Western mathematics. Culture implies a shared meaning system manifested in the daily lives of its members, so that, according to D'Ambrosio (2001a, p. 24):

> …in the same culture, individuals provide the same explanations use the same material and intellectual instruments in their everyday activities. The set of these instruments is manifested in the manners, modes, abilities, arts, techniques in dealing with the ***tics*** of dealing with the environment, of understanding and explaining facts

> and phenomena, of teaching and sharing all this, which is ***mathema*** of the group, of the community, of the ***ethno***. That is, it is their ethnomathematics

This viewpoint is expressed succinctly in his diagram (Fig. 2).

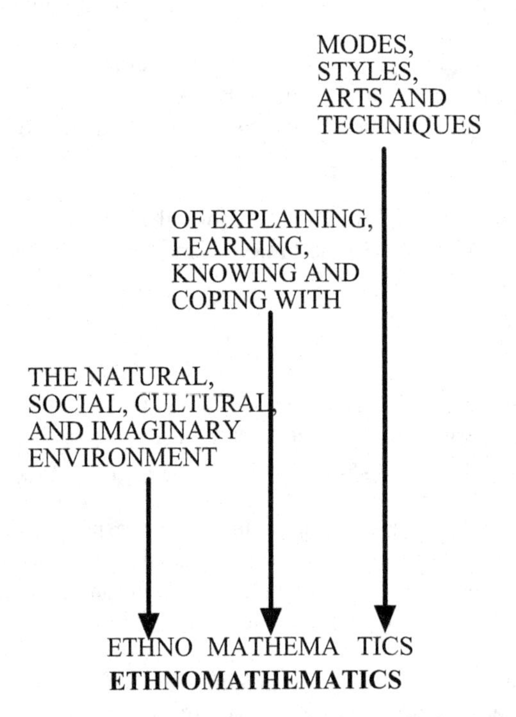

Figure 2. Conceptual underpinnings of ethnomathematics (D'Ambrosio, 2001a)

Ethnomathematics, accordingly, provides a framework that can accommodate the artistic/decorative impulse that we see manifested in cultural artifacts such as baskets and pottery, as well as the delight in abstract patterns that characterizes mathematicians. As a theory of cognition, it is clearly in line with recent perspectives of cognitive science around the general principle that the knowledge used by "just plain folks" (a term coined by Jean Lave (1988) in discussing cognitive characteristics of everyday practice) is situated, while the knowledge that previously might have been characterized as "abstract" is also situated, albeit in a different sense.

By employing D'Ambrosio's ideas of ethnomathematics and focusing on historiography as a methodology, my focus for this paper is to reinforce a deeper understanding of human cognition vis-à-vis epistemology, history and sociology. I rely on my core assumption that *knowledge is situated* (Lave, 1988; Lave & Wenger, 1991), as a result, learning is viewed both as a process and a product that resides, in its duality, in a relationship to the community – a community of practice. Situated in this paradigm, one realizes that a doer's perspective and motivation might not match that of a researcher/observer, often creating misinterpretation, followed by misrepresentation, of the doer. There are plenty of examples in both psychology and anthropology that remind us of the implications of the damages of these mismatches resulting from professional irresponsibility.

One major element in Ethnomathematics has been the deconstruction of the Eurocentric view of the history of mathematics (D'Ambrosio, 1991b; Joseph, 2000; Powell & Frankenstein, 1997; Zaslavsky, 2005). Bishop (1988), in investigating how cultural practices are interwoven with mathematical thinking, developed a taxonomy of six categories to describe salient pan-human activities, namely counting, locating, explaining, designing, locating, and playing. Inspired by the scholars of ethnomathematics and informed by Bishop's six categories, I put forward a study that looks at the cultural practices of basket making of the Tlingit tribe, which, as for many other tribes, declined almost to the point of extinction. Fortunately, there is a renewal of interest and members of various tribes along with the elders are reviving much of their "lost"/"silenced?" crafts.

Tlingit Indians and their basketry

The Tlingit Indians are indigenous to south-eastern Alaska, particularly in Sitka and Juneau (Fig. 3).

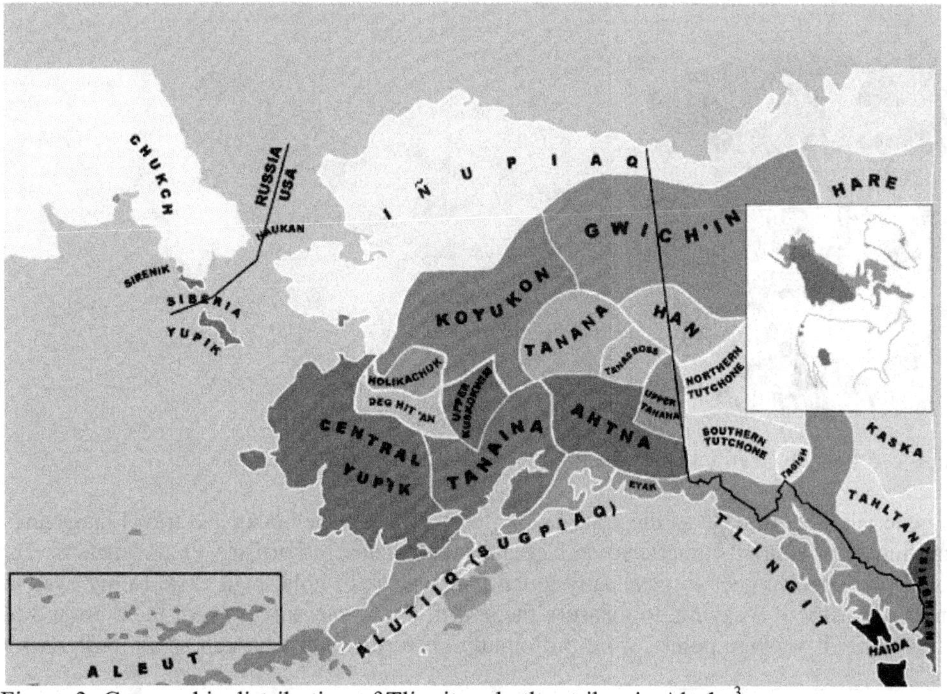

Figure 3: Geographic distribution of Tlingit and other tribes in Alaska[3].

Living on islands and narrow strips of coastline surrounded by fjords, their general livelihood is from fishing. They are well known for their dug-out cedar canoes that they have used for fishing expeditions, although these days they use motor-boats more often. Besides the canoes,

[3] Courtesy: Benny Boyd, Alaska Native Language Network, http://www.ankn.uaf.edu/IKS/subsistence/languagemap.html

the Tlingits are also known for their extraordinary carvings of totem poles. Tlingit is one of the major languages of the Northwest tribes.

Tlingit baskets are among the oldest artifacts that drew the attention of others. The baskets that had purely everyday, utilitarian purposes were perceived differently for their aesthetic values after contact with European explorers and traders. Emmons, one of the earliest scholars to study these baskets, conjectured that perhaps these baskets were the most important element in the Tlingit economy (Emmons, 1903/1993). The baskets, made of spruce roots, came in different sizes and served various functions, such as storage, food gathering, food preparation, garments, and for ceremonial purposes. The baskets were woven very tightly in a technique called twining[4]. Most baskets had complex geometric adornments. While the Northwest coast Indians, especially the Tlingit and Haida, are noted for their realistic woodcarvings depicting living things, their baskets mostly depict geometric patterns – which might appear purely abstract but, as seen below, did represent, in a highly stylized fashion, natural forms. For example, a smaller berry basket (Fig. 4) is about 13 cm high, and the top and the bottom diameters are 16.7 cm and 15.1 cm respectively. The basket has two identical bands – about 2.5 cm high, with a 1 cm space in between. Also according to the museum records, the basket was received around 1932, which gives an idea of its age.

Figure 4: A berry basket with two identical bands (ID: 1-404)

One generally sees baskets as illustrations in glossy coffee table books and travel magazines promoting local tourist attractions, and greatly appreciates the artistry of the makers. By contrast, the opportunities to physically touch the baskets, to hold them close to my eyes to count the density of weaving, to measure them with a tape measure, to turn them around to look at from all vantage points, were profoundly powerful. Some baskets were darker with

[4] Twining is a basketry technique in which two wefts cross over each other between warps. There are variations to twining that result in different surface features of the weave. For an illustration, see http://www.washington.edu/burkemuseum/baskets/Teachersguideforbasketry.htm

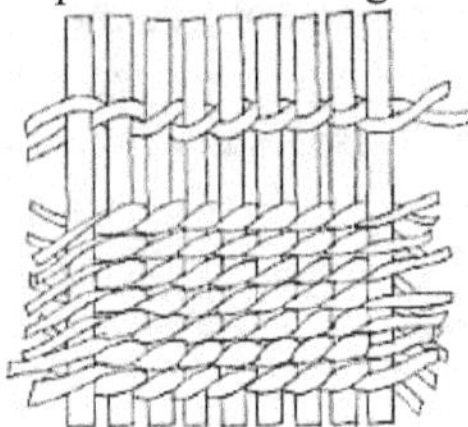

stains, distorted with use, with fraying handles – these were the telltale characteristics of artifacts made for local consumption. However, in a formal collection of native artifacts, as in the case of museums, one more often sees objects that were designed and created for outsiders – the tourists and the collectors. This is particularly true for the native northwest American artifacts. With the advent of European contact, both the forms and functions of the Tlingit baskets started diversifying. The interests expressed by the collectors not only had an economic impact on the tribes, but also on innovation in designs of baskets. For example, one sees basketry teapots, a definite effect of post-contact intervention. Nine cm high, a basketry teapot (Fig. 5) is 13.5 cm wide at the base and 8 cm at the top. It is, like others, a twined spruce root basket, with two rows of false embroidered design called "shaman's hat" alternating in red and dark maroon. There is a spout and a lid but no handle. The spout and the lid also carry the same design in same color as the body of the teapot. The object was made around 1910.

Figure 5: A basketry teapot, made around 1910. (ID: 1-10723)

Making baskets: Tlingit style

For this study, I studied artifacts that were made more than a generation ago. With a loss of participation, there was less documentation of the process of basket making. As a result, I relied primarily on the writings of George T. Emmons, a U.S. Navy officer who took a keen interest in the ethnography of Alaskan tribes, particularly the Tlingit. His treatise *The basketry of the Tlingit*, was first published in 1903 for the American Museum of Natural History. I also relied heavily on Frances Paul's 1944 publication *Spruce root basketry of the Alaska Tlingit*. These anthropological documentations served as guides in this ethnohistoric investigation.

The main materials for twined Tlingit baskets were spruce roots. Young and strong spruce roots of "reasonable diameter (not exceeding two centimeters)" (Emmons, 1903/1993, p. 234), and between one and six meters in length, were collected by women and girls in early spring. The crude root-lengths were prepared for taking the bark off with utmost care. The bark was charred by placing coiled roots on the embers of a small fire. Since both the color and pliability of roots were prized for weaving, the process of debarking was considered highly specialized and was usually undertaken by elderly women. The roots were then straightened out by passing then through a tool – a stick split through the center – and allowed to dry for the next step of splitting to finer strands. Splitting the roots required keeping the root bundles moistened by soaking in water and keeping them weighted down. Since the characteristics of the strands varied greatly with the thickness, the outermost strands closest to the bark were highly valued because of the sheen. Women used knives, made either of traditional shells or metals, for dividing the roots into strands. The strands once prepared to the desired thickness was carefully coiled and stored.

Besides spruce roots that serves as the main material for weaving baskets, the Tlingit women also collected various kinds of grass and plant stems to use for patterns. The technique of weaving to adorn the spruce root baskets with patterns in various colors is called "false embroidery"[5]. Various grasses and maidenhair fern that were popular were collected in early summer when the stalks were still tender and pliable. In earlier days, the dyes used were natural (vegetable and mineral) but with the contact of the Europeans, there was a tendency to use colorants[6]. Besides the natural color of spruce roots and grasses, the colors used were black, red, yellow, and bluish green.

Since the Tlingit used the baskets for various ceremonial and everyday uses, their baskets came in different sizes and forms, specific to their purposes. The most important purpose was, of course, food preparation, which included gathering, preparation, and storage. Emmons conjectured that the earlier baskets were cylindrical in shape and that flare added to the sides was a later idea. However, it is hard to evaluate this conjecture as many of the baskets prior to Emmons' era were already destroyed and, with wear and tear of daily use, the baskets often lost their original shape. The measurement of the baskets is always a challenging task: Many of the baskets, even with the best museum care, have become too brittle for handling.

Emmons described twenty-three types of baskets in his treatise, some of which were very specialized and not available for my analysis. The names of the baskets referred to their purpose and sometimes to their shapes, as the purpose and the shapes were intimately connected.

For berry picking, (the largest number in my sample), there were three types of baskets:
1. The smallest berry baskets (capacity 1- 2 quarts). The Tlingit name for these translates as "to hang from around the neck in front of the body" and "the noise made by throwing the berry into the basket as it bounced from side to side". These baskets were made proportionately smaller for the children, who were also responsible for collecting berries. The

---

5

False embroidery (the inappropriateness of this term is indicative of ethnocentrism) is a weaving technique of overlaying to introduce different colors in the design. Usually a third color as a weft is introduced and twined in such a way that the design only appears on the outer surface of the object. In contrast to *embroidery* which is done after completing an object,
false embroidery is a part of making the object and continues as the weaving of the object progresses. A characteristic of false embroidery is that it is at a slant angle and opposite orientation to the twining. The Tlingit basket makers used moss, maidenhair fern, grass, etc. in their false embroidering of designs.

[6] Emmons wrote that often the Tlingit women would take a European textile and boil it to extract dyes, then steep the material in the resulting liquid (Emmons, 1903/1993, p. 238). However, in later days when aniline dyes were available, the use of colors in baskets was radically impacted.

heights of these baskets were more or less equal to the diameter of the base. These baskets were not precisely cylindrical, having a flare at the top.

2. Dsu-na or yan-nah ("to pack on the back") or "on the top of the back". These are bigger, with a capacity of one-quarter to three-quarters of a bushel. The proportion of height to diameter of the base was the same as for the smaller baskets described above. The contents of these baskets were dumped into the next bigger basket.

3. The largest category of baskets had the descriptive names "to empty into from the back basket" or "to sit in one place" (i.e., stationary). They were placed at some central location so that the berry pickers could easily empty their relatively small baskets. These baskets were often used for festive occasions, for large gatherings when food was served in large quantities.

4. Called the "vessel between" or "basket bucket", these baskets were used for carrying water, and were smaller than the previous category. Usually without decoration as constant wetting impacted the designs, the shape tend to be almost regular cylindrical with less flare. Metal containers that were brought in after White contact replaced these baskets. This type of basket was also used as a storage container for clothes and household goods during traveling.

5. The smallest kind of basket, "diminutive basket", was used as an eating container for children or as their play object.

6. Also called "half-basket", this is a "primitive" vessel for cooking. Generally, the height of a basket was half the diameter of the base. Cooking was done by dropping preheated rocks into the food and water held in the basket and then covering with seaweed.

7. Small cylindrical baskets with very tight weave, called "small buckets" or "salt-water cups", were often used as drinking cups (drinking salt water for men was a part of the early morning activity). The height of the basket was usually one and a half times the diameter of the base. These baskets also claimed to have the finest weave, with adornment of foreign objects (such as, a feather or shell) added to the designs.

Besides the above categories, the baskets were made to hold spoons, to screen berries, as rattles (called "noise inside"), etc. Later on, as trade developed with the European settlers, covered baskets were more popular. Although sometimes they were used as storage baskets, these seldom show signs of wear and tear. While highly decorated, their weaving was less tight in texture and they were perhaps made for sale. It is not at all uncommon to find Tlingit baskets made in the shape of a teapot or a narrow neck jar.

Method of inquiry

This study relied on ethnohistory. By using photographic slides of baskets from a museum with a highly-regarded collection and documentation from earlier scholars, I am studying the mathematical nature of the representations by the Tlingit. Granted much of the documentation on Tlingit artifacts, especially the baskets, is more than fifty years old, my intention has been to connect to the emic perspective of the indigenous people as close as possible, and further the conversation (on basket designs) in future with members of the Tlingit tribes to reconstruct their mathematical meaning making. Since the contact with the Europeans, which subsequently resulted in taking over the land and colonizing the indigenous population, the Native American tribes have undergone oppression no different from cultural genocide. Their languages, religion, cultural practices have been seriously interrupted to the point that much of the cultural activities, such as weaving, hunting, etc. have become close to extinction. They have been subjected to schooling practices of an alien culture. In the past few years, there have been genuine efforts aimed at revival of activities that are closely associated with Native identities. At the forefront of this movement is the Alaska Native Knowledge Network of University of Alaska, Fairbanks (http://ankn.uaf.edu/index.html).

Although ethnohistory as an interdisciplinary approach provides a rich perspective to know the social and cultural history of groups of people from multiple standpoints, it is not unproblematic. We need to realize the circumstances in which cultural artifacts are procured for museum's collections. There is no doubt that museums play a critical role in public education and that the contributions of the donors and collectors are decidedly responsible in making art and cultures of people accessible to many. However, besides questions around ownership of cultural artifacts, one is also confronted with the anonymity of the artist. Unlike artwork from Europe, for example, displayed cultural artifacts fail to acknowledge the names of their creators, the only names they carry being those of the collectors. While recognizing that the names of the makers are generally unknown, we need to ask why this is the case. Indeed, the situation bears an uncanny parallel to the naming of slaves after their owners.

This study was based on analysis of secondary material, artifacts from a museum renowned for its collection of Northwest Indian artifacts. I purchased about 120 slides from the Ethnology department of the Burke Museum, University of Washington, Seattle, WA, after going over their online catalog. Most of the slides were baskets by the Tlingit Indians, with a few from the geographically adjoining tribes such as Haida and Tsimshian. Initially I chose the baskets by reviewing their design elements and later on had special permission to measure the dimensions of the baskets that I decided to work on. My intent was to add the dimensions – height, top and bottom diameters – to the descriptions of the baskets. Since most of these baskets were in a fragile state, one has to be very careful in physically handling them. In measuring the diameter of the base and the top, I had to be particularly careful and thus had to rely on some degree of approximation.

My analysis focused on 113 slides, of which 90 are Tlingit baskets, with 23 from other tribes such as Haida and Tsimshian. As stated earlier, there were baskets of varying sizes and all had geometric patterns woven on them. Except for three items, a Tlingit mat and two Tsimshian baskets[7], the patterns were geometric in nature. In my sample, despite the crudeness of measurement of some brittle specimens, the baskets are categorized by size in the following groups (Table 1)

[7] The Tlingit mat had a painted motif, similar to the wood-carved items. One Tsimshian basket had the word SISTER woven on it, the other had two stylized faces of a bear.

Table 1: Categorization of baskets by size

Category by height	Number
Flat baskets, lids, rattles, mat (height < 4cm)	6
Baskets (height 5 -9 cm)	17
Baskets (height 10 – 20 cm)	60
Baskets (height 21 – 30 cm)	17
Baskets (30 cm and higher)	5
No measurement data available	8

Most of these baskets are not truly cylindrical in shape, but have a flared top, with the ratio bottom diameter/top diameter ranging from 0.87 to 0.91. It was evident that a slight flare in a basket helped in emptying the contents into a larger basket.

Analysis of the patterns

As noted earlier, the patterns in basketry were geometric designs, and constituted a notable exception to the characteristic figurative nature of Tlingit artifacts. It is also worth mentioning that whereas wood carving was a male activity, basket weaving was a female activity.

Emmons (1903/1993) and others scholars (Boas, 1927/1955; Paul, 1944) wrote at length about the highly impressive craftsmanship exhibited in the fine weaves of the baskets and the complexities of the designs that adorned them. For example, Emmons (p. 258) reported:

> Ornamentation in baskets is universally practiced by the Tlingit…excepting fish-trap and oil-strainer, few pieces of weaves are considered unworthy of some slight decoration, even should it be a single line of color or a different weave along the border.

Before the European contact, the colors used in the baskets were strictly organic – from plants and other natural sources. The colors were subtle and were not fade resistant. Aniline dyes, introduced by the European contacts, were brighter in color, durable and reasonably affordable and they naturally replaced the indigenous dyes[8].

Weaving techniques employed were twinning and false embroidery, as described earlier. False embroidery, which, as explained above, is a misnomer – and actually is a weaving technique of adding material other than spruce roots – was described in the early twentieth century[9]

In a patterned weave, the pattern is generated by working one row at a time –like stacking layers of the disembedded patterns, one row at a time (Figure 6).

---

[8] This phenomenon, common all over the world, has made the knowledge of ethnobotany and dye chemistry vulnerable and gradually disappear.

[9] Attributed by Willoughby (1905).

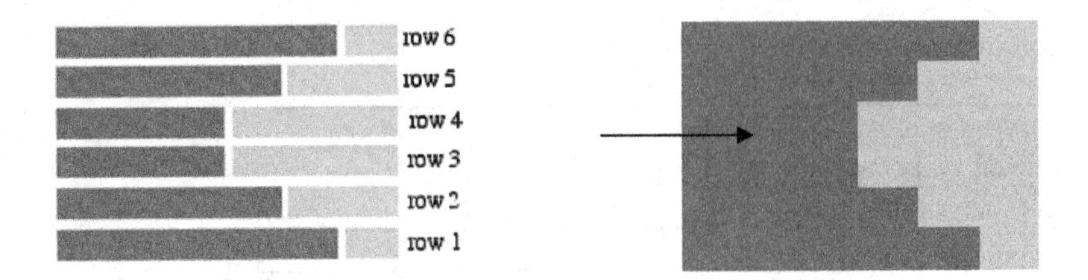

Figure 6: A schematic and enlarged version of a weaving sequence

Without a routinized algorithm, the weaver relies heavily on her capacity of visualizing the entire pattern, breaking down each layer of it, updating a counting sequence as well as visually predicting the entire sequence of pattern and self-correcting counting mistakes made.

According to Emmons, and then Paul, the origins of the Tlingit patterns are unknown. As Emmons (Emmons, 1903/1993, p. 262) put it "...the designs are [neither] tribal [nor] individual. They are common to all." Although geometric in style, they were inspired by nature, and take their descriptive names accordingly. Recall the basket from Figure 1. The wide pattern in two bands was called "the leaves of fireweed". The leaves of the plant fireweed are lance shaped and arranged alternately. Some patterns were influenced and inspired by European contact and adopted latter.

Emmons described forty-nine distinct patterns, some with variations within a class. Most of these patterns exhibit a strong sense of symmetry. My sample cannot be claimed to be representative, for two reasons. First, any museum collection of this sort, however rich and broad, cannot be assumed to constitute a representative sample from the artifacts of the culture. For example, many baskets those were for local consumption only deteriorated before coming into the hands of any collectors. Second, from the museum's collection, I choose baskets with a wide variety of patterns for their aesthetic qualities. Some baskets, for example, have more than one band, some are lidded, with the lids being rattles, and have patterns with rotational symmetry. Table 2 below captures the variability in patterns in the sample by categories such as number of bands, number of headers, etc.

Table 2: Categories of patterns in the sample

Patterns on the body of the baskets

Figura-tive, Words	All over grid	1 band	2 bands	3 bands	4 bands	5 bands	6 bands	7 bands	Header	Drop-per	Mono-chrom-atic
4	3	30	14	37	5	4	1	1	5	6	4

Patterns on the lids of the baskets, rattles, flat trays

Rotational Symmetry

2-fold	3-fold	4-fold	5-fold	6-fold	7-fold	8 fold
3	8	4	4	1	0	4

Examples of common patterns follow.

(i) “The leaves of the fireweed” (Figure 1), as discussed above.

(ii) “”The double around the cross” (Figure 1). This pattern is a variation of a “half-cross pattern” with a contour drawn around it. The “half-cross” pattern is a half of a cross, split along the vertical axis. Paul (1944) claims that this was one of the more recent patterns, credited to the influence of Christianity via the Russian Orthodox church.

(iii, iv) The “war club” pattern and “butterfly pattern” (Figure 7). A basket with three bands, the top and bottom being the same width (5 cm) and the middle band is narrower (2 cm) in red. The top band consists of “the war club” with a “cross” in a complementary form was called “the double khá-tu” (Emmons, 1903/1993, p. 273). The middle band in red is the “butterfly”. It is an old and a common design (Paul, 1944) that represents “the outlines of the expanded wings” or “the path a butterfly makes in flight”. Always placed horizontally, it is a symmetric pattern with its vertical line of symmetry aligned with that of the top and the bottom pattern.

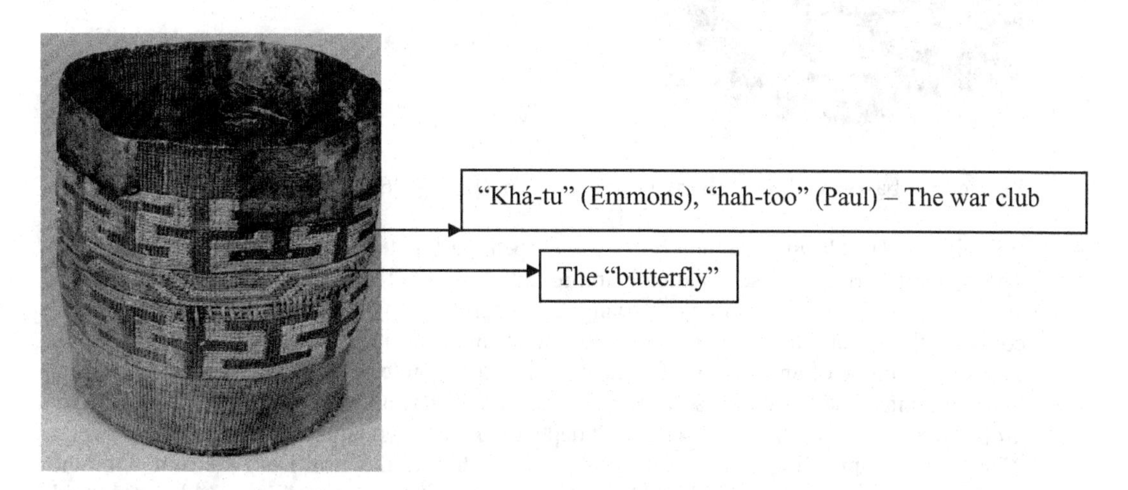

Figure 7: A basket with the “war club”, “cross” and the “butterfly” pattern. (ID: 1-1361)

(v) The "winding around" or "tying" is a pattern consists of rectangles in an echelon – stacked up to connect only at the corner (Figure 8). The pattern represents tying around with a piece of string.

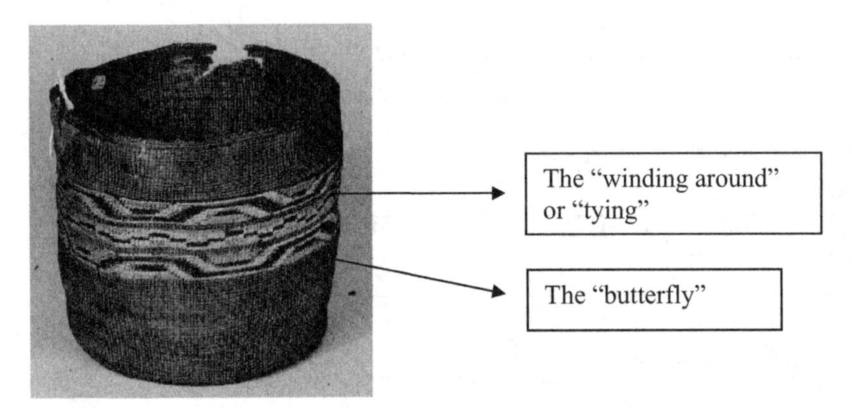

Figure 8: A basket with the "butterfly" and the "winding around" patterns. (ID:1-1293)

(vi) The "flying goose" (Figure 9). The Tlingit term for this literally means "the flight of a flock of brant" (Paul, 1944; p. 65). It represents bird migration and is claimed to be an adaptation of the simpler "tying" pattern. The symmetry properties are obvious.

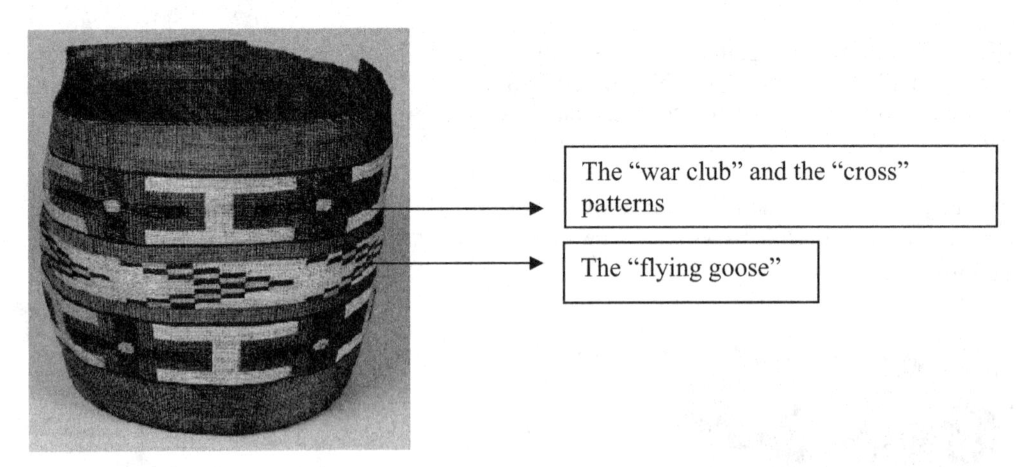

Figure 9: A basket with the "flying goose" pattern. (ID: 1-406)

(vii, viii, ix) The "head of salmon berry", the "spear barb pattern", the "rainbow" (Figure 10). The salmon berries are schematically represented as a set of rhombi, and "the half of a salmon berry" as an isosceles triangle. Patterns using such triangles, in various configurations, particularly as echelons, generate different names. The "spear barb" (Figure 11) is an example of an echelon of isosceles triangles. Single rows of such triangles between two horizontal lines, giving rise to units in reverse order, are common and are called "the drop" (Figure 12) or "the small sand hill" depending on the geographical origin of the basket. "The drop" conjures the image of droplets of water before they fall from the roof. Emmons (p. 277) writes that although "the half of the head of a salmon berry" is an old pattern with wide variations, the "head of a salmon berry" originated from copying "oil-cloth and cheap prints". He attributed the amalgamation of patterns picked up from "carpets, oil-cloth, china" as an effect of the increase in demand for basketry. He derides this economically driven adaptive behavior as a way to "destroy their individuality completely" (p. 277). Emmons points out that the "rainbow" pattern literally means "the wings of different colors" (Emmons, 1903/1993, p. 267). It consists of an arrangement of equally spaced rhombi.

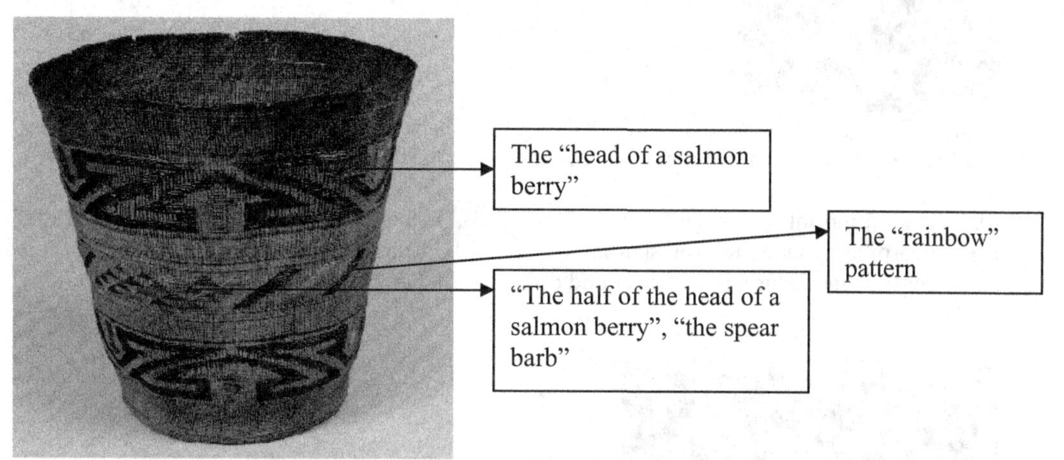

Figure 10: A basket with two bands of the "head of salmon berry" and "the spear barb" (ID: 1-2195)

Figure 11: The "spear barb' (ID:1990-120/1)

Figure 12: The "drop" (ID: 1-11372)

(x, xi) The "shaman's hat" pattern and "beaver-skin stretched on a frame" pattern (Figure 13; see also Figure 5). As one of the oldest and most popular patterns (Paul, 1944), the "shaman's hat" is a step-like design with rectangles utilizing parallel lines. The name comes from a descriptor "the work or embroidery around the head…exclusively worn by the shaman" (Emmons, 1903/1993, p. 274). The "beaver-skin stretched on a frame" pattern is a chain of non-square rhombi or lozenge-shapes.

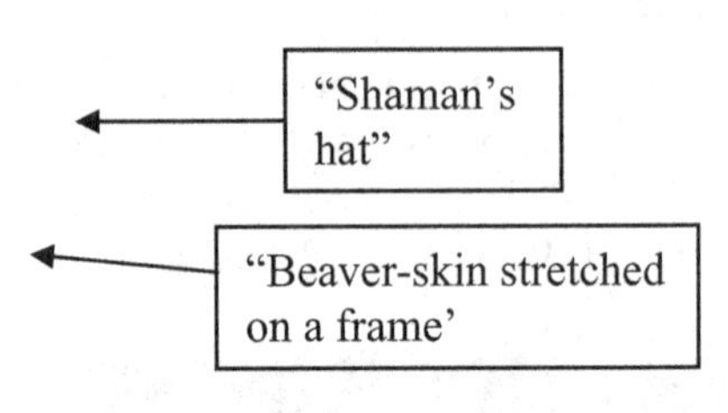

Figure 13: A basket with two wider bands of the “shaman’s hat” pattern and two smaller bands of the “beaver-skin stretched on a frame” pattern (ID: 1-1526)

(xii) The “footprint of a brown bear” (Figure 14). This pattern uses stacked-up horizontal bars to form trapeziums to represent the tracks of the bear. As one of the older patterns, it is an example of the nature, both ornate and complex, of Tlingit pattern.

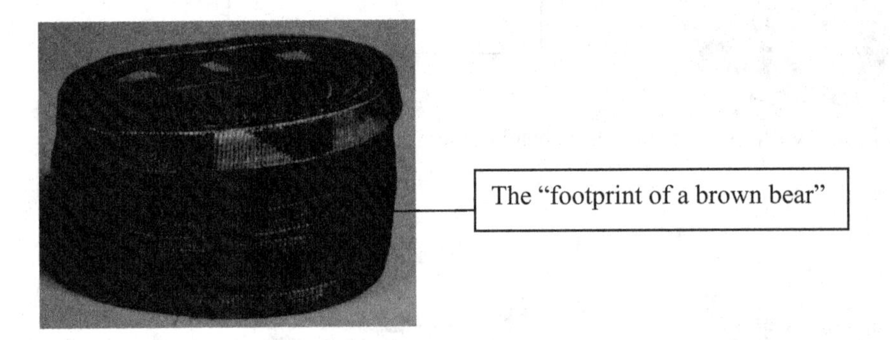

Figure 14: The “footprint of a brown bear” pattern on a lidded basket (ID: 1-782)

Implications for teaching

Early in this paper I cited D’Ambrosio’s characterization of ethnomathematics. To me, ethnomathematics carries a wider meaning of how mathematics is embedded in people’s lives, particularly in the *others*, who do not fall within the Western mainstream norm. They are not only the people from non-Western cultures, they are also the economically depressed within Western societies. Taking ethnomathematics as an approach to teaching establishes respect by honoring the diversity and complexity of cultural practices of *all* in a multicultural world.

For some years, I have been engaged in teaching mathematics through ethnomathematics. As a teacher educator, I teach courses on ethnomathematics and also have organized my courses on teacher preparation following Bishop’s six categories of mathematics-as-pancultural-activities. One of the activities that I incorporate is learning mathematics by studying the cultural artifacts at local museums (Mukhopadhyay & Best, in preparation). We start by participating in field trips to museums to disembed the mathematics situated in constructing the artifacts. We conclude by a hands-on activity of creating artifacts, which to many is “making art”, within strict mathematical constraints.

As an educational outreach, I have been organizing annual seminar series on Alternative Forms of Knowledge Construction in Mathematics for 2006 and 2007. These lectures are available as streaming videos at free of cost.

Becoming a cultural broker: Constructing two-way bridges

The role of a multicultural educator of mathematics is to assume the position of a cultural broker. As an agent of sustainable change, cultural brokers are opposed to cultural imperialists. Their work is mindful in creating two-way bridges. Stressing the importance of two-way communications and interactions as essential to preserving cultural identities, this holistic perspective honors embedded mathematics as a part of the culture and presents the whole, instead of disembedding from the cultural matrix what fits the standardized curriculum of schools. Thus, this approach of learning and teaching mathematics goes beyond teaching traditional school mathematics that is too often viewed as boring, unconnected to people's lives, and even oppressive in its Eurocentric portrayal of mathematics as an academic practice. This perspective undoubtedly addresses diversity as a complex and ongoing critical process that looks beyond tokenism and simplistic assumptions.

Looking forward

Incorporating ethnomathematics as a perspective, as described above, demands an agenda for ethnomathematics as a rigorous research program. My next step for this study needs an authentic human interface. I hope to pursue an ethnographic investigation in Tlingit communities for which I need to work with them to examine the following questions:

- What do the community members make out of their (ancestral) basketry?
- In the revival of basketry, what does pattern making mean?
- How does one learn to make these patterns?
- How does one teach others to make these patterns?
- How does school mathematics, which is currently designed and intended to establish universal uniformity, connect to authentically grounded cultural practices such as making these patterns?

A rallying-cry often heard these days is "Mathematics for all". I would suggest that a better vision is "Mathematics of all".

References

Bishop, A. J. (1988). *Mathematical enculteration. A cultural perspective on mathematics.* Dordrecht, Netherlands: Kluwer.

Boas, F. (1927/1955). *Primitive art.* New York: Dover.

Emmons, G. T. (1903/1993). *The basketry of the Tlingit and the Chilkat blanket.* Sitka, AS: Sheldon Jackson Museum. (Originally published in July, 1903, Memoirs of the American Museum of Natural History, volume III)

D'Ambrosio, U. (1985). Ethnomathematics and its place in the history and pedagogy of mathematics. *For the Learning of Mathematics, 5*, 44 – 48.

D'Ambrosio, U. (2001a). *Ethnomathematics. Link between traditions and modernity.* Rotterdam, The Netherlands: Sense Publishers.

D'Ambrosio, U. (2001b). What is ethnomathematics and how can it help children in schools? *Teaching children mathematics, 7,* 308-310.

Joseph, G. G. (2000). *The Crest of the Peacock: Non-European Roots of Mathematics* (2nd. ed.). London: Penguin, Princeton: Princeton University Press.

Lave, J. (1988). *Cognition in practice.* Cambridge: Cambridge University Press.

Lave, J. & Wenger, E. (1991). *Situated learning. Legitimate peripheral participation.* Cambridge: Cambridge University Press.

Lomawaima, K. T. (1999). The un-natural history of American Indian education. In K. G. Swisher & J. Tippeconnic III (Eds.), *Next steps: Research and practice to advance Indian education* (pp. 3- 31). Charleston, WV: ERIC Clearinghouse on rural education and small schools.

Mukhopadhyay, S. & Best, K. (in preparation). E*thnomathematics: Learning mathematics through cultural artifacts.*

Paul, F. (1944). *Spruce root basketry of the Alaska Tlingit.* Sitka, AS: Sheldon Jackson Museum.

Powell, A. B. & Frankenstein, M. (Eds.) (1997). *Ethnomathematics. Challenging Eurocentrisn in mathematics education.* Albany: SUNY Press.

Sawyer, W. W. (1955). *Prelude to mathematics*. London: Penguin.

Willoughby, C. C. (1905). Textile fabrics of New England Indians *American Anthropologist*, New Series, 7, 85-93.

Zaslavsky, C. (2005). Multicultural math: One road to the goal of mathematics for all. In E. Gutstein and B. Peterson (eds.), *Rethinking Mathematics: Teaching Social Justice by the Numbers*, pp. 124-129. Milwaukee, WI: Rethinking Schools.

MATHEMATICS EDUCATION RESEARCH EMBRACING ARTS AND SCIENCES

Norma Presmeg
Illinois State University

As a young field in its own right (unlike the ancient discipline of mathematics), mathematics education research has been eclectic in drawing upon the established knowledge bases and methodologies of other fields. Psychology served as an early model for a paradigm that valorized psychometric research, largely based in the theoretical frameworks of cognitive science. More recently, with the recognition of the need for sociocultural theories, because mathematics is generally learned in social groups, sociology and anthropology have contributed to methodologies that gradually moved away from psychometrics towards qualitative methods that sought a deeper understanding of issues involved. The emergent perspective struck a balance between research on individual learning (including learners' beliefs and affect) and the dynamics of classroom mathematical practices. Now, as the field matures, the value of both quantitative and qualitative methods is acknowledged, and these are frequently combined in research that uses mixed methods, sometimes taking the form of design experiments or multi-tiered teaching experiments. Creativity and rigor are required in all mathematics education research, thus it is argued in this paper, using examples, that characteristics of both the arts and the sciences are implicated in this work.

A vignette

In the 1990s, I taught a course on informal geometry—a content course—to students at The Florida State University who are prospective high school mathematics teachers. In the first week I asked them to bring or wear to the next class, something that had geometry in it, and to come to class prepared to tell why they had chosen that particular item and to talk about its geometry. In an interview, one of the students, Dena (who wanted to teach algebra rather than geometry), told me about her reactions to this task, as follows (Presmeg, 1998a).

Dena: I noticed when you said, for us to bring something to class or wear something that had geometry in it, for a little while I was having a difficult time, because, everything I picked up had geometry in it. And, I said, maybe there's something I misunderstood about the directions. Y'know.

Interviewer: In fact, even just the shape of a piece of clothing, any clothing.

Dena: Yeah. Anything, has geometry in it. So, for a little while I was confused. I didn't know what to bring to class, until, until I realized that, everything is going to have. I said to myself, everything, of course everything is going to have geometry to it because, y'know, anytime ... You're going to make a desk. I mean, you draw, y'know. Your plans, for making the desk, involves geometry. And everything, that is, just everywhere. I think that geometry is taught as something abstract, sketching things with proofs and rules and, not as very, everyday.

Dena's recollections of her high school geometry experiences were negative ones. "I didn't like it at all!" she concluded. Implicit in this episode are several points which are relevant to the emergence of mathematics education as a discipline in its own right, separate from but not unrelated to other disciplines such as mathematics, psychology, sociology, philosophy, linguistics, history, and (relatively recently) anthropology. It is significant that in coming of

B.Sriraman, C.Michelsen, A. Beckmann & V. Freiman (Eds). (2008). *Proceedings of the Second International Symposium on Mathematics and its Connections to the Arts and Sciences* (MACAS2). Information Age Publishing, Charlotte: NC, pp.75-88

age, mathematics education research broke away from its primary reliance on psychometric research and emulation of the hard sciences. After all, in the complex worlds of human beings learning mathematics in group settings, all aspects of the arts and the sciences that might have bearing on the improvement of this learning are relevant.

Firstly, the disciplines of mathematics and mathematics education are related by their common interest in mathematics. However, these fields differ substantially because their subject matters are different. In mathematics education it is the complex "inner" and "outer" worlds of human beings (Bruner, 1986), as they engage in activities associated with learning of mathematics, that form a primary focus of the enterprise, and therefore also of its research. Dena's agonizing over the nature and boundaries of geometry is fruitful and provocative subject matter to a mathematics education researcher interested in the teaching and learning of geometry. The avenues along which this research may lead depend not only on the data, but also on the interests and interpretations of the researcher. The tendency of such hermeneutic research to use progressive focusing rather than pre-ordinate design (Hartnett, 1982) makes this kind of research as interesting as a mystery story, even if the mystery is to some extent self-created. In this respect, mathematics education research may have elements in common with mathematics research. Certainly, the arts as well as the sciences are implicated in both.

A second point is that the inner and outer worlds of a student—while in this context specific to learning geometry—relate to the disciplines of psychology and sociology respectively, and to the interactions between their elements, as they concern an individual such as Dena. A balance between elements of these two disciplines is required in mathematics education, as witnessed in 1990s debates on the necessity of steering a course between Piaget and Vygotsky, representing individual and social aspects of learning respectively, in constructing theory in mathematics education (Confrey, 1991, 1992, 1994/5; Ontiveros, 1991). It is significant that Confrey believed that neither Piaget's nor Vygotsky's theory alone was adequate to model the complex processes of human learning. She elaborated as follows.

> "What I argue is that proposing an interaction between the two strands will constitute a significant change in both theories, and will require a theory which is neither Piagetian nor Vygotskian, but draws heavily on both. I argue this due to Vygotsky's rejection of the possibility of the development of many of the basic processes of Piaget, such as the development and awareness of schemes, of operations and of reflective abstraction, until social interaction is established. An alternative theory will have to propose a much stronger and more detailed description of how the 'natural' and the 'socio-cultural' activities of the child are linked, allowing for the complexity of each and probably requiring a renaming of the natural strand to reflect a more constructivist view" (Confrey, 1992, pp. 5-6).

Confrey's analysis prefigures the point—well expressed by Cobb (2007)—that in the face of incommensurable theories one way of proceeding is to find out how practitioners in the discipline of the parent theory view the canons of their research. This perspective enables the mathematics education researcher to bring a broader vision to the construction of home-grown theories that will be useful in addressing problems of mathematics education. Cobb (2007) explained the benefit of this attitude as follows.

> The openness inherent in this stance to incommensurability has the benefit that in coming to understand what adherents of an alternative perspective think they are doing, we develop a more sensitive and critical understanding of some of the taken-for-granted aspects of our own perspective. (p. 32)

In the creativity literature it has long been a well-accepted principle that new views may be garnered by *making the familiar strange*, and by *making the strange familiar* (e.g., De Bono, 1970). However, Cobb (2007) went much further than that. He compared four theoretical perspectives that have been influential in mathematics education research, namely, those of experimental psychology (whose methodologies have been advocated again—as in the 1950s and 1960s—by funding agencies in the U.S.A. recently as the only form of *scientific* research in mathematics education: see U.S. Congress, 2001), cognitive psychology (from the *actor*'s perspective rather than the *subject*'s), Vygotskian sociocultural theory, and distributed cognition. In comparing these four perspectives with regard to their characterization of the individual learner, and in their usefulness for design research in mathematics classrooms, Cobb came to the balanced conclusion that each perspective has merits *for certain purposes*, but not necessarily for designing effective mathematics teaching. In his view, scientific randomized experiments are useful to and serve the administrative and political purposes of policy makers. He makes a strong case that insistence on the hegemony of scientific research in the form of randomized statistical experiments would be short-changing the community of classroom teachers of mathematics. As he shows clearly, all theories are based on philosophical premises, although those advocating a particular stance may not acknowledge the limiting effect of these choices.

A third point implicit in Dena's pondering in the initial vignette is that philosophy is ubiquitous also in all questions which are of concern to mathematics education researchers. The nature of geometry is an ontological issue, while how it was taught in Dena's school experience relates to issues of epistemology. Both components are essential in mathematics education theory building, since one's beliefs about the nature of mathematics and mathematical knowledge are the 'spectacles' through which one looks at its teaching and learning.

Tension between the view that "Everything is mathematics" (as Dena expressed it, "Everything is going to have geometry to it"), and the rigorous mathematical position that "Only formal mathematics is valid", was well expressed by Millroy (1992) in her monograph on the mathematical ideas of a group of carpenters, as follows.

> [I]t became clear to me that in order to proceed with the exploration of the mathematics of an unfamiliar culture, I would have to navigate a passage between two dangerous areas. The foundering point on the left represents the overwhelming notion that 'everything is mathematics' (like being swept away by a tidal wave!) while the foundering point on the right represents the constricting notion that 'formal academic mathematics is the only valid representation of people's mathematical ideas' (like being stranded on a desert island!). Part of the way in which to ensure a safe passage seemed to be to openly acknowledge that when I examined the mathematizing engaged in by the carpenters there would be examples of mathematical ideas and practices that I would recognize and that I would be able to describe in terms of the vocabulary of conventional Western mathematics. However, it was likely that there would also be mathematics that I could not recognize and for which I would have no familiar descriptive words. (pp. 11-13)

On the basis of her research results, Millroy argued strongly for the broadening of traditional ideas of what constitutes mathematics. She wrote, "We need to bring nonconventional mathematics into classrooms, to value and to build on the mathematical ideas that students already have through their experiences in their homes and in their communities" (ibid., p. 192). Steen's (1990) view of mathematics as the science of pattern and order opens the door to this lifting of the limiting boundaries of mathematics. Millroy's recommendation is consonant with those in the National Council of Teachers of Mathematics (NCTM)'s recent calls for connected knowledge in mathematics education (2000). A related point is that a "mathematical cast of mind" may be a characteristic of students who are gifted in mathematics (Krutetskii, 1976). This mathematical cast of mind enables these students to identify and reason about mathematical elements in all their experiences; they construct their worlds with mathematical eyes, as it were. But unless teachers are aware of the necessity of encouraging students to recognize mathematics in diverse areas of their experience, only a few students will develop this mathematical cast of mind on their own. Many more will continue to regard mathematics as "a bunch of formulas" to be committed to short term memory for a specific purpose such as an examination, and thereafter forgotten (Presmeg, 1993).

The foregoing sets the scene for a fourth point which emerges from these considerations, namely, the links which mathematics education research has been building with various branches of anthropology, particularly with regard to methodology and construction of theory. Millroy's (1992) study was ethnographic. Entering to some extent into the worlds of Cape Town carpenters in order to experience their "mathematizing" required that Millroy become an apprentice carpenter for what she called an extended period, although the four-and-a-quarter months of this experience might still seem scant to an anthropologist (Eisenhart, 1988). But the point is that the ethnographic methodology of anthropological research is peculiarly facilitative of the kinds of interpreted knowledge which are valuable to mathematics education researchers and practitioners. After all, each mathematics classroom may be considered to have its own culture (Nickson, 1992). In order to understand the learning, or, sadly, the prevention of learning which may take place there, the ethnographic mathematics education researcher needs to be part of this world, interpreting its events for an extended period, and then documenting the culture of this world, making the familiar strange and the strange familiar while walking the tightrope of being in but not totally of the world that is observed.

The Arts and the Sciences—at War?

I shall start this section with a brief intellectual autobiography, which resonates with the recent history of research in the field in which I work, mathematics education. When I was a teenager, a senior in high school, I was reading Sir James Jeans' books about the universe, and I was also particularly inspired by the life and work of Marie Curie, who was a dedicated woman in the man's world of the hard sciences at the end of the 19th century, and by the incomparable life and work of Albert Einstein (1970, 1973, 1976, 1979). As valedictorian of my high school in South Africa, I was "good at" all subjects—but it seemed that the arts and the sciences were at war in me, because I was drawn to both. I chose the path of the sciences, doing a B.Sc. degree with majors in mathematics and physics, intending to become a nuclear physicist. The reasoning was that if I chose the sciences, I could learn languages, read poetry and philosophy, play a musical instrument, and appreciate art all on my own, whereas if I chose the arts it was unlikely that I would study the hard sciences independently. This reasoning has been justified to some extent. But for a career, I soon realized that I wanted not a lab coat, but a profession that dealt with people in all their complexity; so the best compromise was to become a high school teacher of mathematics, and that is the career path that I followed for twelve years.

My master's degree was in educational psychology, and included a thesis examining the creative thought of Albert Einstein, to investigate whether there were implications for the classroom learning of mathematics. And there were! Albert Einstein was a visualizer, and his mental imagery was the rich source of his creative insights (Holton, 1973; Schilpp, 1959). But there were students in high school mathematics classes who were visualizers, as I knew from the exceptionally high spatial scores they were achieving on the battery of tests they were doing for vocational guidance—and they were achieving poorly in mathematics, as had Einstein in the military environment of the *gymnasium* he attended in Munich before moving to Switzerland. The question of *why* demanded further investigation, and I was privileged to spend three years doing a Ph.D. at Cambridge University in England (Presmeg, 1985), with the following central research goal as it concerned mathematics education:

> *To understand more about the circumstances which affect the visual pupil's operating in his or her preferred mode, and how the mathematics teacher facilitates this or otherwise.*

The research was exciting, full of surprises, as absorbing as a mystery novel, and my three years at Cambridge from 1982 to 1985 remain a highlight of my life. Some of the results of this research will be outlined later, but first let me draw comparisons with the field of mathematics education, which was starting to emerge as a field of study in its own right.

Complex human worlds: mathematics education as an emergent field

While the history of mathematics goes back several millennia, mathematics education as a field of study in its own right is barely half a century old (Sierpinska & Kilpatrick, 1998). The foremost and oldest fully international journal in this field, *Educational Studies in Mathematic*, a few years ago celebrated its 50th anniversary. As suggested in the opening section, initially the study of problems in the learning of mathematics was a small subset of the wider realm of the concerns of psychology. With respect and admiration for the relative certainty of results obtained by researchers in the hard sciences, in which empirical investigation was used to confirm or disconfirm theory, early researchers in mathematics education (especially in the 1960s and 1970s) tried to emulate this research. Psychometric research was the only genre of research in mathematics education that was considered worthy of the name. Of this period, the Soviet psychologist Krutetskii (1976) wrote as follows:

> It is hard to understand how theory or practice can be enriched by, for instance, the research of Kennedy [in 1963], who compared, for 130 mathematically gifted adolescents, their scores on different kinds of tests and studied the correlation between them, finding that in some cases it was significant and in others not. The process of solution did not interest the investigator. But what rich material could be provided by a study of the process of mathematical thinking in 130 mathematically able adolescents! (p. 14)

Indeed, it was lamented that mathematics education research was having little impact, in fact appeared to be irrelevant, in mathematics teachers' classroom practices. Research as epitomized in "Aptitude-Treatment Interaction" studies (ATIs) seemed to have little impact or relevance in mathematics classrooms. The question of relevance is still an issue in mathematics education research, but more recent developments in this growing field have embraced diverse methods from the human sciences, and even more recently from the social sciences, taking into account some of the complexity that is endemic to research involving human beings.

In the early 1980s, when I was engaged in my doctoral research, qualitative, hermeneutic research under banners such as "illuminative evaluation" (McCormick, Bynner, Clift, James, & Brown, 1977) was starting to be viewed as legitimate in mathematics education because it could address questions about details of teaching and learning that were inaccessible to purely statistical research. My study involved both quantitative and qualitative methods. At about the same time, research carried out by teachers in their own classrooms (now called "action research", e.g., Ball, 2000) was gaining currency. It was recognized that methods from other disciplines might need adaptation to the particular requirements of mathematics education research, but that there was a rich variety of methodologies that could be valuable. In the last three decades, mathematics education journals and conferences have proliferated, and universities internationally have established programs in mathematics education, housed either in schools of education or more rarely in mathematics departments. These changes accelerated in the 1990s. In a search for identity in its own right (Sierpinska & Kilpatrick, 1998), mathematics education became recognized as a legitimate field, distinct from, yet informed by, the disciplines of mathematics, psychology, sociology, anthropology, philosophy, and even linguistics (Sfard, 2000; Dörfler, 2000). Mathematics education, as a human science, embraces the arts as well as the sciences, and various qualitative research methodologies adapted from the humanities became recognized as legitimate in addition to the previously dominant psychometric paradigms. In particular, following Bishop's (1988, 2004) seminal work, there was increasing recognition of cultural and social aspects of the classroom learning of mathematics, complementing the psychological emphasis of cognitive theories of learning. In this field there is no need for war between the arts and the sciences – both are important. I have come home!

Creativity in the arts and in the sciences: mathematics education creativity spanning both

As mentioned, the heart of Albert Einstein's immensely creative thought was his capacity to visualize (Schilpp, 1959). Mathematics has an obvious visual component, not only overtly, as in geometry or trigonometry, but also in the mental imagery that by self-report enhances the thinking of many creative mathematicians (Sfard, 1994). Why, then, were there visualizers in high school mathematics classes who were finding this subject so difficult that they were obtaining failing grades on examinations (Presmeg, 1985)?

The purpose of my doctoral research was to investigate the strengths and limitations of visual processing in mathematics in a classroom context at senior high school level, and to investigate the effect on learners who are visualizers of the preferred cognitive modes, attitudes, and actions of their mathematics teachers. (For a fuller account, see Presmeg, 2006a&b.) Selection of students and teachers required the development of a new mathematical processing instrument to measure preference for visual thinking in mathematics. I still use this instrument to understand more about the visualization styles of students in my classes. On the basis of the preference for mathematical visualization (MV) scores obtained using this instrument, 13 mathematics teachers were chosen to represent the full range of scores available. In the senior classes of these teachers, 54 visualizers (23 boys and 31 girls) were chosen from 277 high school students. Visualizers were taken to be those who scored above the median score for this population, on the preference test.

The research methodology included participant observation in the classes of the teachers over an eight-month period, and tape-recorded interviews with teachers and students, as well as sparing use of non-parametric statistics to identify trends in the data from the visualization instrument. As a framework for observation in lessons, 17 classroom aspects (CAs) were identified that the literature suggested were facilitative of formation and use of visual imagery in mathematics. The teaching visuality scores obtained by triangulation of viewpoints (teacher's, students', and researcher's) on the basis of the CAs were only weakly correlated with the teachers' MV scores from the preference instrument. It made sense that a good teacher who feels little need of visual supports might recognize the need of mathematics learners for more of these supports. After item analysis and refinement of the CAs, teaching visuality scores divided the teachers neatly into three groups, namely, a nonvisual, a middle, and a visual group. Analysis of 108 transcripts of lessons revealed 45 further classroom aspects that differentiated the three groups of teachers, and that suggested that the visual teachers manifested traits associated with creativity, such as use of humor in their teaching. (Einstein had a marvelous sense of humor—see Dukas & Hoffmann, 1979.)

One of the biggest surprises in this research was that it was the teaching of the middle group of teachers, not the visual group, which was optimal for the visualizers in the study. All the difficulties experienced by the visualizers in their learning of mathematics related in one way or another to the generality of mathematical principles. An image or a diagram, by its nature, is one concrete case, and students need to learn how to distinguish the general elements from the specific ones in learning mathematics. Visual teachers, who had mastered these distinctions, were not cognizant of the difficulties experienced by their students. In my data, there were two ways in which a mental image or related diagram could represent generalized mathematical information. Firstly, the image itself could be of a more general form, which I designated *pattern imagery*. Secondly, a concrete picture (mental or represented on paper or a computer screen) could be used *metaphorically* to stand for a general principle. This latter result of this research led me to the fascinating study of the use of metaphor and metonymy in mathematics education, during the decade of the nineties (Presmeg, 1992, 1997a&b, 1998b). However, I also became involved in another compelling research agenda, which I shall describe in the next section.

Different bridges: semiotic chaining linking mathematics in and out of school

In the last two decades, two strands of significance have been developing in the mathematics education research community. On the one hand, there have been increasing calls that teachers facilitate the construction of *connected* knowledge in mathematics classrooms (National Council of Teachers of Mathematics, 1989, 2000). These connections entail not only the linking of various branches of mathematics that have been taught as separate courses at high school level, but also the linking of classroom mathematics with other subjects in the curriculum. And particularly, the importance is stressed of linking school mathematics with the experiential realities of learners. On the other hand, the importance of symbolizing and discourse in the teaching and learning of mathematics has come to the fore (Cobb, Yackel, & McClain, 2000), along with recognition of the significance of sociocultural aspects of the learning of mathematics (Bishop, 1988).

I set out to link these two significant strands by exploring answers to the following question: *How can teachers use semiotic theories to help them facilitate the construction of connections in the classroom learning of mathematics?* In particular, semiotic chaining presented a fruitful method of bridging the formal mathematics of the classroom and the informal out-of-school mathematical experiences of learners. The significance for mathematics education of theories originating in linguistics was becoming apparent to me. At first in this research I used chaining of signifiers based on Lacan's inversion of Saussure's dyadic model of semiosis (Saussure, 1959). I investigated how teachers and graduate students could use these chains to link the cultural activities of learners with mathematical principles. Working with two research assistants and a doctoral student, Matthew Hall, we interviewed students and taught teachers to build such chains and use them in the mathematics classroom (see Hall, 2000). There was the potential for the celebration of diversity and equity. We had some success, but the research suggested the need for a more complex model, because not just signifiers and signifieds, but *interpretation*, were endemic in the activities. Thus I was led to development a nested model of chaining based on the triadic theory of Charles Sanders Peirce (1992, 1998). Some of his many constructs illuminated the research, like searchlights, and I am still excited and involved in the exploration of the repercussions of this work. Many instances of the potential of semiotic chaining to foster connected knowledge of mathematics illustrated its significance (e.g., Presmeg, 2006c), and the research is ongoing. Recently, I have been using a triadic Peircean lens to investigate ways that students connect, or fail to connect, the various registers (Duval, 1999) of school trigonometry (Presmeg, 2006b).

There are clearly elements of the arts and the sciences in this mathematics education research. In the next section I invoke Habermas's (1978) *knowledge-constitutive interests* to argue this case further.

Knowledge-constitutive interests invoking arts and sciences

Using Ewert (1991) and Grundy (1990) as sources, in figure 1 I have summarized the three types of knowledge and their philosophical bases posited by Habermas (1978). This triad comprises not merely three different ways of looking at knowledge, but three different ways of characterizing what *counts* as knowledge. It is beyond the scope of this paper to discuss Habermas's theory in depth. (Interested readers should consult the original sources.) In this paper I shall use this summary to argue that there is room in mathematics education research for all three kinds of knowledge.

Technical	**Practical**	**Emancipatory**	
Social media:			
	labour	*interaction*	*power*
Conditions for the three sciences:			
	empirical-analytic	*hermeneutic*	*critical*
→ procedures for basic activities:			
	control of external conditions	*communication*	*reflection*
Trichotomous division between sciences:			
	natural science	*cultural science*	*critical science*
Forms of knowledge:			
	instrumental rationality	*subjective meaning*	*critical theory*
Philosophical basis:			
	positivism	*phenomenology*	*critical theory*

****

Eidos and disposition:			
	specific, definable ideas - techne (skill)	*the Good - phronesis (judgement)*	*liberation - critique (critical community)*
Action and outcome:			
	poietike → product	*practical action → interaction*	*emancipatory action → praxis*

****

Figure 1. Three Knowledge-constitutive Interests

Of Habermas's three types of interests that constitute knowledge, it is obviously the technical one that epitomizes knowledge in the hard sciences. Literary creativity and research are examples of the seeking for knowledge of the second type, in which interpretation of the human condition is paramount. The enterprise seeks to understand that condition, but not necessarily to change it. The critical reflection called for in the third category, by way of contrast, has the goal of changing the human condition in some way—hence its designation as emancipatory. In contemporary mathematics education research, examples are to found of all three types of interests. In broad categories, the *technical* interest is ongoing in large-scale statistical studies, the *practical* interest is evident in hermeneutic studies that aim for understanding of the mathematical thinking of individual students or small groups of students, and the *emancipatory* interest is apparent in studies that address issues of social justice and critical issues such as access to the study of mathematics. It is beyond the scope of this paper to characterize the landscape of mathematics education research in detail, but the following are examples of research in each of these three categories.

As an example of research in the first category, the investigations of Gagatsis and his co-researchers at the University of Cyprus, Nicosia seek new knowledge of issues in the teaching and learning of mathematics through the statistical investigation, using large samples, of such topics as "Students' improper proportional reasoning" (Modestou and

Gagatsis, 2007), or "Exploring young children's geometrical strategies" (Gagatsis, Sriraman, Elia, & Modestou, 2006). Because it is not feasible to assign children randomly to the classes in these studies, the studies may be characterized as of pseudo-experimental design. The methodology enables group trends and relationships to be uncovered, without seeking to ascertain the reasons *why* these trends and relationships are significant. In-depth investigation of the question of "Why?" would entail research in the second category. In my own research on visualization, the construction and validation of an instrument for preference for visualization involved interests in the technical category: validity and reliability were established using non-parametric statistics (Presmeg, 1985). Large samples showed that there was no statistically significant difference between the boys and the girls with regard to their preference for visual thinking in mathematics; however, there was a significant difference between the preference for visualization of the teachers in this part of the study, and their students, who needed far more visual supports than they did.

Again, the question of *why* was deferred to Habermas's second category. Insights into the difficulties and strengths of visualization in teaching and learning mathematics came from interpretive research involving a whole school year of classroom observation and interviews with 54 high school "visualizers" and their 13 mathematics teachers. All of the problems experienced by these learners related in one way or another to the need for mathematical abstraction and generalization, as indicated in an earlier section of this paper. Whereas this kind of research provided insights, it did not have the overt goal of changing classroom practice, although teacher awareness of the results might in fact result in "practical action"—*praxis*— in the classroom (Grundy, 1990). Emancipatory interests, in contrast, have the goal of praxis.

Examples of research involving emancipatory interests can be found in the chapters of the monograph on *International perspectives on social justice in mathematics education* (Sriraman, 2007). After a useful historical introduction to issues of social justice by the editor, Sriraman, several of the chapters describe projects that in one way or another attempt to address the issues of equity that are implicit in social justice applied to mathematics education. For instance, Merrilyn Goos, Tom Lowrie, and Lesley Jolly describe a framework for analyzing key features of partnerships amongst families, schools, and communities in Australian numeracy education. Iben Maj Christiansen contributes a thoughtful and exploratory chapter based on her experiences introducing mathematical ideas to university students in South Africa and Denmark, through social data that highlight inequity. Her analysis leads her to the startling question, "Does our insistence on these 'critical examples' end up being 'imposition of emancipation'?" Tod Shockey contributes the positive influence of a culturally appropriate curriculum for Native Peoples in Maine, USA. Libby Knott explores issues of status and values in the professional development of mathematics teachers in Montana, USA. Eric Gutstein provides a companion piece to his recent influential book on social justice in a Chicago school classroom (Gutstein, 2006). These chapters and others have the more or less explicit goal of changing praxis in mathematics education. Although the monograph also contributes useful empirical and theoretical ideas to the ongoing conversation about social justice in mathematics education (practical interest), its emancipatory interest places it squarely in Habermas's third category. My own research on ways that teachers may incorporate the cultural practices of students in their classes into the praxis of school teaching and learning of mathematics also embraces this category to some extent.

Final thoughts

Of the three categories of Habermas's (1978) knowledge-constitutive interests, the technical one pertains to the sciences, whereas the practical and emancipatory belong to the concerns and complexities of human life and its interpretation, to the thoughts and feelings of human beings. The discipline of mathematics itself, with its inexorable logic and *instrumental rationality*, resides as a content domain in the technical category, although the creative domain of mathematicians doing research in mathematics might arguably relate better to the *subjective meaning* of the practical category. In contrast, because the teaching and learning of mathematics are practices engaged in by human beings, subjective meaning is all-important if mathematics is to be learned meaningfully, and *critical theory* relates to the improvement of this teaching and learning in mathematics classrooms. However, the content of mathematics with its historically-constituted canons is the subject of this teaching and learning. Thus I would argue that both the sciences and the arts are implicated in mathematics education, whose research also requires the full gamut of methodologies available in the arts and the sciences.

References

Ball, D. L. (2000). Working on the inside: Using one's own practice as a site for studying teaching and learning. In A. E. Kelly & R. A. Lesh (Eds.), *Handbook of research design in mathematics and science education* (pp. 365-402). Mahwah, NJ: Lawrence Erlbaum Associates.

Bishop, A. J. (1988). *Mathematical enculturation: A cultural perspective on mathematics education.* Dordrecht: Kluwer Academic Publishers.

Bishop, A. J. (2004). Mathematics education in its cultural context. In T. P. Carpenter, J. A. Dossey, & J. L. Koehler (Eds.), *Classics in mathematics education research* (pp. 201-207). Reston, VA: National Council of Teachers of Mathematics. Reprinted from *Educational Studies in Mathematics, 19* (1988), 179-191.

Bruner, J. (1986). *Actual minds, possible worlds.* Cambridge, MA: Harvard University Press.

Cobb, P. (2007). Putting philosophy to work. In F. K. Lester, Jr. (Ed.), *Second handbook of research in mathematics teaching and learning* (pp. 3-38). Charlotte, NC: Information Age Publishing.

Cobb, P. Yackel, E., & McClain, K. (Eds.) (2000). *Symbolizing and communicating in mathematics classrooms: Perspectives on discourse, tools, and instructional design.* Mahwah, NJ: Lawrence Erlbaum Associates.

Confrey, J. (1991). Steering a course between Piaget and Vygotsky. *Educational Researcher 20*(8), 28-32.

Confrey, J. (1992). *How compatible are radical constructivism, social-cultural approaches and social constructivism?* Paper presented at a conference on Alternative Epistemologies in Education, University of Georgia, Athens, GA, February 20-23, 1992.

Confrey, J. (1994). A theory of intellectual development: Part 1. *For the Learning of Mathematics 14*(3), 2-8.

Confrey, J. (1995). A theory of intellectual development: Part 2. *For the Learning of Mathematics 15*(1), 38-48.

Confrey, J. (1995). A theory of intellectual development: Part 3. *For the Learning of Mathematics 15*(2), 36-45.

De Bono, E. (1970). *Lateral thinking: A textbook of creativity.* London: Pelican.

Dörfler, W. (2000). Means for meaning. In P. Cobb, E. Yackel, & K. McClain (Eds.), *Symbolizing and communicating in mathematics classrooms: Perspectives on discourse, tools, and instructional design* (pp. 99-131). Mahwah, NJ: Lawrence Erlbaum Associates.

Dukas, H. & Hoffmann, B. (Eds.) (1979). *Albert Einstein: The human side*. Princeton, NJ: Princeton University Press.

Duval, R. (1999). Representations, vision and visualization: Cognitive functions in mathematical thinking. Basic issues for learning. In F. Hitt & M. Santos (Eds.), *Proceedings of the 21st Annual Meeting of the North American Chapter of the International Group for the Psychology of Mathematics Education* (Vol. 1, pp. 3-26). Columbus, OH: ERIC Clearinghouse for Science, Mathematics, & Environmental Education.

Einstein, A. (1970). *Out of my later years*. New York: Greenwood Press.

Einstein, A. (1973). *Ideas and opinions*. London: Souvenir Press.

Einstein, A. (1976). *Relativity: The special and the general theory*. London: Methuen.

Einstein, A. (1979). *The world as I see it*. New York: Citadel Press.

Eisenhart, M.A. (1988). The ethnographic research tradition and mathematics education research. *Journal for Research in Mathematics Education 19*(2), 99-114.

Ewert, G. D. (1991). Habermas and education: A comprehensive overview. *Review of Educational Research, 61*(3), 345-378.

Gagatsis, A., Sriraman, B., Elia, I, & Modestou, M. (2006). Exploring young children's geometrical strategies. *Nordic Studies in Mathematics Education, 11*(2), 23-50.

Grundy, S. (1990). *Curriculum: Product of praxis*. New York: The Falmer Press

Hall, M. (2000). *Bridging the gap between everyday and classroom mathematics: An investigation of two teachers' intentional use of semiotic chains*. Unpublished doctoral dissertation, The Florida State University.

Hartnett, A. E. (1982). *The social sciences in educational studies: A selective guide to the literature*. London: Heinemann.

Habermas, J. (1978). *Knowledge and human interests*. London: Heinemann.

Holton, G. (1973). *Thematic origins of scientific thought: Kepler to Einstein*. Cambridge, MA: Harvard University Press.

Krutetskii, V. A. (1976). *The psychology of mathematical abilities in schoolchildren*. Chicago: University of Chicago Press.

McCormick, R., Bynner, J., Clift, P., James, M., & Brown, C. M. (Eds.) (1977).*Calling education to account*. London: Heinemann.

Modestou, M & Gagatsis, A. (2007). Students' improper proportional reasoning: A result of the epistemological obstacle of "linearity". *Educational psychology, 27*(1), 75-92.

Millroy, W.L. (1992). *An ethnographic study of the mathematics of a group of carpenters*. Reston, VA: National Council of Teachers of Mathematics, Monograph 5.

National Council of Teachers of Mathematics (1989). *Curriculum and evaluation standards for school mathematics*. Reston, VA: The Council.

National Council of Teachers of Mathematics (2000). *Principles and standards for school mathematics*. Reston, Virginia: The Council.

Nickson, M. (1992). The culture of the mathematics classroom: An unknown quantity? In D. A. Grouws (Ed.) *Handbook of research on mathematics teaching and learning* (pp. 101-114). New York: Macmillan.

Ontiveros, J. Q. (1991). *Piaget and Vygotsky: Two interactionist perspectives in the construction of knowledge,* paper presented at the 5th International Conference on Theory of Mathematics Education, Paderno Del Grappa, June 20-27, 1991.

Peirce, C. S. (1992). *The essential Peirce*. Volume 1, edited by N. Houser & C. Kloesel, Bloomington: Indiana University Press.

Peirce, C. S. (1998). *The essential Peirce.* Volume 2, edited by the Peirce Edition Project. Bloomington: Indiana University Press.

Presmeg, N. C. (1985), *The role of visually mediated processes in high school mathematics: A classroom investigation.* Unpublished Ph.D. dissertation, University of Cambridge.

Presmeg, N. C. (1992). Prototypes, metaphors, metonymies, and imaginative rationality in high school mathematics. *Educational Studies in Mathematics 23*(6), 595-610.

Presmeg, N. C. (1993). Mathematics - 'A bunch of formulas'? Interplay of beliefs and problem solving styles. In I. Hirabayashi, N. Nohda, K. Shigematsu, & F.-L. Lin (Eds.), *Proceedings of the 17th Annual Meeting of the International Group for the Psychology of Mathematics Education*, Tsukuba, Japan, July 18-23, Vol. 3, 57-64.

Presmeg, N. C. (1997a). Reasoning with metaphors and metonymies in mathematics learning. In L. D. English (Ed.), *Mathematical reasoning: Analogies, metaphors, and images* (pp. 267-279). Hillsdale, NJ: Lawrence Erlbaum Associates.

Presmeg, N. C. (1997b). Generalization using imagery in mathematics. In L. D. English (Ed.), *Mathematical reasoning: Analogies, metaphors, and images* (pp. 299-312). Hillsdale, NJ: Lawrence Erlbaum Associates.

Presmeg, N. C. (1998a). Balancing complex human worlds: Mathematics education as an emergent discipline in its own right. In A. Sierpinska & J. Kilpatrick (Eds.), *Mathematics education as a research domain: A search for identity* (Vol. 1, pp. 57-70). International Commission on Mathematical Instruction Study Publication. Dordrecht: Kluwer Academic Publishers.

Presmeg, N. C. (1998b). Metaphoric and metonymic signification in mathematics. *The Journal of Mathematical Behavior, 17*(1), 25-32.

Presmeg, N. C. (2006a). Research on visualization in learning and teaching mathematics. In A. Gutiérrez & P. Boero (Eds.), *Handbook of research on the psychology of mathematics education: Past, present and future* (pp. 205-235). Rotterdam, The Netherlands: Sense Publishers.

Presmeg, N. C. (2006b). A semiotic view of the role of imagery and inscriptions in mathematics teaching and learning. Plenary Paper. In J. Novotna, H. Moraova, M. Kratka, & N. Stehlikova (Eds.), *Proceedings of the 30th Annual Meeting of the International Group for the Psychology of Mathematics Education,* Vol. 1, pp. 19-34. Prague, July 16-21, 2006.

Presmeg, N. C. (2006c). Semiotics and the "connections" standard: Significance of semiotics for teachers of mathematics. *Educational Studies in Mathematics, 61*(1-2), 163-182.

Saussure, F. de (1959). *Course in general linguistics.* New York: McGraw-Hill.

Schilpp, P. A. (Ed.) (1959). *Albert Einstein: Philosopher-scientist.* Vol. I & II, Library of Living Philosophers. London: Harper & Rowe.

Sfard, A. (1994). Reification as the birth of metaphor. *For the Learning of Mathematics, 14*(1), 44-55.

Sfard, A. (2000). Symbolizing mathematical reality into being – or how mathematical discourse and mathematical objects create each other. In P. Cobb, E. Yackel, & K. McClain (Eds.), *Symbolizing and communicating in mathematics classrooms: Perspectives on discourse, tools, and instructional design* (pp. 37-98). Mahwah, NJ: Lawrence Erlbaum Associates.

Sierpinska, A. and Kilpatrick, J. (Eds.) (1998). *Mathematics education as a research domain: A search for identity.* Dordrecht: Kluwer Academic Publishers.

Sriraman, B. (Ed.). (2007). *International perspectives on social justice in mathematics education.* The University of Montana Press: Monograph 1, The Montana Mathematics Enthusiast (http://www.math.umt.edu/TMME/Monograph1/).
(also published as)

Sriraman, B. (Ed). (2007). *International perspectives on social justice in mathematics education.* Charlotte, NC: Information Age Publishing.

Steen, L.A.: 1990, *On the Shoulders of Giants: New Approaches to Numeracy,* National Academy Press, Washington, DC.

U.S. Congress (2001). *No child left behind Act of 2001.* Washington, DC: Author.

THE GEOMETRY OF 17th CENTURY DUTCH PERSPECTIVE BOXES

Claus Jensen(Retired since July 2003)
Hasseris Gymnasium, Aalborg, Denmark

Representing space *in the plane*: The Central Projection.

An ordinary perspective painting like that of fig. 1 gives the impression of space in the plane. In the foreground of the painting there is a dog, whose lower part is missing. Behind the dog a tiled floor stretches towards the background where there is a door. Through the door we see another room with a tiled floor, and so on – almost to infinity. Space is indeed overwhelming.

Fig. 1. Left-hand end panel of Samuel van Hoogstraten's perspective box, National Gallery in London. Reproduced with the gallery's permission after its cardboard model.

B.Sriraman, C.Michelsen, A. Beckmann & V. Freiman (Eds). (2008). *Proceedings of the Second International Symposium on Mathematics and its Connections to the Arts and Sciences* (MACAS2). Copyright held by Claus Jensen, pp.89-106

Fig. 2. Right-hand end panel of Samuel van Hoogstraten's perspective box, National Gallery in London. Reproduced with the gallery's permission after its cardboard model.

Fig. 2 shows another example of a perspective painting. In the foreground we see a broom whose lower part is missing. Mathematically, perspective paintings like those of figs. 1 and 2 are based on the central projection, illustrated by Albrecht Dürer in one of his famous woodcuts of 1525 (fig. 3). Here, the eye firmly screwed into the wall represents the eye point of the projection, and the wooden frame its picture plane. A canvas is mounted on the frame by means of two hinges, allowing the canvas to turn so as to coincide with the picture plane.

The task is to draw the lute in perspective. This is done point by point. On the lute any point P is chosen and connected with the eye point by a chord representing the ray of vision. It's intersection with the picture plane defines the image P' of P, cf. fig. A1. The position of P' is marked by means of two movable intersecting strings attached to the frame. After removing the chord, the canvas may be folded towards the frame, and the crossing of the strings may be marked as a point on the canvas.

Then the canvas is opened, a new point P is chosen and the entire procedure repeated. On the canvas the image of the lute drawn in perspective does actually appear point by point.

The central projection has some important properties to be used in what follows. These may be found in any standard exposition of perspective theory, but for easy reference and for clarity of the arguments figures A1-A5 in the appendix below briefly outline some of these properties.

Fig. 3. Albrecht Dürer's method of drawing a lute in perspective, *Unterweisung der Messung*, Nürnberg 1525.

Representing space *in space*: Perspective boxes.
As already mentioned a perspective painting gives the viewer an illusion of space *in the plane*. This illusion may be extended to one of space *in space*, viz. when a few of these paintings are combined in a proper way: Indeed, in fig. 4 the paintings of figs. 1 and 2, together with a third painting, surround a real, tiled floor. In fig. 5 the three surrounding paintings are raised so that the main part of a box is formed. This is actually part of a cardboard model[1] of a so-called perspective box, the subject of this paper. Fig. 6 shows the original, i.e. the very perspective box[2] made around 1655 by the Dutch artist Samuel van Hoogstraten (1627-1678), a pupil of Rembrandt, and the famous maker of numerous persuasive illusionistic paintings[3].

Fig. 4. Samuel van Hoogstraten's perspective box, National Gallery in London: Floor panel surrounded by three panels laid flat. Reproduced with the gallery's permission after its cardboard model.

Fig. 5. The floor and the three surrounding panels of fig. 4 folded to form part of a perspective box. Reproduced with London National Gallery's permission after its cardboard model.

Fig. 6. Samuel van Hoogstraten's perspective box, National Gallery in London. Reproduced from plate V in Brusati (1995) with permission from the author and the gallery.

Looking through the rectangular opening in the front panel of this box gives a strange impression of distorted floors, doors, and furniture. Actually we should not look into the interior through this opening, which was probably originally covered by a piece of paper[4] that admitted light into the box but prevented peeping. Instead, each of the two opposing end panels has a peephole through which one should look into the box. And when so doing one actually gets an exciting impression of space in space: From the right-hand peephole, for instance, three intersecting panels are seen (fig. 7), although their joins are hardly noticed. And the tiles painted on the left-hand end panel seem to lie in continuation of those of the real floor. Furthermore, the above-mentioned dog, now complete, seems to be sitting in the middle of this partly real, partly illusionistic floor. Thus, the perspective box seems to be longer than it actually is, and similar illusionistic tiles painted on the back panel make the box seem wider than it actually is. The ceiling is painted, too, so that the box also seems higher than it really is. Moreover, a perspective box that is actually triangular or pentagonal may even appear rectangular!

Fig. 7. Samuel van Hoogstraten's perspective box: the dog panel seen from the right-hand peephole. Reproduced from plate 12 in Brown et al. (1987) with permission from National Gallery in London.

Constructing perspective boxes was a popular activity among Dutch painters of the 17th century, but only for the relatively short period of time about 1650-1675. People were absolutely fascinated by these boxes as we may learn from the British 17th century author John Evelyn who saw a perspective box in London in 1656 and described his experience in the following way:

Was shown me a pretty perspective and well represented in a triangular box, the Great Church of Harlem in Holland, to be seen through a small hole at one of the corners and contrived into a handsome cabinet. It was so rarely done that all the artists and painters in town flocked to see and admire it.[5]

Unfortunately, today only six of these boxes are preserved. But even though limited in number, the preserved boxes exhibit a considerable variety in shape and size, as is evident from fig. 8. Photos of these boxes are available in print[6] as well as on web sites[7].
Geometrical analyses of the perspective constructions of the six preserved boxes – with the exception of that of the Copenhagen rectangular box – are found in Andersen (2007)[8], Jensen (2004), and Verweij (2001). Koslow (1967) has rudiments of an analysis of perspective box geometry, being repeated in Bomford (1987)[9]. Bomford (1998) recycles some of the ideas

and also attempts an unsuccessful[10] perspectival analysis by introducing lines of convergence onto the floor patterns.

Place	Date	Horizontal section through peephole(s)	Number of peepholes	Motif	Artist
Nationalmuseum Copenhagen	1655-1660		1	A reformed church	Anonymous
Nationalmuseum Copenhagen	ca. 1660	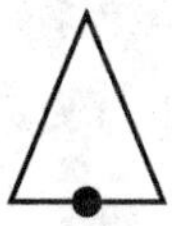	1	A catholic church	Anonymous
Nationalmuseum Copenhagen	1665-1670	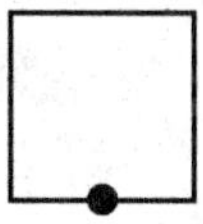	1	A Dutch *voorhuis*	Anonymous
Bredius Museum The Hague	1670-1675	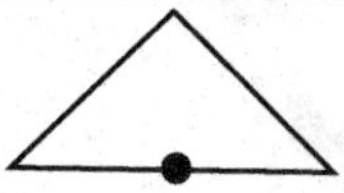	1	A Dutch *voorhuis*	Anonymous
Detroit Institute of Arts	1663	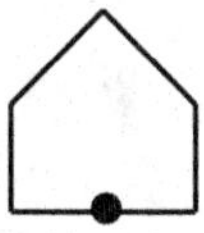	1	A hall	Anonymous
National Gallery London	1655-1660	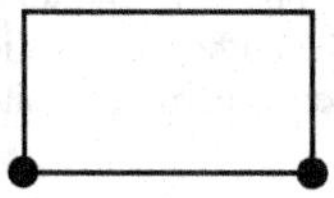	2	A Dutch interior	Samuel van Hoogstraten

Fig. 8. Preserved Dutch 17th Century perspective boxes.

Perspectival analysis of Samuel van Hoogstraten's perspective box.

We will carry out a perspectival analysis of Samuel van Hoogstraten's perspective box in order to find out how the artist obtained his illusion of space in space:

Let us begin with the left-hand end panel – the dog panel – as seen from the right-hand peephole. The peephole represents the eye point E of a central projection having the dog panel as its picture plane (fig. 9, bottom left). The point A is the corner of the box situated immediately below E. In the real floor the tiles are turned so that their diagonals are perpendicular to the end panel. So, when the dog panel and the real floor are combined and laid flat (fig. 9) the four dotted lines on the real floor represent orthogonals (fig. A3). In a virtual prolongation of the real floor these orthogonals prolonged would be projected into the dotted lines of the dog panel. Therefore, these lines passing through the tile vertices meet at the dog panel's principal vanishing point *V*. Indeed, *V* is situated at the peephole opposite to *E*, thus complying with perspective theory (fig. A3).

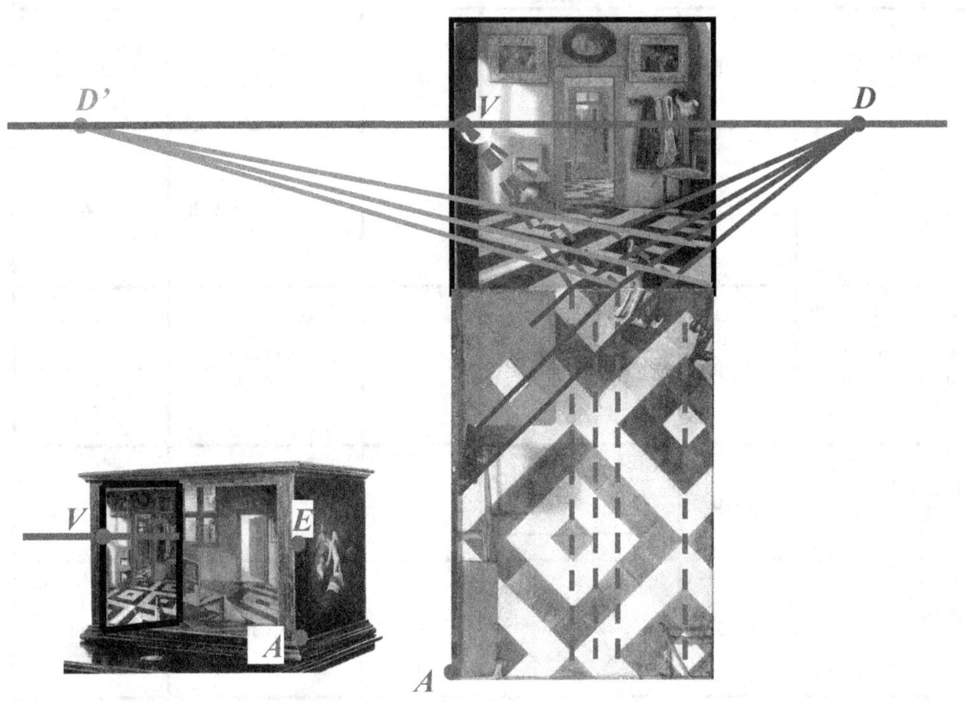

Fig. 9. Perspectival analysis of the dog panel. The box in the bottom left part is reproduced from plate V in Brusati (1995) with permission from the author and National Gallery. The floor and the dog panel laid flat are reproduced with the gallery's permission after its cardboard model.

Fig. 10. Perspectival analysis of the back panel seen from the left-hand peephole. The box in the bottom left part is reproduced from plate IV in Brusati (1995) with permission from the author and National Gallery. The three panels laid flat are reproduced with the gallery's permission after its cardboard model.

Furthermore, on the real floor each set of tile sides make an angle of 45° with the dog panel. So, the tile sides of the dog panel meet at this panel's two distance points D and D', situated on the horizon line, symmetrically around V – though outside the panel (figs. 9, A4, and A5). Measurement shows that $VD = VD'$ = the length of the box, in accordance with the general theory (fig. A5).

The back panel seen from the left-hand peephole may be treated similarly. Now, the left-hand peephole represents the eye point E of another central projection having the back panel as its picture plane (fig. 10, bottom left) (in practice only the right-hand half of the back panel is visible from E). The point B is the corner of the box situated immediately below E. The real floor, the back panel, and the dog panel are laid flat, and four dotted orthogonals are shown on the real floor. Their continuations on the back panel – vertex to vertex to vertex … – meet at the principal vanishing point V of this projection. At first sight it may seem surprising that V is situated at the join between the back and end panels, but from the three-dimensional situation (fig. 10, bottom left) it becomes clear that this is in compliance with the fact that the principal vanishing point is obtained by projecting the eye point perpendicularly onto the picture plane (fig. A3). Actually a cluster of pinholes is seen in the London box around this point[11] thus testifying that strings have been stretched from here in the process of drawing the lines of the illusionistic floor tiles. The two distance points are obtained in the same way as above, by means of the two sets of tile sides. In this situation it is evident without measurement that the distance of the projection equals the width *VD'* of the box (fig. A5).

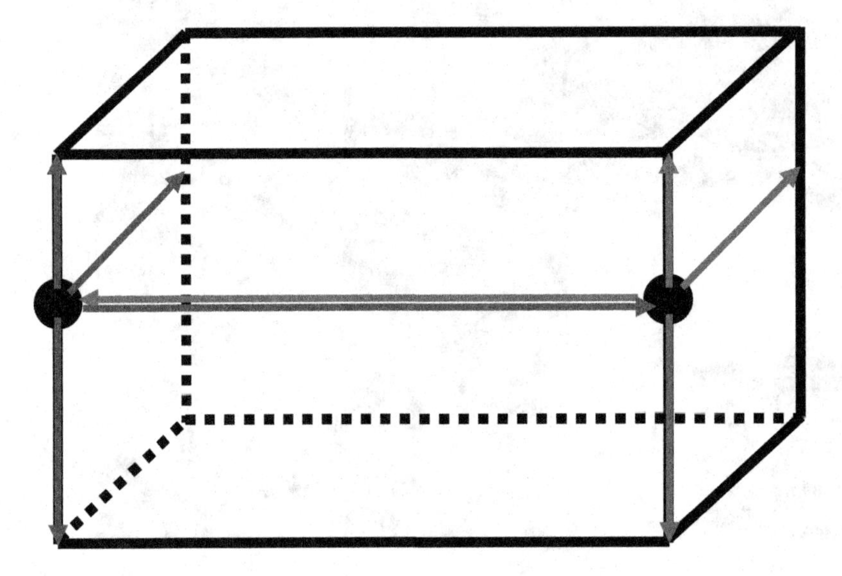

Fig. 11. The London box: The 8 central projections and their principal vanishing points.

The above analysis makes use of two different central projections, and we may proceed similarly with every combination of any peephole and any panel (including the floor and the ceiling) visible from the peephole in question. So, evidently van Hoogstraten used a total of 8 central projections (fig. 11) giving rise to 8 principal vanishing points, i.e. the arrowheads of fig. 11, and 16 distance points. What makes the illusion of space in space so perfect is that these projections come in fours having a common eye point, i.e. one of the peepholes.

Construct your own perspective box.
From the section above it is evident that illusionistic tiles may be constructed on the vertical panels of any box as soon as the following facts are known:

- the size and shape of the box
- the position of the peephole
- the tiles of the real floor.

The reader might wish to try it out in the situation described in fig. 12. Enjoy!

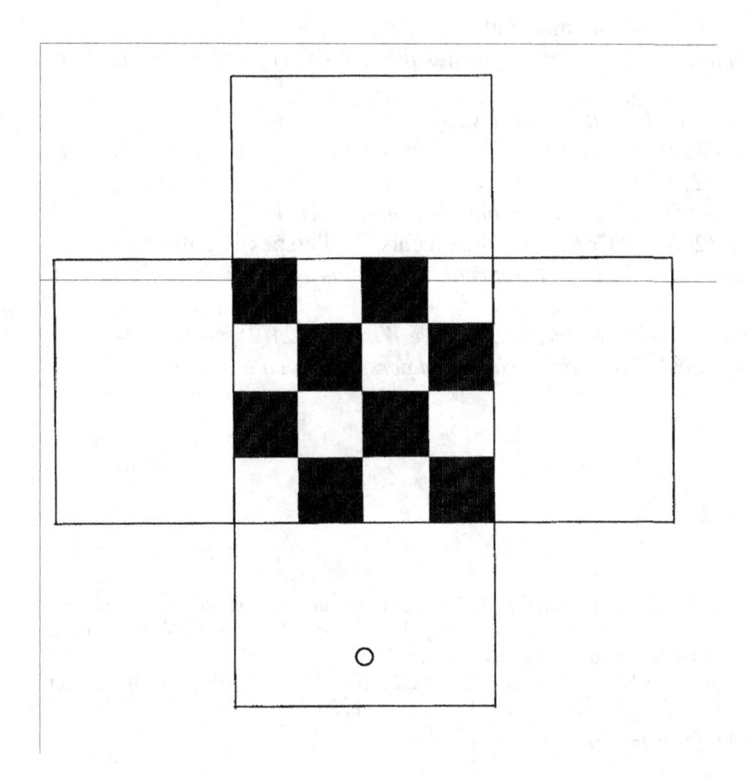

Fig. 12. A floor of squared tiles surrounded by four panels that may be folded to form a box. Light is admitted into the box through the missing lid. There is a peephole in one of the panels.

REFERENCES:

Andersen, Kirsti (2007). *The Geometry of an Art. The History of the Mathematical Theory of Perspective from Alberti to Monge.*

Bomford, David (1987). "Perspective and peepshow construction", *National Gallery Technical Bulletin*, 11, 65-77.

Bomford, David (1998). "Perspective, Anamorphosis, and Illusion: Seventeenth-Century Dutch Peep Shows", *Vermeer Studies*, ed. Ivan Gaskell and Michiel Jonker, *Studies in the History of Art*, vol. 55, National Gallery of Art Washington, 124-135.

Brown, Christopher; Bomford, David; Plesters, Joyce; Mills, John (1987). "Samuel van Hoogstraten: Perspective and Painting", *National Gallery Technical Bulletin*, 11, 60-85.

Brusati, Celeste (1995). *Artifice and Illusion. The Art and Writing of Samuel van Hoogstraten.*

Cole, Alison (1992). *Perspective.*

Dürer, Albrecht (1525). *Unterweisung der Messung*. Facsimile edition 1983.

Ebert-Schifferer, Sybille (2002). *Deceptions and Illusions. Five Centuries of Trompe l'Oeil Painting.*

Elffers, Joost; Schuyt Michael; Leeman Fred (1981). *Anamorphosen. Ein Spiel mit der Wahrnehmung, dem Schein und der Wirklichkeit.*

Gundestrup, Bente (1991). *Det kongelige danske Kunstkammer 1737/The Royal Danish Kunstkammer 1737*, vols. I-II, *Register/Index*, 1995.

Jensen, Claus (2004). "Perspektivkasser og matematik", *Normat*, 52(4), 160-171.
Kemp, Martin (1990). *The Science of Art. Optical themes in western art from Brunelleschi to Seurat.*
Koester, Olaf (1999). *Illusions: Gijsbrechts - Royal Master of Deception.*
Koslow, S. (1967). "De wonderlijke Perspectyfkas". An Aspect of Seventeenth Century Dutch Painting, *Oud Holland 82*, 35-56.
Liedtke, Walter (2000). *A View of Delft. Vermeer and his Contemporaries.*
Pedersen, Eva de la Fuente (2005). "Cornelius Gijsbrechts og Perspektivkammeret i Det Kongelige Danske Kunstkammer", *SMK Art Journal 2003-2004*, 84-107 (translated into English at pp.152-160).
Verweij, Agnes (2001). "Perspectief in een kastje", *Nieuwe Wiskrant*, 21(2), 6-16.
Zuvillaga, Javier Navarro de (2000). *Mirando a través. La perspectiva en las artes.*

---

[1] The cardboard model was produced by National Gallery, London, but unfortunately it is no longer available. However, a slightly smaller model comes with Zuvillaga (2000). I am grateful to Agnes Verweij, Delft University of Technology, The Netherlands, for supplying this information.

[2] Now in National Gallery, London. The box is 77,9 cm long, 51,2 cm wide, and 52,4 cm high (interior measures). A full-scale copy of the box is housed in Göteborgs Konstmuseum, Sweden.

[3] For Samuel van Hoogstraten's career and oeuvre, see Brusati (1995).

[4] Brown et al. (1987) p. 70.

[5] Koslow (1967) p. 37, note 8.

[6] For instance Andersen (2007), Brown et al. (1987), Bomford (1987), Brusati (1995), Cole (1992), Ebert-Schifferer (2002), Elffers et al. (1981), Gundestrup (1995), Kemp (1990), Koester (1999), Koslow (1967), Liedtke (2000), Pedersen (2005), Verweij (2004).

[7] Copenhagen: http://www.kunstkammer.dk/Schilderi/genstande_schilderi_lit.asp?ID=132
http://www.kunstkammer.dk/Schilderi/genstande_schilderi.asp?ID=247

The Hague: http://www.museumbredius.nl/schilders/elinga.htm
Detroit: http://www.dia.org/the_collection/overview/full.asp?objectID=48296&image=1
London: http://www.nationalgallery.org.uk/cgibin/WebObjects.dll/CollectionPublisher.woa/wa/work?workNumber=NG3832

[8] The present paper is a revised version of a talk given in 1995 at The Steno Department for Studies of Science and Science Education, University of Aarhus, Denmark. I am grateful to Kirsti Andersen of that department for fruitful discussions we had about the geometry behind van Hoogstraten's box.

[9] Some of Koslow's phrases are even repeated verbatim, without acknowledgement.

[10] The text mixes up, among other things, principal vanishing points with distance points.

[11] I am grateful to David Bomford for this information.

APPENDICES: Elements of Perspective Theory

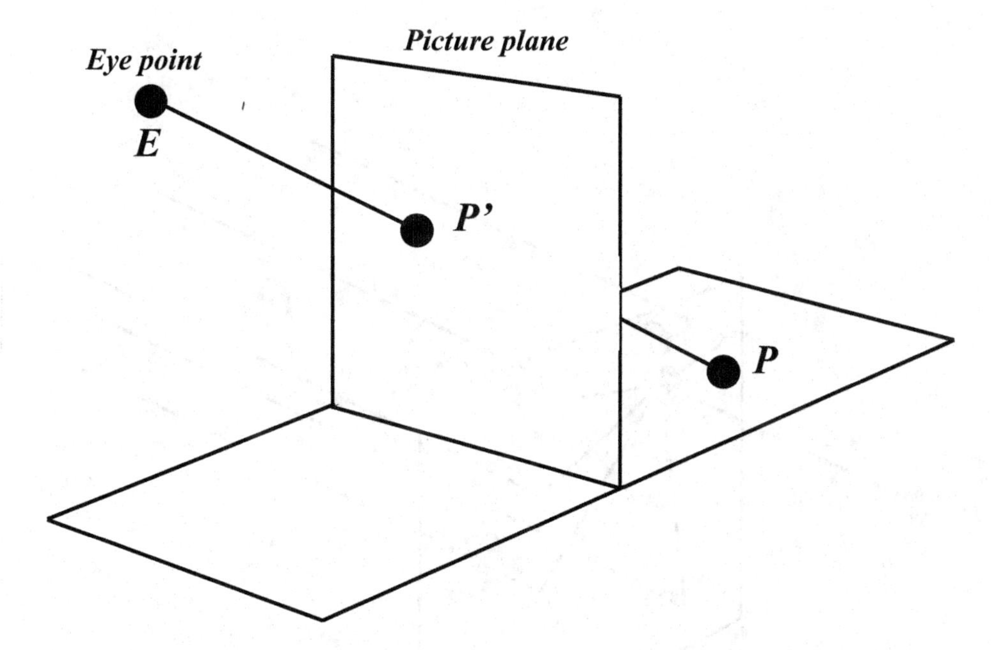

Fig. A1. The central projection with *eye point E*: A point *P* is projected into the point *P'* of the two-dimensional picture plane

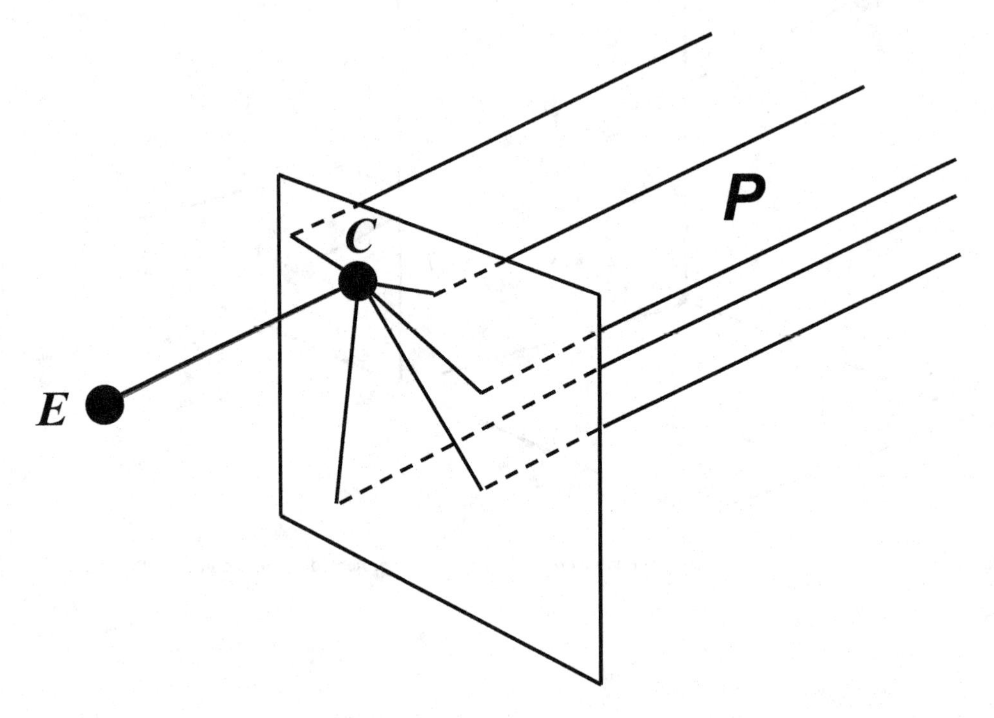

Fig. A2. A pencil *P* of parallel lines in space is projected into a pencil of concurrent lines in the picture plane. The point *C* in common is *the vanishing point* of *P*. This vanishing point is obtained by intersecting the picture plane with the line passing through *E*, parallel to any line of *P*.

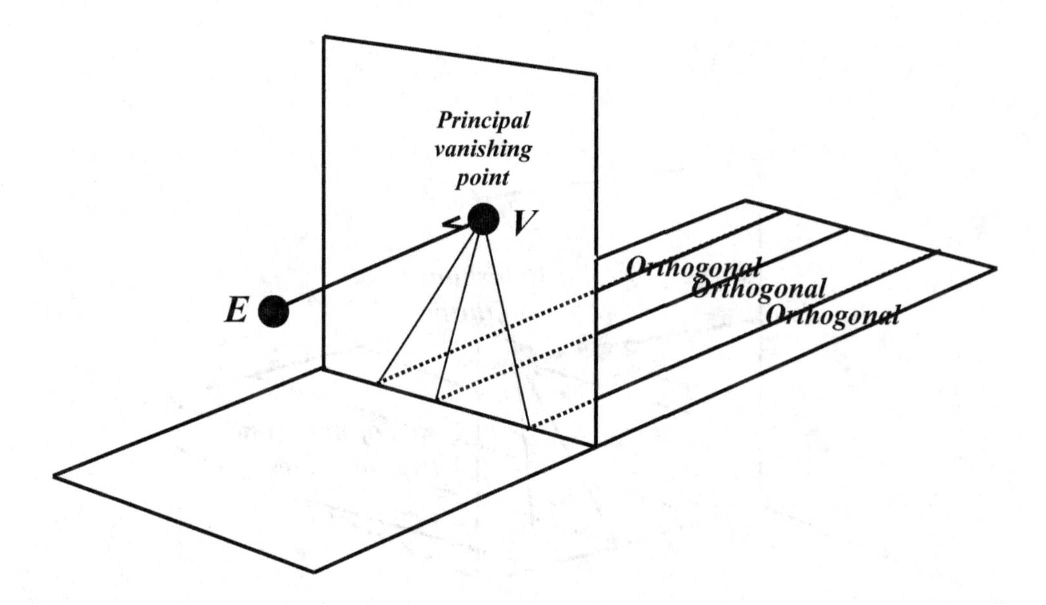

Fig. A3. *The orthogonals* are lines in space perpendicular to the picture plane. Their vanishing point V is *the principal vanishing point* of the central projection. According to fig. A2, V is obtained as the orthogonal projection of E onto the picture plane.

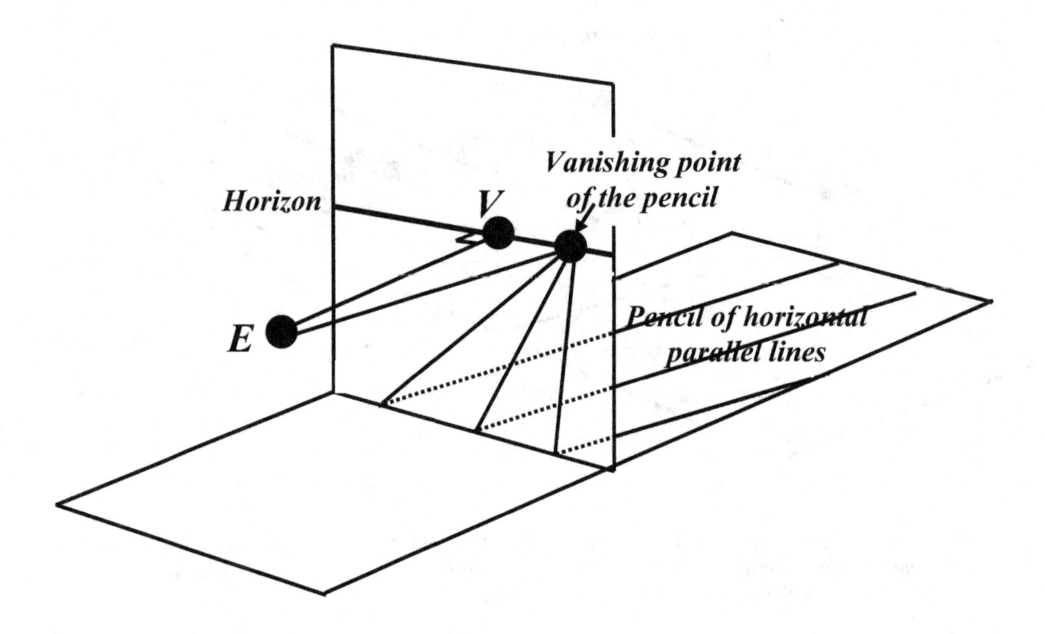

Fig. A4. The vanishing point of any *horizontal* pencil of parallel lines in space is situated on the *horizon*, i.e. on the horizontal line of the picture plane, passing through the principal vanishing point.

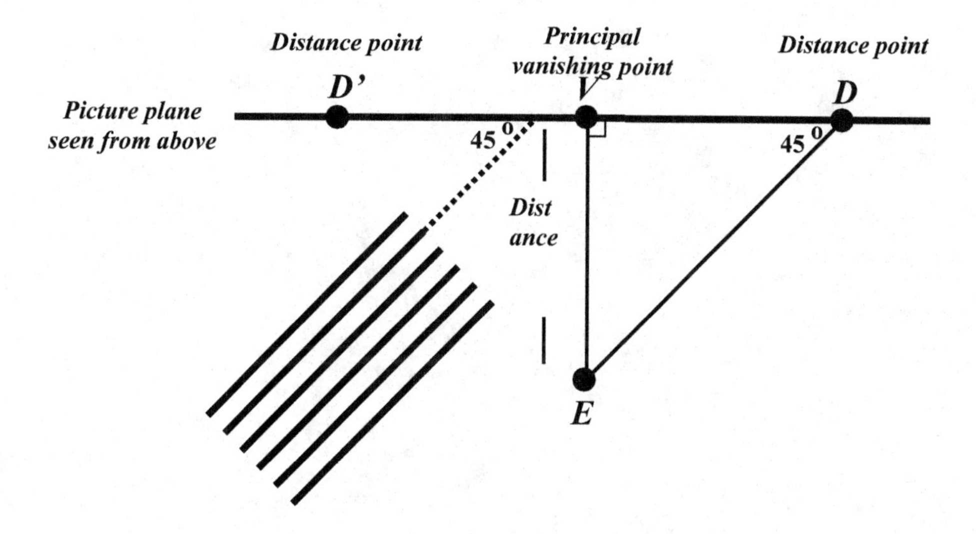

Fig. A5. Horizontal section through the eye point: The *distance* of the central projection indicates how far its eye point is removed from its picture plane. In space there are two horizontal pencils of parallel lines each making an angle of 45° with the picture plane. *The distance points* are the two corresponding vanishing points. They are situated on the horizon, lying symmetrically around the principal vanishing point. The distance equals the length of the segment from any distance point to the principal vanishing point.

THE ARITHMETIC MEAN AND CAR DIFFERENTIAL

Damjan Kobal
Department of Mathematics
University of Ljubljana, Slovenia

The mechanism of a car differential (differential gear) has been known for about two thousand years. It represents an ingenious technical invention, which is nothing else than a realisation of a simple mathematical idea of the arithmetic mean. The study of a car differential provides a practical and intuitive insight into an otherwise abstract concept of variable dependency in simple mathematical equations.

INTRODUCTION

All of us daily take advantage of the comfort, which is provided by technology and the car differential is an important technical device, which we all use regularly. But very few are aware of it or have ever contemplated this simple technical idea. And even fewer have ever thought about its natural connection to the very simple mathematical ideas. For anyone trying to understand mathematics, it can be of a great help if abstract mathematical ideas are given deeper meaning or its mechanical realisations that provide for our comfort living.

We all know, that a car is powered by a motor. But how? How is the power (the rotation) of the motor transferred to the wheels that make the car move. On a bicycle, we use a chain that transfers the rotation of the pedal to the back wheel. Is it not done very similarly for a car, just that the source of power is a motor and not our muscles? Well, the very first cars were truly done that way. By the use of a chain the rotation was transferred from the motor to the (back) axle. So both, right and left wheel rotated simultaneously and made the vehicle to move.

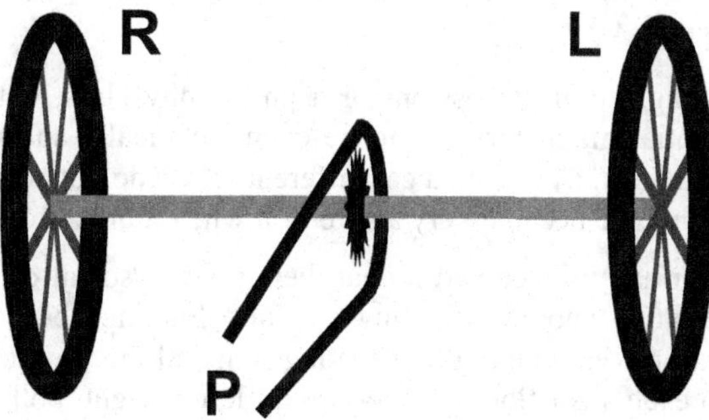

Figure 1: Right and left wheels are attached to the same axes of rotation

B.Sriraman, C.Michelsen, A. Beckmann & V. Freiman (Eds). (2008). *Proceedings of the Second International Symposium on Mathematics and its Connections to the Arts and Sciences* (MACAS2). Information Age Publishing, Charlotte: NC , pp.107-116

With a look on the picture, disregarding possible transmission ratio and denoting 'power' (engine rotation) by P, right wheel rotation by R and left wheel rotation by L, we get a very simple equation (system of equations):

$$P = R = L$$

But does a car work like that? Well, the very first cars did function like that and as a consequence, the steering was very hard. Namely, in a left turn, the right wheel travels longer route than the left and in a right turn, the situation is reversed. Theoretically, with a mechanism like on the above picture, the steering is impossible as both wheels rotate identically and thus, travel the path of the very same length. In reality, the steering is done while both wheels must slide slightly on the ground. That makes steering with such a mechanism physically quite hard. Today, one can experience this effect while driving a tractor or a jeep with a 'blocked differential'. On a rough terrain, for example farmers must use their tractors with 'blocked differential' to increase the puling power. But while driving with 'blocked differential' it is very (physically) hard to move a steering wheel to either right or left turn position. In reality, several 'tractor accidents' are caused by weakness of the driver to turn the tractor while in a 'blocked differential' position. Sometimes a steering wheel and steering (usually front) wheels might even be turned in the right position, but the tractor with a 'blocked differential' might just push straight.

How is this problem solved in reality? A simple solution would be to transmit the engine rotation only to one of the two (right or left) wheels. That way steering would be 'easy' but for today's standards of comfortable driving, driving and steering would be truly bizarre and dangerous. For example with the power on the right wheel, driving into a left turn would feel like really slowing down, while a turn to the right would be a stunt (a car would feel like speeding up into a right turn).

SOCIAL DIFERENTIAL

It might be an interesting moral discussion for a philosophy class, but thinking about our 'build in' social differential might turn out to be even technically and mathematically very intuitive and easy to understand. As with a car differential, we do not acknowledge we have it until it works properly, but we became very aware of it when it brakes down.

We could say every normally developed human being has a 'social differential'. Imagine a couple walking one next to another, chatting and not thinking about the path they walk possibly in a nice park, or even walking home using stairs all the way to the fourth floor and making sharp turns on each semi-floor. While turning left or right, both promenaders would (subconsciously) adopt their pace as to remain lined with a friend. If we think about what happens while turning, it is quite obvious that for example in a right turn, the right promenader would slow down a bit and the left would speed up a bit and their average speed would remain unchanged.

As mentioned above, malfunctioning of social differential can be noticed easily. Usually it happens with 'very important' people, who are too 'important' to adopt their speed to their subordinate and maintain their constant walking pace also when turning ... It is really funny to observe such a scène when for example 'a subordinate' student on the right side of 'an important' professor with a broken social differential runs in a left turn and remains almost still in a right turn.

THE ARITHMETIC MEAN

A (normal) promenading couple turns to the left so that the left promenader slows down and the right speeds up a bit. Their average speed remains unchanged. If P is their average speed, R is the speed of the right and L is the speed of the left promenader, then their 'social differential' is described by a simple equation:

$$P = \frac{R+L}{2}$$

The equation nicely and fully describes the relation between their speeds and average speed throughout the promenade, when their path is straight and their speeds are equal (in this case we have $P=R=L$) as well as when their speeds differ in left or right turns. Could this simple formula be mechanically realized for powering right and left wheels of a car? But how? It might be surprising, but a positive answer to this question has been known for over 2000 years. In fact, a mechanical realisation of the formula for arithmetic mean is surprisingly simple.

DIFFERENTIAL GEAR

Imagine first, that equal powering of the right and left wheels is achieved by a 'rotating handle on a disc', which is attached to the right and left disc, that are welded at the end of the right and left wheel axes, as shown on the bellow picture.

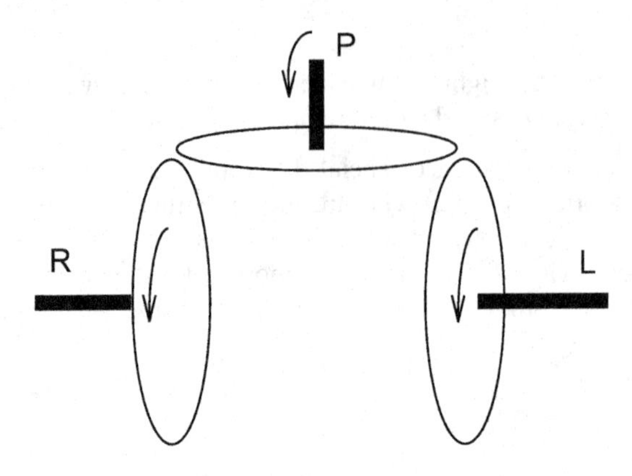

Figure 2: Right, left and the powering discs

Instead of discs we can imagine cogs. It is obvious, that with a help of such a mechanism, a rotation of 'the handle P' would imply an equal rotation of the left and right wheel, thus $P=R=L$. This seems like still far away from the desired equation:

But it is not. Starting with this formula, $$P=\frac{R+L}{2}$$ we can easily check, that

$$P-R=L-P=\frac{L-R}{2}.$$

Denoting

$$\frac{L-R}{2}=X,$$

we have

$R=P-X$ and $L=P+X$.

The value of X can be understood as a free parameter in the relation of three variables within a single equation

$$P=\frac{R+L}{2}.$$

The variable X has such an important role in the mechanic realisation of the arithmetic mean, that its understanding completely resolves the dilemma of the powering of the car wheels.

Namely, if we allow that our 'power disc' in the above picture, is freely rotatable (free variable X) around the 'handle', as shown in the bellow picture, we already have a model of a differential gear.

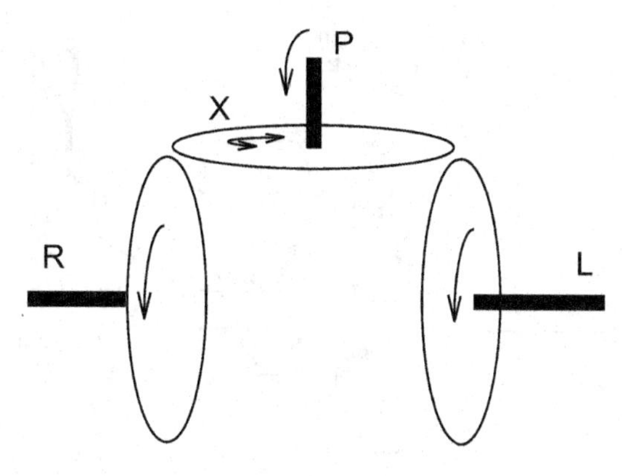

Figure 3: Right, left and freely rotate-able powering discs

With a look on the above picture, let us think again about the formula

$$P = \frac{R + L}{2}.$$

If both, right and left discs are freely rotate-able, the push (rotation) of our handle will cause both discs to rotate evenly and the powering disc will not rotate (*X=0*). If either right or left discs is stopped (or only partially braked), the powering disc will start rotating as we push (rotate) the handle and the opposite disc will rotate even faster. Thus, 'variable *X*' will exactly transform for example slowing down of the right disc into speeding up of the left disc.

The above simple sketch illustrates the essence of a differential gear and gives a mechanic realisation of the simple but abstract *arithmetic mean* mathematical idea. Freely revolvable powering disc takes care of *differentiating* the resistance on the left and right half shafts. As much as one of the wheel torques (left or right) is diminished because of the resistance, as much the other is increased. The question of how to transmit the engine torque through the (cardan) drive shaft to our powering disc is not trivial, but regarding the described ingenious idea, this question is only technical.

On the picture below one can see a sketch of a true (classical) differential gear.

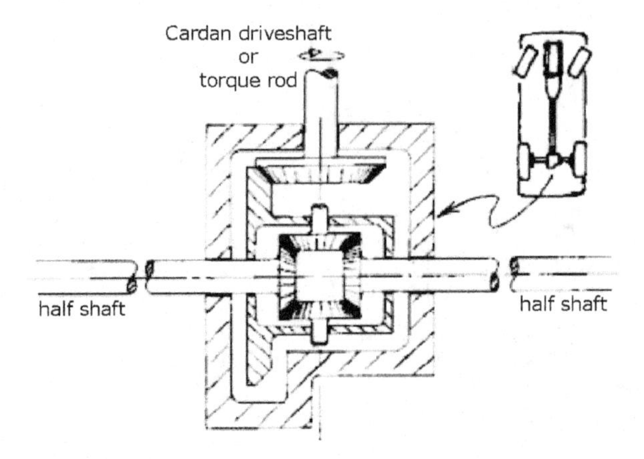

Figure 4: Differential gear profile

Engine torque is transmitted over the *cardan driveshaft* named also *torque rod.* It is interesting that the name *cardan* is directly derived from the name of Italian mathematician, physician and inventor *Girolamo Cardano* (1501-1576), who invented the *universal joint. Universal joint* is an essential part of a usable *torque rod*, but one that is not associated directly to our idea of a differential gear. *Universal joint* is another ingenious idea which provides a simple solution for 'around the corner rotation'. In other words it is a joint joining two simultaneously rotatable rods that are joined under an angle between 90 and 180 degrees.

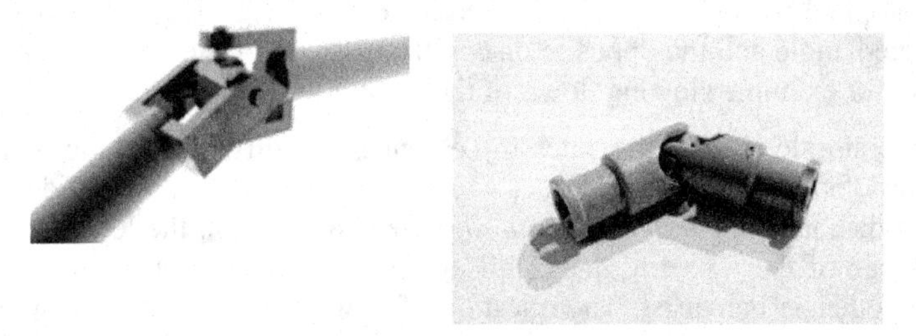

Figure 5: Cardan – a universal joint, drawing and Lego swivel

We believe the idea of a differential gear can be a useful didactical motivational tool. It provides a useful, complex and yet simple technical and intuitive idea from where we can derive and contemplate deeper meanings of otherwise only abstract mathematical ideas. Namely, even to a mathematics expert, this simple idea of arithmetic mean can pose several quite nontrivial questions. Furthermore, abstract ideas can be given intuitive and technical meanings via well known questions related to common experience of driving, turning. One can easily experiment as today, even Lego (Technics) provides sophisticated but yet simple models of devices like differential gear.

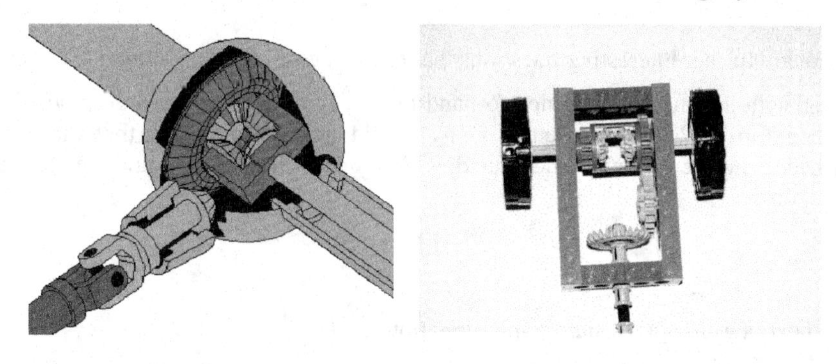

Figure 6: Drawing of a differential gear with universal joint and Lego model of differential

ARITHMETIC MEAN AND SNOW DRIVING

Many people have experiences related to the functioning of a differential gear mechanism. That might be tractor or jeep driving as mentioned at the begging of this article, but more often and unfortunately unpleasant are experiences of a car driving in a snow. It happens very easily that driving in a snow leaves us powerless on the road, when we are unable to move the car. The engine would just helplessly rotate one of the wheels, which would freely slide on the smooth snow. Usually this happens when a car leans to one side and whatever we tray, the car just sinks deeper into the snow. Always it is only one wheel that rotates, and even that is the wrong one. If the wheel on the side where the car leans would rotate, the force might be strong enough to move the car forward... It even happens that we get some strong help and powerful boys try to lift the side of the car that is burdened because of the lean. At the same time they might push down the other side of the car to put pressure on the spinning wheel... Sometimes it might help but even more common is the situation that shifting the leaning of the car to the other side only causes the shift from one freely spinning wheel to the other. Well, this is the situation when a differential gear is doing right the opposite of what would be productive. Namely, differential gear makes always the easier rotatable wheel to spin. Of course all this wrestling can be easily explained by the arithmetic mean formula

$$P = \frac{R+L}{2}.$$

These thoughts can be an insightful start of other mathematical chapters, like for example system of equations. Namely, considering P as given (constant – engine power), this is really a good and simple example, which tells us that one equation can only tell us the relation but not the absolute values of the two variables it connects. Thus a natural need for two equations to determine two variables is given.

There are further interesting questions that one can consider with students. For example:

A car with a turned off engine, no hand brakes applied in forward gear position: Can it be pushed forward? By experience, many would answer correctly, that the car can not be pushed..., but few would understand, how this could be implied from the equation

$$P = \frac{R+L}{2}.$$

The next question gives simple interpretation of that.

Let a car be in exactly the same position as above, but rather than on the ground, let us imagine the car is lifted up as in a garage. Can a powering wheel be turned around by hand?

It is very interesting that usually only 'technically inclined professionals' answer correctly to this question. Even mathematics teachers after workshops of work on this idea, are deceived by misunderstanding of their experience. Namely, in the above described position, powering wheel can easily be turned around..., while the opposite wheel turns to the opposite direction. Of course, since the car engine is off and the car is in forward gear position P in our equation

$$P = \frac{R+L}{2}$$

is forced to be 0. Thus $R = -L$. So how come that forward (or backward) gear position works as a break of a car standing on a road? Well, of course, moving a car forward or backward would mean that both left and right wheel turn to the same direction, while when $P=0$, R and L can only be of the same sign if they are both 0.

DIFFERENTIAL, THE STRAIGHTNESS CONCEPT AND GEODESICS

It is interesting that such a simple idea can be developed further into a wonderful intuitive understanding of a complex and abstract differential geometry concept of geodesics. Before that high school students can be engaged into a debate over what *straight* means in reality. In mathematics we know the idea of a straight line, but in real life, do we know anything that is 'more straight' than the 'equator circle'. From here it is easy to derive a concept of '*straightness*' as the shortest distance... Of course, on a plane that is a straight line. On a sphere *straightness* is best described by great circles. This also explains why long distance airplanes fly 'strange arced paths' on our usual maps. It took a long time to mankind to comprehend that a straight edge of a table is no straighter to a man than a highway loop around a small town is to an ant. Formal definition of a straight line is quite complicated and abstract. In a surface, which is by our experience 'flat' but by our limited understanding in fact its curvature remains unknown to us, we define straightness as *the shortest distance*. The shortest paths are called geodesics. And what has this to do with our differential gear? As described in the very beginning of this article, a tractor or a jeep with a blocked differential would only drive straight. As would a simple Lego model with two wheels attached to the same axes, if carefully pushed, only go straight. But what if the 'driving ground' is not flat? Well, then our vehicle would travel wonderful intuitive paths of geodesics... posing many further questions, inspiring our imagination and challenging our understanding.

HISTORIC REMARKS

It is not known who invented differential gear mechanism. It seems obvious that the idea is much older than many of Leonardo da Vinci's (1452-1519) inventions. British inventor James Starley (1830-1881), known as ' Father of the Bicycle Industry', used a differential gear mechanism in a special sewing machine in about 1850. In 1877 he used the differential gear in a road vehicle. Supposedly, differential gear was used in a road vehicle for the first time by German Rudolph Ackerman in 1810. Several sophisticated mechanical devices that included differential gear mechanism are much, much older. Findings in China prove the existence of this mechanism dating back to about 300. In the year 1900 an extremely sophisticated **Antikythera mechanism** (named after nearby Antikythera island of Greece, where the ship was discovered) was found in a ship wreck. The mechanism was a carefully designed and crafted in bronze and wood. It was a sort of astrological computer to calculate the position of planets and stars. And what is the most amazing, a device has been dated to about 125 BC and it had a differential gear mechanism. The device is displayed in the Bronze Collection of the National Archaeological Museum of Athens, while several exact reconstructions have been made and are on display around the globe, usually in computer museums (like for example in American Computer Museum in Bozeman, Montana).

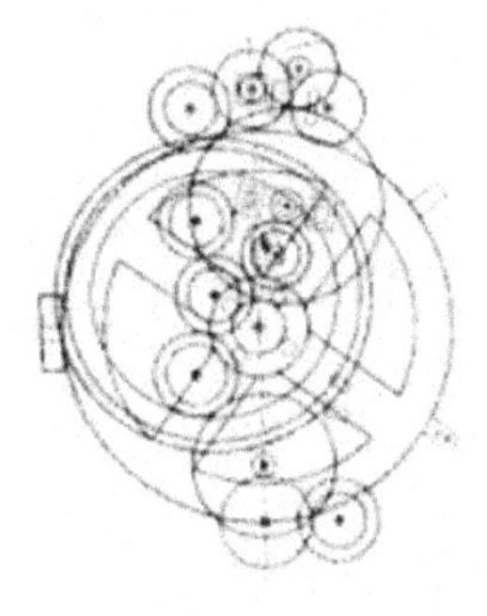

Figure 7: The main fragment of The Antikythera mechanism (~125 BC) and its reconstructed plan

GRAPH THEORY AS A METHOD OF IMPROVING CHEMISTRY AND MATHEMATICS CURRICULA

Franka Miriam Brückler,

Department of Mathematics, University of Zagreb, Croatia

Vladimir Stilinović,

Department of Chemistry, University of Zagreb, Croatia

Very often various classes in schools are presented independently. An important part of schooling should be that pupils learn how the various subjects relate to each other. This paper deals with one possible way of showing such a relationship between mathematics and chemistry using the mathematical discipline of graph theory. The discipline is mostly ignored on pre-university level, although the basic principles are easy to understand and have very natural applications in chemistry.

INTRODUCTION

Mathematics students usually encounter graph theory in their undergraduate studies. As applications, usually computer science applications are explained or at least mentioned. Rarely, and this mostly only as short examples, chemical applications are shown. On the other side, chemical applications of graph theory are usually approached rather late in chemistry studies, in graduate chemistry courses or undergraduate at earliest. A large number of articles, and even books, about graph theory in chemistry already exist. Most of them are of a too advanced level to be understood by non-specialists, and are often too complicated to be understood even by students of chemistry or mathematics and thus unsuitable for any possible presentations in primary or secondary schools. The purpose of this paper is to show how graph theory can be introduced in math and chemistry education at a much earlier time. This connection is particularly useful since an important aspect of modern education should be to show the connections between various subjects taught in school. Besides, mathematics and chemistry are often named as the two most difficult school subjects (often accompanied by physics). The didactical benefit here for both subjects is that pupils tend to get a more positive feeling for them if they see that the subjects are not separate entities but have connecting topics. The best effect should be achieved when the topic that shows the connection is at the same time discussed in both the classrooms or, when possible, presented as a joint class.

Historically, the first appearance of graph theory dates back to the 18th century, when the great Swiss mathematician Leonhard Euler solved the famous Königsberg bridges problem and gave a general method for solving similar problems. In the Königsberg bridges problem,

B.Sriraman, C.Michelsen, A. Beckmann & V. Freiman (Eds). (2008). *Proceedings of the Second International Symposium on Mathematics and its Connections to the Arts and Sciences* (MACAS2). Information Age Publishing, Charlotte: NC , pp.117-126

the question is if there is a tour visiting the four city parts of Königsberg in such a way that each of the seven bridges is visited only once and ending the walk in the same city part where it started. Euler noticed that such questions don't depend on the distances involved, but just on the number of connections between the objects city parts. He constructed the corresponding graph replacing city parts with points ("vertices"), and representing bridges as connecting lines ("edges") between the vertices. As an answer to the question is if there is a tour on the graph traversing every edge exactly once, Euler gave the theorem that such a tour (called Eulerian tour) exists if and only if every vertex is of even degree (the degree of an vertex being the number of edges meeting in the vertex). Except for some smaller results, graph theory didn't much evolve after Euler until mid 19th century, when the English mathematician Arthur Cayley obtained some of the most important new results. Remarkably, many of his results are not only applicable in chemistry, but arose from chemical problems. Nowadays, the mathematical discipline of graph theory is highly developed. One of its applications still is chemistry, and these applications have advanced to quite complicated notions like various topological indices in quantitative structure-activity and structure-property relationship studies.

INTRODUCING BASIC GRAPH THEORY CONCEPTS

When first dealt with, graph theory is most entertainingly introduced by recreational problems. Equivalent to the Euler Königsberg bridges problem are the diagram-tracing puzzles: "Can you draw the following picture in one line without tracing any part of it twice?"

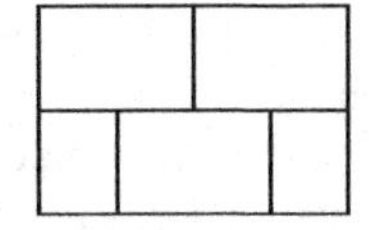

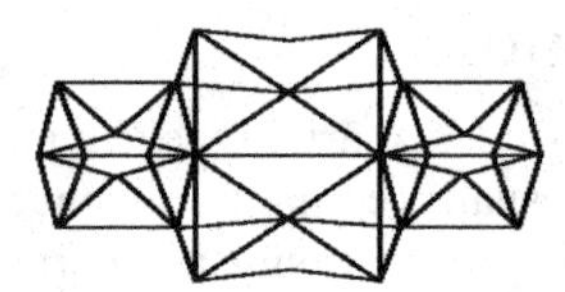

Fig. 4. One diagram that can and one that cannot be traced without tracing any part of it twice – which is which?

Another type of a recreational graph theory problem is the following: Consider a chessboard. Is it possible to cover all the fields of it by traversing it with a knight so that no field is visited twice? Is such a tour possible so that the knight returns to its starting field? What about other field sizes besides the 8×8 field? Remember that knight-moves are L-shaped. If the field is small, say 3×4, it's easy to check the solution by the trial-and-error method. We get there is a solution shown in Fig.1. (in fact, there are three different solutions, but there is no closed tour).

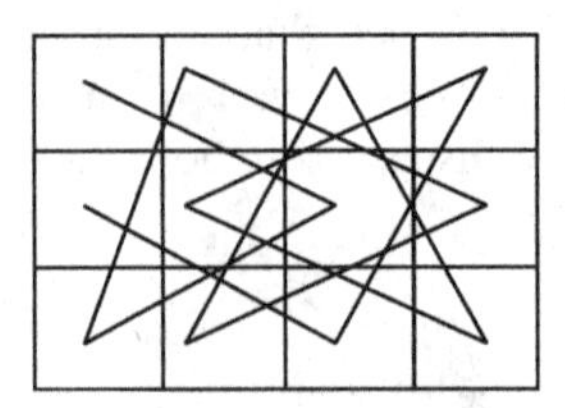

Fig.1. 3×4-chessboard with a knight tour

But how can one find a systematic way to check if there is such a tour on a board of given size (and to find one or more if it exists)? To find an efficient way to solve the problem it is a good idea to consider the fields as points such that any two are connected by a line if one can come from one to the other by a regular knight's move. Formulated as a graph theory problem, the question is to find a sequence of points of the model such that each point is visited exactly once and one may move from one to another point only if they are connected by a line (such sequences are called Hamiltonian circuits); additionally, we can require that the sequence ends where it starts. The graph for the 3×4 problem is shown in Fig.2. It turns out that there is no knight's tour for $n \times n$ chessboards if n is odd or if n=2 or n=4.

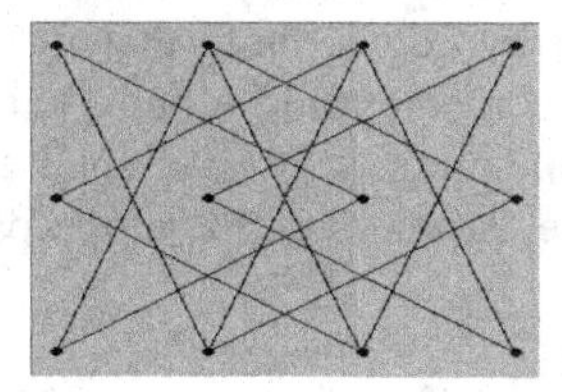

Fig.2. Graph of the 3×4-board knight tour problem

Another recreational problem which is about finding Hamiltonian circuits is "Can you visit all the vertices of a given polyhedron travelling over the edges so that each vertex is visited exactly once?". In contrast to Eulerian tours, there is no simple theorem that decides if one can find a Hamiltionian circuit. Still, it is known that such a circuit is possible on any Platonic or, more generally, Archimedean solid. Most recreational problems connected to graph theory require finding Eulerian tours or Hamiltonian circuits.

Let us introduce basic graph-theoretic terminology. A *graph* consists of two sets. One is called *vertices* and is represented by points in the plane. The other is called *edges* and is represented as connecting lines between vertices. It is not important how long and of what form the edges are, but only which two vertices an edge connects. A *loop* is an edge connecting a vertex to itself. The intersections of edges are a part of the graph only if they are vertices. For example, in the above graph of the knight's tour problem only the emphasized points are vertices, while the other intersections of edges could be imagined as if there one edge passes above the other. A graph is called *simple* if it has no loops and no multiple edges (i.e. no two vertices are connected by more than one edge). The *degree of a vertex* is the number of edges meeting in that vertex. A *path* is an alternating sequence of the type vertex-edge-vertex-edge-...-edge-vertex such that every two consecutive vertices are connected by the edge between them; additionally it is required that no vertex appears twice in the path. A *circuit* is defined as a path, only that the starting vertex is the same as the ending one. A graph is *connected* if there is a path from each of its vertices to any other vertex. An edge in a connected graph is called a *bridge* if by deleting it from the graph we would obtain a disconnected graph. A *tree* is a graph that is connected and has no circuits. The most important theorem about trees states: if a tree has n vertices, then it has n-1 edges. And vice versa: if the graph is connected and the number of its edges is for 1 less than the number of its vertices, then the graph is a tree. In a tree every edge is a bridge and a path between any two vertices is unique.

GRAPH THEORY IN CHEMISTRY

Graph theory can be introduced in chemistry education together with the concepts of valence and molecular structure. The concept of valence, as the property of an atom that determines its capability if binding with other atoms, was firstly discovered in mid 19th century. It is an essential concept in chemistry, and an important step in teaching in a beginner's course. When first introduced, the concept of valence swiftly led to the concept of molecular structure, with atoms connected with one another in various ways, always in a manner which doesn't require change of an atoms valence. Very soon after the concept was introduced, first molecular structural formulas (graphs) appeared.

Graphs commonly used in chemistry can be classified in one of two groups: *reaction graphs*, representing reaction schemes (mechanisms) and *molecular (*or: *structural= graphs*, being graphical representations of molecules. In this paper we shall deal only with the latter group. A molecular graph can be constructed by representing each atom of a molecule by a vertex and bonds between atoms by edges. Note that in such a graph, the degree of each vertex equals the valence of the corresponding atom. (This procedure can be modified so that the degree of a vertex equals the number of outer shell electrons of the corresponding atom.) For example, propane, propene, propine, and cyclopropane are represented by graphs shown in Fig. 3.

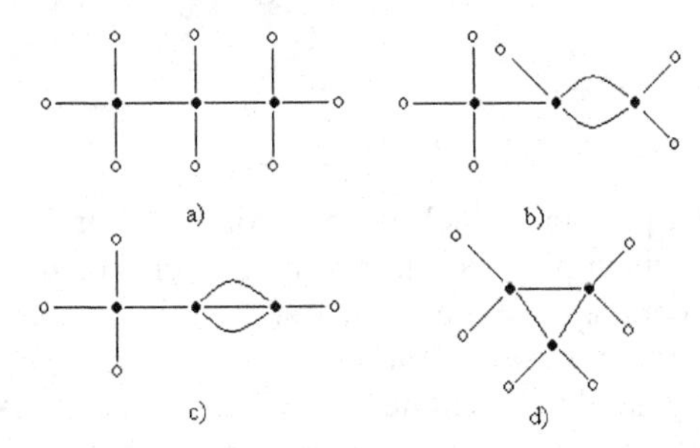

Fig.3. Molecular graphs of hydrogencarbons: a) propane, b) propene, c) propine, d) cyclopropane

In essence, molecular graphs are line formulas: two-dimensional representations of molecules (molecular entities) in which atoms are shown joined by lines representing single or multiple bonds without any indication or implication concerning the spatial direction of bonds. As we see, graph a) is a tree, while other graphs contain circuits. A circuit containing two vertices corresponds to a double bond, while circuits containing n vertices ($n>2$) correspond to rings containing n atoms. Whereas most chemical graphs have no loops, many have multiple edges (bonds) so chemical graphs are often not simple graphs.

The oldest chemical application of graph theory, due to A. Cayley, was isomer enumeration. Two compounds are isomers if they have the same empirical formula, but different line and/or stereochemical formula. Isomerism is a relation between isomers. Graph theory can be used to find the number of possible isomers of a given empirical formula. Although Cayley's enumeration of alkanes strongly used graph theory (trees), more modern results rest on molecular graphs, but use more combinatorics and group theory, as shall be explained in Application 2 below.

Although it is possible to construct molecular graphs for many inorganic compounds, the majority of uses of graph theory are for organic substances. A very important use of graph theory in organic chemistry is in conceiving descriptors (usually referred to as *topological indices*) of molecular shape, size and branching in particular. Since properties of substances depend of not only their chemical composition, but also the shape of their molecules, correlations of descriptors of molecular shape to certain properties of substances (physical properties, chemical reactivity, biological activity...) exist. The first such descriptor, the Wiener number, was introduced in 1947. by H. Wiener who called it path number. In later years several modifications of the Wiener index as well as different topological indices, such as Randić branching index, Balaban index, Hosoya index and molecular ID number were introduced. Molecular graphs used for construction of these topological indices are *hydrogen suppressed graphs*, i.e. vertices of the graph correspond to non-hydrogen atoms. Let us define three of the topological indices. The Wiener number W, was originally defined as the sum of the distances between any two carbon atoms in the molecule, in terms of carbon-carbon bonds, but this definition was generalised to the sum of distances (i.e. edges) between all pairs of vertices of the hydrogen suppressed molecular graph. So the Wiener number is

$$W = \frac{1}{2}\sum_{i,j=1}^{N} d_{ij}$$

where d_{ij} is the number of edges in the shortest path connecting the pair of vertices i and j and N the number of vertices. Note that W can be defined only for molecules with molecular graphs that are trees. The Randić branching index R, for a hydrogen suppressed graph, is defined as

$$R = \sum_{\text{all edges}} \sqrt{\frac{1}{mn}}$$

where for any edge in the summation term, m and n stand for degrees of adjacent vertices joined by that edge. The Hosoya index is defined as

$$Z = \sum_{k=0}^{[M/2]} p(k)$$

where M is the number of edges in the graph, $p(k)$ the number of ways of choosing k non-adjacent edges from the graph. Note that $p(k)$ is zero for $k > [M/2]$ (the symbol $[x]$ denotes the largest integer smaller or equal to x) since there is no set of k non-adjacent edges in a graph of M edges if $k > [M/2]$, and $p(1) = M$ i.e. the number of ways of choosing one non adjacent edge is equal to the number of edges. By definition, $p(0) = 1$. The correlation of various topological indices has been studied extensively and one can find many exercises suitable for pupils here. The students' task here, for instance, could be to find an expected value of the boiling point of a compound of some class (e.g. amine) not listed in tables, and comparing it to an experimental value. Such an exercise gives the student a perfect view of how a property of a substance may depend on its molecular structure.

COMBINING CHEMISTRY AND MATH LESSONS THROUGH GRAPH THEORY

After introducing basic graph theory through recreational problems, one can turn to applications. We present three of the possible graph theory applications in chemistry suitable for incorporating in school curricula.

Application 1: Trees

An alkane is a chemical compound with a molecular formula C_mH_n such that there are no circuits in the molecule and no multiple bonds. That means that the graph of an alkane molecule is a tree. This could be an appropriate moment to introduce the notion of a tree, for example by contrasting trees with graphs that have cycles. When the pupils have grasped the idea i.e. are able to tell if a graph is a tree, or is not, the theorem that in a tree the number of edges if for 1 less than the number of vertices could be stated or pupils induced to discover it (without a proof, of course, but since the proof is not too complicated, it could be a task for the more advanced students or group work in the mathematical class). The pupils at this point should know that a carbon atom has valence 4, and a hydrogen atom has valence 1. A few molecular graphs could be shown and pupils asked to find out which ones could represent an alkane, and which could not. Note that it is not necessary to label the vertices of the graph by names of corresponding elements since the only two elements involved have different valences so the corresponding vertices have different degrees.

Now the pupils are prepared to solve the task of discovering how the number m of C-atoms and the number n of H-atoms can relate in an alkane molecule. Altogether we have $m+n$ atoms i.e. vertices in the graph. Every C-vertex has degree 4 and every H-vertex has degree 1, so the double number of edges (bonds) is $4m+n$. It is the double number of edges because every edge connects two vertices so we have counted each edge twice. By the mentioned theorem about trees, it must be true that $4m+n=2(m+n-1)$ so $n=2m+2$. In other words, a molecule C_mH_n is an alkane only if $n=2m+2$ (the number of H-atoms is twice the number of C-atoms plus 1).

Note that the principle “the sum of the degrees of its vertices is double the number of edges in a graph” is a general principle for graphs and it gives a quick answer if a formula cannot represent a molecule. All one has to know are the valences of the atoms in the formula. Say we have a formula $X_kY_mZ_n$... (X, Y, Z, ... are mutually different element-symbols corresponding to elements of valences $a,b,c,...$) Then the graph of the molecule has $k+m+n+...$ vertices and $(ka+mb+nc+...)/2$ edges i.e. if the formula $X_kY_mZ_n$... represents a molecule, then it must be true that $ka+mb+nc+...=2(k+m+n+...)$, i.e. if the equality doesn’t hold, the formula doesn’t represent a molecule. One should however note that if the equality holds, it doesn’t necessarily mean that the formula represents a molecule.

Application 2: Enumeration problems

A benzene molecule consists of 12 atoms: six carbon atoms forming a ring and six hydrogen atoms. One or more of the hydrogen atoms may be replaced by chlorine to form chloro-derivatives from chlorobenzene (C_6H_5Cl) to hexachlorobenzene (C_6Cl_6). The number of these derivatives can be obtained purely mathematically using a general enumeration theorem that is too advanced to be presented on pre-university education level. Still, it could be motivating for the pupils who have mastered the notion of isomerism to see how some mathematical principles based on group theory, combinatorics and graph theory relatively simply give answers to enumeration problems. The mathematical prerequisites for this topic are familiarity with working with algebraic expressions, and basic knowledge about symmetries (in the geometric sense), permutations and inverse functions. Although groups are mentioned in this example, it is not necessary that the pupils learn many details on groups.

First one should draw a general graph of the molecule and decide which part of it is fixed (in the benzene-example, that is the carbon ring) and what are the free positions (in our example, there is one free 1-valent position on each of the carbon atoms). Let the vertices of the carbon ring be labeled from 1 to 6. To each of the carbon atoms at the vertices one hydrogen or one chlorine atom can be connected, so a benzene chloro-derivative can be viewed as a function f from the set $D=\{1,...,6\}$ to the set {H, Cl} (e.g., $f(2)=H$ if an H-atom is connected to the carbon on position 2). Basic combinatorics gives that there are altogether $|R|^{|D|} = 2^6 = 64$ such functions, but some are counted twice because benzene and its chloro-derivatives are invariant under rotations of the carbon ring, and under reflections of the carbon ring through the axis connecting two opposite vertices or midpoints of opposite edges. More precisely formulated: the benzene isomers are invariant under the group of symmetries of the regular hexagon - the dihedral group D_{6h}.

Generally, the set D consists of labels 1 to n for the free positions, the set R consists of symbols of the possible elements to fill the free positions and the group G consists of all permutations (symmetries) which give an equivalent molecule. Two functions $f,g:D\rightarrow R$ are G-equivalent if $f(p(x))=g(x)$ for some p from G and all x from D. In our example, an element p from $G=D_{6h}$ is the rotation of the benzene ring for 120º, and it gives an equivalent graph-representation of the same molecule (we can imagine that we exchanged the labels of the vertices from 1,2,3,4,5,6 to 3,4,5,6,1,2). It is the same if to the vertices (before the rotation) we apply the function g: $g(1)= g(3)=H$, $g(2)=g(4)=g(5)=g(6)=Cl$ or we rotate and then apply the function f: $f(1)= f(5)=H, f(2)=f(3)=f(4)=f(6)=Cl$:

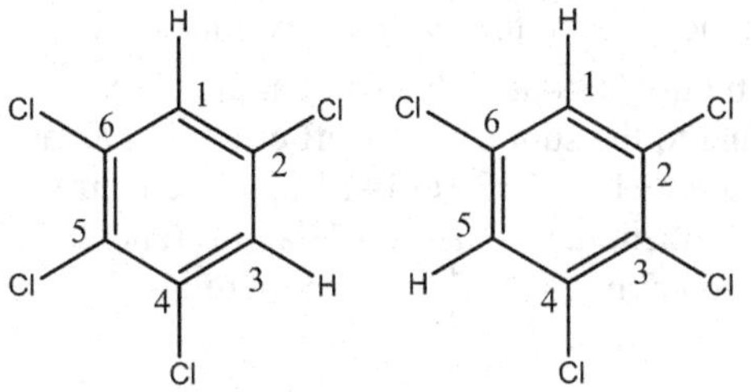

Fig. 5. Equivalent graph-representations of 1,2,3,5-tetrachlorobenzene

Therefore two functions from {1,...,6} to {H,Cl} correspond to the same isomer if and only if they are D_{6h}-equivalent. The enumeration problem is to count the number of functions from D to R that are not G-equivalent (i.e. the number of different isomers taking into account that rotated or axial-symmetric structural formulas represent the same compound). A more precise formulation would be that we want to count the number of equivalence classes under the relation of G-equivalence. To do this, we have to represent all permutations in G as products of cycles. Every permutation can be represented as a product of cycles. A 1-cycle (i) is a fixed point i.e. an element i from D that "stays put" after the permutation. The 120º-rotation has no 1-cycles because all vertices change their positions. If we choose the reflection about the axis joining vertices 1 and 4, then (1) and (4) are 1-cycles because the two vertices don't move in this permutation. A 2-cycle (i,j) is an interchanging of two elements from D i.e. the permutation sends i to j and vice versa. In the reflection example, 2 goes to 6, 6 to 2, 3 to 5 and 5 to 3 so there are two 2-cycles (2,6) and (3,5). A 3-cycle (i,j,k) means that the permutation moves i to j, j to k, and k to position i; and so forth. The 120º-rotation is a 6-cycle of the form (1,3,5,2,4,6). When we write all the disjoint cycles (disjoint means: no two cycles contain the same element from D) of the permutations next to each other, we talk about the permutation being represented as a product of cycles. For example, the reflection about the 1-4-axis of the benzene molecule is represented as (1)(4)(2,6)(3,5).

The cycle index of a group G is the sum of all cycle types of elements in G divided by the number of elements in G (in our example, the number of elements in D_{6h} is 12), where the cycle type of an element is represented by a term of the form $x_1^a \cdot x_2^b \cdot x_3^c \cdot ...$ where a is the number of 1-cycles, b is the number of 2-cycles, c is the number of 3-cycles, ... in the element (the last x in the product has lower index equal to the number of elements to be permuted; in D_{6h} we consider permutations of 6 elements, so we have terms of the form $x_1^a \cdot x_2^b \cdot x_3^c \cdot x_4^d \cdot x_5^e \cdot x_6^f$). So to the 120º-rotation corresponds the term x_6^1 (only 1 6-cycle) and to the reflection about the 1-4-axis the term $x_1^2 \cdot x_2^2$ (2 1-cycles and 2 2-cycles) etc. It turns out that the cycle index of D_{6h} is

$$Z(G) = \frac{1}{12}x_1^6 + \frac{1}{6}x_3^2 + \frac{1}{3}x_2^3 + \frac{1}{4}x_1^2x_2^2 + \frac{1}{6}x_6^1.$$

Now, replace every occurence of x_i in the cycle index with H^i+Cl^i. We get:

$$\frac{1}{12}(H+Cl)^6 + \frac{1}{6}(H^3+Cl^3)^2 + \frac{1}{3}(H^2+Cl^2)^3 + \frac{1}{4}(H+Cl)^2(H^2+Cl^2)^2 + \frac{1}{6}(H^6+Cl^6) =$$

$$H^6 + H^5\cdot Cl + 3H^4\cdot Cl^2 + 3H^3\cdot Cl^3 + 3H^2\cdot Cl^4 + H\cdot Cl^5 + Cl^6.$$

Generally, if $R=\{X,Y,\ldots\}$, we replace every occurrence of x_i in the cycle index with $X^i+Y^i+\ldots$ The solution to the enumeration problem is given by the famous Polya enumeration theorem: the number of all not G-equivalent functions is equal to the sum of the coefficients in the last polynomial expression. Thus in our example ther are 1+1+3+3+3+1+1=13 possible benzene chloro-derivatives. If we want to check how many of them have exactly 2 H-atoms (and 4 Cl-atoms), we just check the coefficient of the term containing H^2Cl^4 and see there are 3 such isomers.

Example 5: Determination of topological indices of molecules

As stated earlier, we can use molecular graphs to construct descriptors of molecular shape and size that correlate to physical or chemical properties of substances. Let us determine Wiener, Randić and Hosoya index (which were defined earlier in this paper) for three molecules: 2-methylbutane, isoprene and cyclohexane. Their hydrogen suppressed molecular graphs are shown in Fig.6.:

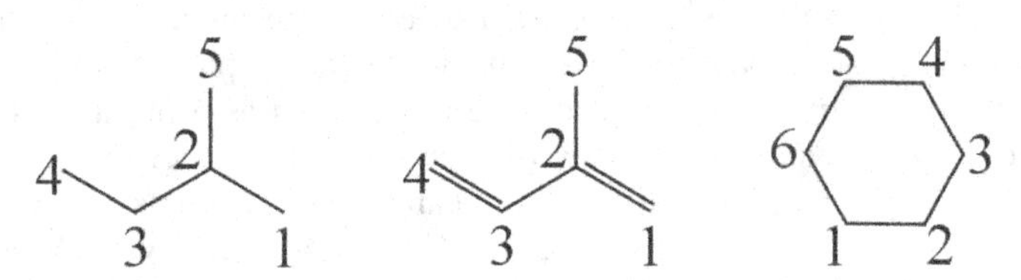

Fig.6. Hydrogen suppressed molecular graphs of 2-methylbutane, isoprene and cyclohexane

The Wiener index of 2-methylbutane is

W = ½ ((1+2+2+3)+(1+1+1+2)+(1+1+2+2)+(1+2+3+3)+(1+2+2+3)) = 18,

since atom 1 is one bond "away" from atom 2, two bonds from atoms 3 and 5, and three bonds from atom 4, atom 2 is one bond "away" from atoms 1,3 and 5, and two bonds from 4 and so on. Its Randić index is

$$R = \sqrt{\frac{1}{1\cdot 3}} + \sqrt{\frac{1}{3\cdot 2}} + \sqrt{\frac{1}{2\cdot 1}} + \sqrt{\frac{1}{1\cdot 3}} \approx 2{,}270$$

The first bond (between atoms 1 and 2) connects vertices of the first and of third degree, the second of third (2) and second (3) degree etc. There are four edges, and two ways of choosing two non adjacent edges Fig. 7.

Fig. 7. Two ways of choosing two non adjacent bonds in 2-methylbutane

So the Hosoya index for 2-methylbutane is Z = $p(0)+p(1)+p(2)=1+4+2=7$.

For isoprene and cyclohexane we cannot compute *W* since the molecular graphs are not trees. The Randić index for isoprene is

$$R=\sqrt{\frac{1}{2\cdot 4}}+\sqrt{\frac{1}{4\cdot 3}}+\sqrt{\frac{1}{3\cdot 2}}+\sqrt{\frac{1}{4\cdot 1}}\approx 2{,}408$$

and Hosoya index is $Z = 1$ + 6 + 6 = 13. The Randić index of cyclohexane is $R=6\sqrt{1/(2\cdot 2)}=3$, and the Hosoya index is $Z = 1 + 6 + 18 + 2 = 27$.

CONCLUSION

Basic graph theory is easily introduced in schools and has several more or less easy applications to chemistry, relying on the identification of graphs with structural formulas. In order to obtain a better understanding of both chemistry and mathematics, lessons about graph theory could (and we think: should) be incorporated in school curricula, particularly in secondary schools. The above mentioned applications are not the only graph theory applications in chemistry that can be introduced in schools; for example, we didn't deal here with a relatively new field of application consisting in analysing (topological) chirality of molecules, where the graph theoretical concepts of (non)planarity find their application. Because of very wide applications of chemistry, introductory graph theory in school curricula is also of benefit for the more advanced students who shall possibly pursue further studies in mathematics or chemistry.

References:

Biggs, N. L., Lloyd, E. K., & Wilson, R. J. (1976). *Graph Theory 1736-1936*. Oxford: Clarendon Press.

Feigelstock, S. (1994). Isomers, groups and combinatorics. *Maple Tech, 1*, 82-85.

Hansen, P. J., & Jurs, P. C. (1988). Chemical Applications of Graph Theory: Part I. Fundamentals and Topological Indices. *Journal of Chemical Education*, 65, 574-580.

Hansen, P. J., & Jurs, P. C. (1988). Chemical Applications of Graph Theory: Part II. Isomer Enumeration. *Journal of Chemical Education*, 65, 661-664.

Hosoya, H. (1971). Topological Index. *Bulletin of Chemical Society, Japan*, 44, 2331-2339.

McNaught, A. D., & Wilkinson, A. (Eds.). (1997). *International Union of Pure and Applied Chemistry: Compendium of Chemical Terminology "The Gold Book"*. Retrieved February 1, 2007. from *http://www.chem.qmul.ac.uk/iupac/bibliog/gold.html*

Randić, M. (1975). On Characterization of Molecular Branching. *Journal of the American Chemical Society*, 97, 6609-6615.

Rappaport, Z. (1974). Handbook *of Tables for Organic Compound Identification*. Cleveland: CRC Press.

Tannenbaum, P. (2004). *Excursions in Modern Mathematics*. Upper Saddle River: Pearson Education, Inc.

Trudeau, R. J. (1994). *Introduction to Graph Theory*. New York: Dover Publications.

Veljan, D. (1989). *Kombinatorika s teorijom grafova*. Zagreb: Školska knjiga.

Weisstein, E. W. (2007). Cycle Index. Retrieved March 25, 2007. from *MathWorld* – A Wolfram Web Resource, *http://mathworld.wolfram.com/CycleIndex.html*

Weisstein, E. W. (2007). Icosian Game. Retrieved March 25, 2007. from *MathWorld* – A Wolfram Web Resource, *http://mathworld.wolfram.com/IcosianGame.html*

Weisstein, E. W. (2007). Knight's Tour. Retrieved March 25, 2007. from *MathWorld* – A Wolfram Web Resource, *http://mathworld.wolfram.com/KnightsTour.html*

MATHEMATICAL PARADOXES AS PATHWAYS INTO BELIEFS AND POLYMATHY

Bharath Sriraman
Dept. of Mathematics
The University of Montana

This paper addresses the role of mathematical paradoxes in fostering polymathy among pre-service elementary teachers. The results of a 3-year study with 120 students are reported with implications for mathematics pre-service education as well as interdisciplinary education. A hermeneutic-phenomenological approach is used to recreate the emotions, voices and struggles of students as they tried to unravel Russell's paradox presented in its linguistic form. Based on the gathered evidence some arguments are made for the benefits and dangers in the use of paradoxes in mathematics pre-service education to foster polymathy, change beliefs, discover structures and open new avenues for interdisciplinary pedagogy.

Keywords: beliefs; interdisciplinarity; paradoxes; pre-service teacher education; polymathy; Russell's paradox;

1. Introduction

Elementary set theory serves as the backbone of mathematics content required by prospective elementary school teachers around the world. This is evident in the content standards of numerous curricular documents (e.g., Australian Research Council, 1990; National Council of Teachers of Mathematics, 2000) which call for both a foundational and contextual understanding of the models for the four arithmetic operations (+, - , ×, ÷) developed for the natural, whole, rational and real numbers. In addition the use of manipulatives such as Dienes base-10 blocks, Cuisenaire rods etc, greatly facilitate the enact-ion of the elementary arithmetic operations for the particular set under consideration. However the set theoretic and philosophical foundations of these operations are typically thought to be beyond the scope of pre-service education. The two fundamental questions explored in this paper are: (1) How can we facilitate the discovery of the mathematical foundations, paradoxes and structures? and (2) How can deeply rooted beliefs about the nature of mathematics be impacted?

2. Motivation and conceptual framework

In the United States, teacher professional development programs typically target in-service teachers and use an interventionist attempt to shift their beliefs and practices about the nature of mathematics. A large body of extant research addresses pre-service and practicing school teachers' beliefs and attitudes (e.g., Thompson, 1992) towards mathematics and describes the affective factors (Leder, Pehkonen & Törner, 2002) which influence mathematical understanding (Ball, 1990)and problem solving (Goldin, 2000, 2002). There is also research which addresses limitations of current research approaches to studying teacher beliefs (Leatham, 2006; Wedege & Skott, 2006). Wedege & Skott (2006) argue that the main-stream trend of "research on belief-practice relationships runs the risk of becoming a self-fulfilling prophecy. It often contains a circular argument of claiming that certain observed mathematical practices are due to beliefs, while at the same time inferring mathematical beliefs from the very same practices." (p.34).

B.Sriraman, C.Michelsen, A. Beckmann & V. Freiman (Eds). (2008). *Proceedings of the Second International Symposium on Mathematics and its Connections to the Arts and Sciences* (MACAS2). Information Age Publishing, Charlotte: NC , pp.127-142

Similarly Leatham (2006) critiques research on teacher beliefs as assuming that "teachers can easily articulate their beliefs and that there is a one-to-one correspondence between what teachers state and what researchers think those statements mean. Research conducted under this paradigm often reports inconsistencies between teachers' beliefs and their actions." (p.91).
Ernest (1989) categorized three philosophies of mathematics, namely the instrumentalist view, the Platonist view, and the problem solving view. The instrumentalist sees mathematics as a collection of facts and procedures which have utility. The Platonist sees mathematics as a static but unified body of knowledge. Mathematics is discovered, not created. The problem solving view looks on mathematics as continually expanding and yet lacking ontological certainty. The problem solving view sees mathematics as a cultural artifact. This implies that what is thought as true today, may not be seen as true tomorrow. (pp. 99-199). Ernest also describes absolutist and fallibilist views of mathematical certainty. The absolutist sees mathematics as completely certain and the fallibilist recognizes that mathematical truth may be challenged and revised (Ernest, 1991, p.3).

Lerman (1990) recognizes that ones philosophy is related to ones preferred teaching style. The absolutist teacher will prefer a direct teaching style whereas a fallibilist is much more likely to engage in exploratory activities and open-ended problems. In what is now one of the seminal studies in the domain of teacher beliefs, Thompson (1984) studied three junior high teachers, all of whom had different beliefs about the nature of mathematics. The first teacher viewed mathematics as a coherent collection of interrelated concepts and procedures. She regarded mathematics as a subject free of ambiguity and emphasized conceptual development in the students. She would fit Ernest's model as a Platonist. The second teacher had a very different perspective of mathematics. Thompson says her teaching reflected more of a process-oriented approach than a content oriented approach. A view of mathematics as a subject that allows for the discovery of properties and relationships through personal inquiry seemed to underlie her instructional approach. This teacher would fit Ernest's model as person with the problem solving view. The third teacher in Thompson's study saw mathematics as a collection of facts and procedures which help students find the answer. She saw no ambiguity in mathematics. She would fit Ernest's model as a person with an instrumentalist view. Thompson sees at least three distinct ways of viewing mathematics, all of which greatly influence the choice of curriculum and its delivery. Thompson (1992) says that research on teachers cognitions and studies of teachers' conceptions have contributed to a conceptual shift in the field of research on teaching, moving away from a behavioral conception of teaching towards "a conception that takes account of teachers as rational beings" (p.142). Our understanding of teaching from teachers' perspectives complements our growth of understanding of learning from learners' perspectives, which in turn, enriches the idea of schooling as the negotiation of norms, practices and meanings (Cobb,1988). Much earlier, Fenstermacher (1978) predicted that the single most important construct in educational research are beliefs (Pajares 1992). Törner & Sriraman (2007) argue that Thompson's (1982, 1992) theory to explain teacher's actions in a mathematics classroom based on their beliefs about mathematics is one instance of the development of a local philosophy based on problematizing research in the domain of beliefs. Thompson (1992) wrote:

> I think we will get further evidence on the role of teachers' views of mathematics when we go into more detail and investigate their understanding of different domains of mathematics, of specific components such as the meaning of mathematical concepts, proof, definition, theorem, conjecture, variable, symbols, rule, formula, axiom, problem, problem solving, application, model, computation, graphical

representation, visualization, metaphor,etc., both with respect to the various sub-domains of mathematics as well as in a more general sense. (p. 142)

Today we usually speak of teachers' beliefs, which are generally formulated as ''views about mathematics'' (e.g. Grigutsch 1996; Pajares 1992). It is assumed that different beliefs about mathematics have different associated philosophies and/or epistemologies (Törner, 2002). Amidst all this important research which increases our understanding of teacher beliefs, in lieu of Thompson's call, one is left wondering about the dearth of studies (recent or otherwise) in describing or analyzing prospective elementary school teachers understanding of the foundational (set-theoretic) concepts of the mathematics they are exposed to. Could it be that the aftermath of New Math has had an "affective" impact (pun-intended) on the focus of mathematics education researchers engaged in teacher education and led to less emphasis of this particular mathematics content. For the author it is important that we attend to the significant role that tasks may have in teacher education, particularly tasks that foster interdisciplinary thinking as well as tasks that shake the dominant views of prospective teachers on the nature of mathematics. In the remainder of this paper the author will report on the use of a set-theoretic task to help prospective elementary school teachers understand and discover paradoxes and structures and foster polymathy. Based on the findings, limitations and dangers in such tasks are also outlined.

3. Polymathy, Paradoxes and Philosophy

The term polymath is in fact quite old and synonymous with the German term "Renaissance-mensch." Although this term occurs abundantly in the literature in the humanities, relatively few (if any) attempts have been made to isolate the qualitative aspects of thinking that adequately describe this term. Most cognitive theorists believe that skills are domain specific and typically non-transferable across domains. This implicitly assumes that "skills" are that which one learns as a student within a particular discipline. However such an assumption begs the question as to why polymathy occurs in the first place? (Sriraman & Dahl, 2007).

Root-Bernstein (1989,1996, 2000, 2001, 2003) has been instrumental in rekindling an interest in mainstream psychology in a systematic investigation of polymathy. That is the study of individuals, both historical and contemporary, and their interdisciplinary thinking traits which enabled them to contribute to a variety of disciplines. Common thinking traits of the thousands of polymaths (historical and contemporary) as analyzed by Root-Bernstein, Robert Sternberg , Dean Simonton and many others are: (1) Visual geometric thinking and/or thinking in terms of geometric principles, (2) Frequent shifts in perspective, (3) thinking in analogies, (4) Nepistemological awareness (that is, an awareness of domain limitations), (5) Interest in investigating paradoxes (which often reveal interplay between language, mathematics and philosophy), (6) Belief in Occam's Razor [Simple ideas are preferable to complicated ones], (7) acknowledgment of Serendipity and the role of chance, and (8) the drive to influence the Agenda of the times. Although the prodigious writings of these researchers (Root-Bernstein, 1989, 1996, 2000,2001, 2003; Sternberg et al., 2004) typically involve eminent individuals, it has been found that polymathy as a thinking trait occurs frequently in non-eminent samples (such as high school students) when presented with the opportunities to engage in trans-disciplinary behavior. In particular the use of unsolved classical problems, paradoxes and mathematics literature has been found to be particularly effective in fostering inter-disciplinary thinking (Sriraman, 2003a, 2003b, 2004, 2005).

4. Methodology

In order to investigate whether prospective elementary school mathematics teachers display some of the thinking traits of polymaths the author conducted a 3 year study with approximately 120 prospective elementary school mathematics teachers in the 2002-2005 time period. These pre-service teachers were enrolled in the mathematics content sequence for elementary teachers at a large university in the western United States. This two semester content sequence is the only required mathematics content for these prospective teachers due to the particular state legislations in this region. The author was also the instructor of these students. Journal reflective writing was an integral part of this course.

During the course of the semester, among the various tasks assigned to the student was the following:

4.1 The Task

> The town barber shaves all those males, and only those males, who do not shave themselves. Assuming the barber is a male who shaves, who shaves the barber?
> Explain in your own words what this question is asking you? When you construct your response to the question, please justify using clear language why you think your answer is valid? If you are unable to answer the question who shaves the barber, again justify using clear language why you think the question cannot be answered.

This task is the well known linguistic version of Russell's paradox, appropriately called the Barber Paradox. The question as formulated here was read out several times in the class in order to clarify what it was asking. Students were given about 10 days to construct a written response to this task. The purpose of this task was to investigate whether students with no prior exposure to the paradox would be able to decipher the contradictions in the linguistic version of Russell's paradox, and whether they would be able to then construct their own set theoretic (mathematical) version of the paradox. All the students were also asked to complete the following affective tasks parallel to the mathematical task. The students were also requested not to consult the worldwide web in search of a solution.

- Write one paragraph (200-300 words) about your impressions of a given question after you have read it, while tackling it (if possible), and after you've finished it.
- In particular record things such as:
- The immediate feeling/mood about the question (confidence, in confidence, ambivalence, happiness, tenseness etc)
- After you've finished the question record the feeling/mood about the question (if you are confident about your solution; why you are confident? Are you satisfied /unsatisfied? Are you elated/not elated? Are you frustrated? If so why?
- Did you refer to the book, notes? Did you spend a lot of time thinking about what you were doing? Or was the solution procedural (and you simply went through the motions so to speak)
- Was the question difficult, if so why? If not, why not?
- Did you experience any sense of beauty in the question and/or your solution?

4.2 Data Collection and Analysis

Out of the 120 students, 52 students were able to unravel the paradox, i.e., understand and explain the contradiction in their writings, 40 students believed there was no contradiction (i.e., they answered that the barber shaved himself), and 28 students gave up on the problem but completed the affective portion. Over the 3 year period, in addition to the written journal responses of the students to the aforementioned tasks, the author interviewed 20 students from the 120 students who were representative of the larger sample.

Of these 10 students had successfully unraveled the paradox (out of the 52), and the remaining 10 were unable to unravel it. 6 of these students came from the subset of 40 who saw no contradiction, and the remaining represented those who gave up on the problem. These students were purposefully selected on the basis of whether they were able to unravel the paradox and formulate its set theoretic or mathematical equivalent and those that were unsuccessful in their attempts to do so. It should be noted that all students were given full credit for the assignment irrespective of whether they were successful or not. The written artifacts (students journal writing/solution and affective responses), and interview data were analyzed using a phenomenological -hermeneutic approach (Merleau-Ponty, 1962; Romme & Escher, 1993) with the purpose of re-creating the voice of the students. Phenomenology has its roots in the philosophical work of Husserl and Heidegger, which was extended into a theory of embodiment by Merleau-Ponty in order to counter reductionism, dualism and to capture the totality of human experience.

During the interview students re-voiced their experience of unraveling the paradox. The author simply sought clarifications on the written solution, their affective responses and asked students for general comments on the nature of the problem and their struggles with it. This led to questions on their beliefs about mathematics. There was no pre-set direction or protocol in which the interview was made to progress. Each interview lasted approximately 60 minutes. The individual transcripts and the author's interpretation of student voices, particularly their self-reported affective and polymathic experiences was discussed with each student to ensure validity and reliability. Student journal writings were coded several times to categorize and determine the affective mood self reported by the students. Similarly the interview transcripts were also coded for affective categories and to determine consistency in the self reported voices. In addition, the constant comparative method from grounded theory was applied for the purposes of triangulating the categories which emerged from the phenomenological approach (e.g., Annells, 2006). Finally, in all classes a de-briefing session occurred in which various students presented their views on the problem and their solutions with a discussion of the contradiction.

5. Results

Examples of affective responses, student solutions, and interview transcripts are now presented in a phenomenological style, which recreate student voices as they struggled through this paradox. Contrasting voices are presented and are representative of the 20 students who participated in the interviews.

5.1 Voices

Note the following abbreviations:
JWAV- Journal writing affective voice (response) ; JWSV- Journal writing solution voice (response); IVS- Interview voice student ; IVA- Interview voice author

JWAV1 Started the question in class (10/19) after you (author) read it. Stopped occasionally for work and classes, finished it 10/25 at 12.01pm. The question is very confusing and I feel very anxious about this question. Why would this question be asked in a math class? Don't get me wrong, I like to think about questions like these, but they are difficult and time consuming. I'm frustrated...[w]hat a strange question.

JWAV2 I painfully came to the conclusion that the question was answerable through reading the question several times and thinking about it for days. It is a beautiful paradox, if one thing were changed in the question I feel there would be a definite answer. The English in it is perfect.

JWSR1 The question is asking who shaves the barber? However the barber is a male and he shaves only those males, who do not shave themselves. Thus he cannot shave himself because he only shaves the males that do not shave themselves...[B]ut the barber only shaves those who do not shave themselves...[T]his is a paradox, he cannot be the barber and shave himself, and he cannot shave himself and be the barber.

JWSR2 The barber shaves himself. I justify this firstly by the opening sentence "who do not shave themselves" which implies there are those who do shave themselves. For example a mechanic will work on all of other peoples cars in town but if his car has a problem, he would work on it!

JWAV3 My immediate opinion was that the answer would be "yes". I didn't really feel anything when doing this problem other than that the answer was obvious.

JWAV4 This question was the death of me...[I] was more than upset with you. Thinking about this question made me so frustrated that I stopped and decided not to waste my time on it. I found this question to have not any beauty in it, all it caused me was a lot of stress and discomfort in my life. It is impossible to put any mathematical equation into it.

The affective response of frustration and curiosity as well as a sense of beauty in the problem was found predominantly in the group that unraveled the paradox (Group 1). Students in this group voiced high levels of frustration, anger, curiosity and beauty as well as reported a sense of accomplishment in unraveling the paradox.

In contrast, in the groups that saw no contraction (group 2) and the group that gave up on the problem (group 3), although the levels of frustration were comparably high, the proportion of students that became curious or saw some beauty in the problem was considerably lower. Approximately half of the students in group1 and group 3 became angry while attempting this problem, however nearly 80% of group 1 students reported curiosity whereas only 10% of group 3 students experienced a sense of curiosity. The proportion of self reported experience of "sadness" was considerably higher in group 3 in comparison to the other groups. Group 1 reported the highest levels of accomplishment. The following table gives a summary of the coded affective voices/responses extracted from student journal writings and interview transcripts.

Table 1: Affective traits in student groups

Affective mood→ Student groups↓	Frustration	Anger	Curiosity	Sense of beauty	Sense of accomplishment	Set-theoretic formulation	Sadness	No affect reported
Group 1 (52) (unraveled paradox)	52	26	41	28	35	16	0	0
Interview Subgroup 1 (10)	10	7	10	4	8	6	0	0
Group 2[1] (40) (Saw no contradiction-claimed the barber shaved himself)	32	11	14	0	3	0	1	8
Interview Subgroup 2 (6)	3	1	1	0	2	0	1	1
Group 3 (28) (Gave up on the problem)	27	13	3	0	0	0	14	1
Interview subgroup3 (4)	4	2	1	0	0	0	3	0

[1] Four of the students in group 2 tried to construct an explanation to unravel the paradox after stating that the barber shaved himself (see Figure 2)

5.2 Student Journal Writings

In this section, sample journal writings with student solutions are presented

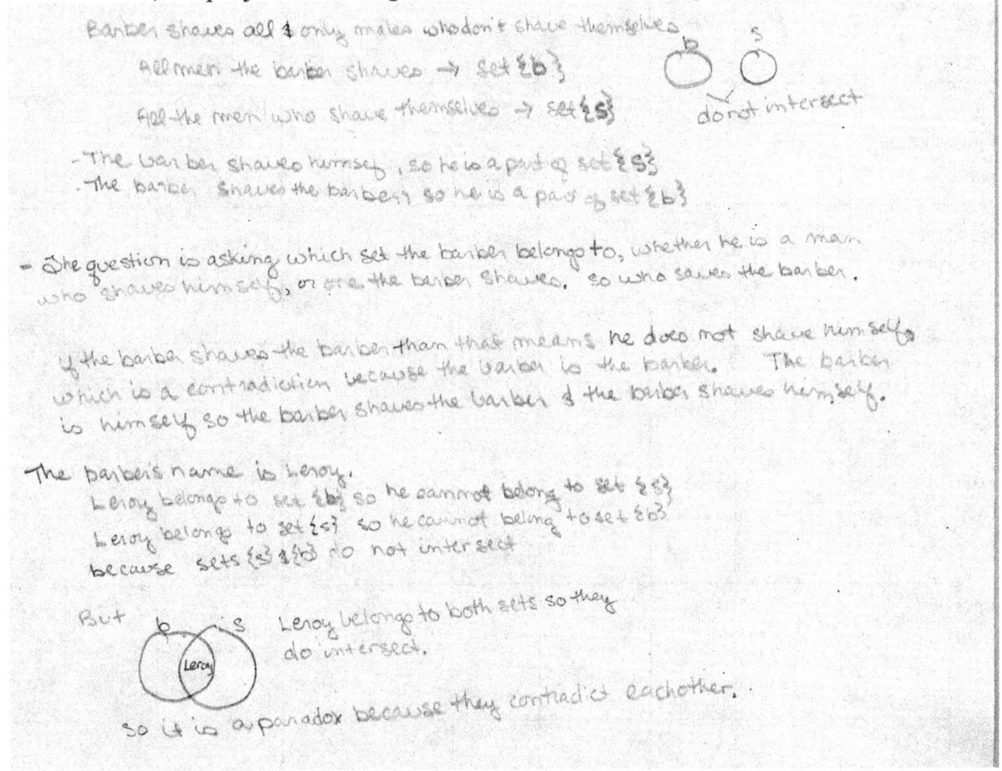

Barber shaves all & only males who don't shave themselves

All men the barber shaves → set {b}

All the men who shave themselves → set {s}

b s do not intersect

- The barber shaves himself, so he is a part of set {s}
- The barber shaves the barber so he is a part of set {b}

- The question is asking which set the barber belongs to, whether he is a man who shaves himself, or one the barber shaves. so who saves the barber.

If the barber shaves the barber than that means he does not shave himself which is a contradiction because the barber is the barber. The barber is himself so the barber shaves the barber & the barber shaves himself.

The barber's name is Leroy.
Leroy belongs to set {b} so he cannot belong to set {s}
Leroy belongs to set {s} so he cannot belong to set {b}
because sets {s} & {b} do not intersect

But b s Leroy belongs to both sets so they do intersect.

Leroy

so it is a paradox because they contradict eachother.

Figure 1: Unraveling the set theoretic version of the paradox (Group 1)

5.3 Commentary 1

This solution represents the numerous solutions obtained from group 1 students where the set-theoretic contradiction became clear to the students. It should be noted that students who discovered and formulated the set theoretic version of the solution spent significantly longer periods of time with the problem in comparison to the other groups. It should be noted that only 16 of the 52 students in group 1 were successful in doing this. It is well known in the foundational mathematics literature that Betrand Russell discovered this paradox and communicated it to Gottlob Frege in a letter right when Frege was about to complete his treatise on the foundations of Arithmetic (*Grundlagen der Arithmetik*), which dealt a devastating blow to his work. The standard way of stating the paradox is if we let R be the set of all sets that are not members of themselves. Then R is neither a member of itself nor not a member of itself.

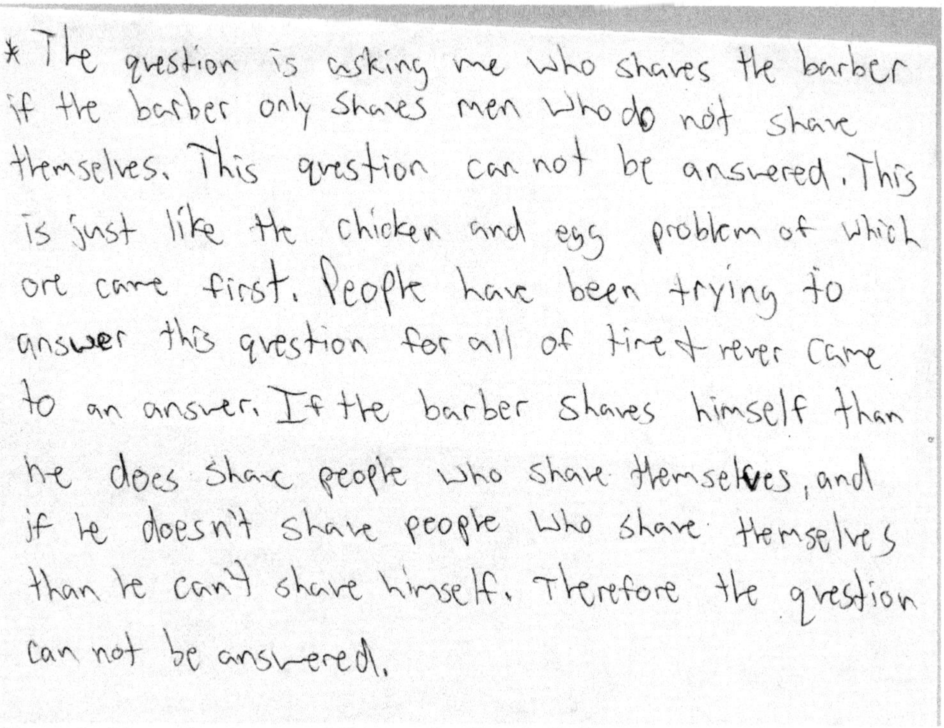

* The question is asking me who shaves the barber if the barber only shaves men who do not shave themselves. This question can not be answered. This is just like the chicken and egg problem of which one came first. People have been trying to answer this question for all of time & never came to an answer. If the barber shaves himself than he does shave people who shave themselves, and if he doesn't shave people who shave themselves than he can't shave himself. Therefore the question can not be answered.

Figure 2: Attempting to unravel the paradox linguistically

5.4 Commentary 2

There were 4 students in group 2 who first claimed that the barber shaved himself but then attempted to construct a linguistic explanation like the one shown in Figure 2, which were incorrect. Two of these students were interviewed during which they voiced the reasons for doing so, which is reported in a following section of this paper.

a. This question is asking me to explain who shaves the barber (or cuts his hair) if he is the one that is shaving (or cutting) everyone that is male and doesn't shave themselves.

b. If the town barber shaves all those males who do not shave themselves, then the males that shave themselves must shave the barber. If the males that don't shave themselves are Bob, Joe, Bill, Jack, and Frank then the Barber (Tom) shaves them. If the males that shave themselves are Jake, George, Stew and Cody, then they must shave Barber Tom. I think my answer is valid because we know that Joe, Bill, Jack, Bob and Frank don't cut Toms hair because they don't even cut their own hair. So, the males that do cut their own hair (shave themselves) must cut Toms hair because there wouldn't be anyone else that could do it. So, Jake, George, Stew or Cody cuts the barbers hair.

Figure 3: Unsuccessful solution to the paradox

5.5 Voices 2

In this section, three interview vignettes are presented representing the three groups.

Interview vignette 1

IVS1: I was tickled by the challenge of this question. I didn't expect to solve a paradox in a math class and had to think a great deal when I was working on other things. I didn't see it as a problem involving sets at least not right away.

IVA1: Did you think it was a mathematical problem?

IVS2: Not really. It made me think of literature or something you would encounter from Zen Buddhism or even Rumi. Math problems don't make me angry and this one did! I was angry but challenged enough that I didn't let my anger get in the way of finding a solution. I also thought about it in terms of art, like the strange tiling pattern posters we have in class, you know what I mean?

IVA2: Yes, I think you're referring to the Escher posters. How long did you spend on the problem?

IVS3: Yeah… that's the one. Days and days… [b]ut when I saw through it and realized it was a paradox, like you know assuming one thing led to another thing which was contradictory, then I thought it was beautiful. I never imagined saying something like that for a math problem.
IVA4: Why not? So is this a math problem?
IVS4: Umm….in math classes, you like use formulas and equations and stuff and get answers. This one made you think really hard. I wish math were like that. Everything is just black and white in math, atleast this is what we are led to believe.
IVA5: Does this make you change your view of what mathematics is?
IVS5: This problem does, but I don't think we can say that for sure.
IVA6: What if I told you there were many such paradoxes in math? Would that make you see math differently?
IVS6: Sure. Look, we're told all through school that you got to memorize equations and formulas and things like that to do math or like there is a set way to do a problem. But with problems like these you kinda get a little from freedom to think for yourself. So definitely, yeah…if we were exposed to more paradoxes like these, we'd change our minds about what math is all about.

Interview vignette 2
IVA1: You wrote this problem caused the "death" of you? I'm sorry you were so frustrated by the problem
IVS1: Yeah, I was really quite upset with it. I really thought about it for a while and became so confused that I didn't want to deal with it anymore.
IVA2: Would you have approached the problem differently if you were this in a non-math class?
IVS2: I actually did think about my philosophy course when I was reading this problem. But I'm not used to this kind of stuff in a math for elementary teachers class. I did look into the book and it said that there were some paradoxes in math in the first chapter. It was easier to give up than go crazy over the problem. Plus I was really under pressure from work and just didn't have the time to think it through. Sorry.
IVA3: No, no, you don't have to apologize. You simply did what you did. Did you really stop thinking about the problem after you gave up?
IVS3: [silence] Yes and no. I didn't really feel like solving it. Part of the reason might have to do with that I didn't really know what a solution is supposed to look like, you know. I did imagine other situations that could be like the one in the problem but it didn't help any.
IVA4: Do you think paradoxes have a place in mathematics?
IVS4: I don't know. I'm going to do my student teaching soon and I was thinking this might be a fun thing for kids to talk about. I don't know how it would go over with parents if the teacher gave problems like these [laughing].
IVA5: You mean they might get upset?
IVS5: Yeah…I mean they'd try to help their kid and get a headache and then blame the teacher for giving this problem. Parents would want the answers to be cut and dry.

Interview Vignette 3
IVA1: I noticed you first created an example with ten people and said it was clear that the barber had to shave himself. Then you expressed doubts after that and said it was like the chicken and the egg problem.

IVS1: My first feeling about the question was complete frustration. I didn't know who or what to relate this question to. How was I supposed to answer the question? So, I did this problem solving thing like making an example. But then I wasn't really using the information given in the problem. Just making stuff up....[s]o I wasn't too happy about it and thought maybe I should do something else. That's when I wrote the stuff about the chicken and the egg problem. Now it seems clear to me that there was a contradiction in the way the problem was given. It was a good brainteaser but very frustrating though.
IVA2: Did you get upset?
IVS2: Honestly, yes. I got pretty worked up. I wasn't satisfied with the solution with people in it. It was all made up. The really upsetting thing was not knowing whether you were right or wrong.
IVA3: Is that how mathematics is supposed to be?
IVS3: [silence]
IVA4: By the way, did the problem make you think of other things?
IVS4: I read and re-read the problem many times and tried to think of it in different ways. I mean I thought of making stuff up like, the barber is married, his wife shaves him. It was tough to stay within the boundaries of the question. It made me think of puzzles or visual tricks that you see in paintings sometimes. Like you see it one way and then you blur your eyes or focus on a different point in the picture and you see something else. You know like those pictures that appear like a duck or a rabbit, or dolphins and people in the same picture. I guess the point of the problem was you couldn't have it both ways. I finally figured out that there were like two sets, one which had all men that shaved themselves, and then another set with all men shaved by the barber, and the barber couldn't belong to either of these sets.
IVA5: Are you happy you figured out the paradox later?
IVS5: I don't know about happy. More like relieved!

5.6 Author Voice Over

These three interview vignettes revealed that students were beginning to make connections between mathematics and other domains of inquiry such as philosophy, language and art. The hermeneutic analysis of student journal writings, affective responses and interview transcripts indicated that nearly half of the students (predominantly in group 1) displayed one or more *polymathic* traits when engaged with the paradox. In particular students reported (1) Frequent shifts in perspective (2) thinking with analogies, (3) tendency towards nepistemology i.e., questioning the validity of the question and its place in the domain of mathematics as well as the fallibility of mathematics. The pre-service teachers also reported an increased interest in the place of paradoxes in mathematics, which they had always believed as an infallible or absolutist science. Some of the students began to connect mathematics with language and voiced the need to engage in discourse as opposed to engaging in such an activity solitarily. One of the dominant and consistently heard student voices over the course of three years was of the deeply held dominant belief of mathematics as infallible or absolutist. Table 2 gives the polymathic traits displayed by the three groups of students.

Table 2. Polymathic themes emerging from journal writing and interview transcript analysis

Polymathic traits→ Student groups↓	shifts in perspective (paradoxes in art and language)	Thinking in analogies	Nepistemological awarness [mathematics as fallible]	Interest in investigating paradoxes	Mathematics as language (need for discourse)	Mathematics as philosophy
Group 1 (52) (unraveled paradox)	16	25	38	52	18	28
Interview Subgroup 1 (10)	6	10	6	10	5	4
Group 2 (40) (Saw no contradiction- claimed the barber shaved himself)	7	12	4	4	2	4
Interview Subgroup 2 (6)	3	2	1	3	2	3
Group 3 (28) (Gave up on the problem)	1	5	1	0	1	0
Interview subgroup3 (4)	0	3	0	0	1	0

6. Implications and Concluding Points

Polymathy and interdisciplinarity are topics on which one finds scant literature in the field of mathematics education, particularly in domain of pre-service elementary teacher education. Although we live in an age where knowledge is increasingly being integrated in emerging domains such as mathematical genetics; bio-informatics; nanotechnology; modeling; ethics in genetics and medicine; ecology and economics in the age of globalization, the curriculum in most parts of the world is typically administered in discrete packages. The analogy of mice in a maze appropriately characterizes a day in the life of students, with mutually exclusive class periods for math, science, literature, languages, social studies etc. Even mathematics is increasingly viewed as a highly specialized field in spite of its intricate connections to the arts and sciences. The thinkers of the Renaissance did not view themselves simply as mathematicians, or inventors or painters, or philosophers or political theorists, but thought of themselves as seekers of Knowledge, Truth and Beauty. In other words there was a Gestalt world-view with polymaths that worked back and forth between multiple domains. The results of this three year study with 120 pre-service students indicate that nearly half of the students displayed polymathic traits – as a result of their attempt to unravel the given paradox. This suggests that interdisciplinary activities can certainly play an important role in the education of these future teachers. By taking a phenomenological approach and trying to understand the first person perspective of these students, deep set hidden beliefs about the nature of mathematics also became apparent through the voices of the students. The pre-service teachers also reported an increased interest in the place of paradoxes in mathematics, which they had believed as an infallible or absolutist science.

However the significance and applicability of these findings come with certain limitations. It should be noted that nearly one fourth of the students (group 3) in the study experienced and reported negative affective experiences as a result of tackling the paradox, which need to be sensitively attended to by the teacher educator. Debriefing sessions conducted during the course of the study were essential to create a positive pedagogical atmosphere on these students and foster a willingness on their part to try interdisciplinary activities in their classroom which integrates mathematics with other subjects. Another missing ingredient in this study voiced by many students in their journal writings and interviews was the need to engage in discourse with other students when confronted with the paradox. It would be of interest to the community of researchers to investigate how pre-service elementary teachers tackle paradoxes in a collaborative group effort. Due to the limitations in the resources available for this study, the author did not pursue this approach but this remains a fertile area for further investigation. Finally, many of these students voiced concern over concrete ways in which interdisciplinary activities could be introduced in the elementary classroom. To this end there have been recent attempts to classify works of mathematics fiction suitable for use by K-12 teachers in conjunction with science and humanities teachers to broaden student learning. Padula (2005) argues that although good elementary teachers have historically known the value of mathematical fiction, mainly picture books, through which children could be engaged in mathematical learning, such an approach also has considerable value at the secondary level. Padula (2005) provides a small classification of books appropriate for use at the middle and high school levels, which integrate paradoxes, art, history, literature and science to "stimulate the interest of reluctant mathematics learners, reinforce the motivation of the student who is already intrigued by mathematics, introduce topics, supply interesting applications, and provide mathematical ideas in a literary and at times, highly visual context" (p. 13)

Mathematical paradoxes played a significant role in the historical development of the field. These paradoxes contain enormous pedagogical potential for pre-service teacher education, particularly in showing that even mathematics can be fallible. As seen in this study, realizing the fallibility within what students believed was an absolutistic and monolithic structure, these prospective teachers experienced both a sense of empowerment and expressed changing views about the nature of mathematics. It is the authors hope that this line of research is further developed by the community of researchers and teacher educators to make a positive impact on the beliefs and practices of future teachers of mathematics.

7. References

Annells, M. (2006). Triangulation of qualitative approaches: hermeneutical phenomenology and grounded theory. Journal of Advanced Nursing, 56(1), 55–61

Australian Education Council (1990). A national statement on mathematics for Australian schools. Melbourne, VC: Australian Educational Council.

Ball, D. L. (1990). The mathematical understandings that pre-service teachers bring to teacher education. Elementary School Journal, 90, 449-466.

Cobb, P. (1988). The tension between theories of learning and theories of instruction in mathematics education. Educational Psychologist, 23, 87- 104.

Ernest, P. (1989). The impact of beliefs on the teaching of mathematics. In C. Keitel, P. Damerow, A. Bishop & P. Gerdes (Eds.), Mathematics, Education and Society (pp. 99- 101). Paris: UNESCO Science and Technology Education Document Series No 35.

Ernest, P. (1991). The philosophy of mathematics education. London: Falmer Press.

Fenstermacher, G.D. (1978). A philosophical consideration of recent research on teacher effectiveness. In L.S. Shulman (Ed.), Review of research in education 6 (pp. 157-185). Ithasca (IL): Peacock.

Goldin, G.A. 2000. Affective pathways and representations in mathematical problem solving. Mathematical Thinking and Learning, 17(2), (pp.209-219).

Goldin, G.A. (2002). Affect, Meta - Affect, and Mathematical Belief Structures. In: G. Leder, E.Pehkonen & G. Törner (Eds), Beliefs: A Hidden Variable in Mathematics Education?(pp. 59 – 72). Dordrecht: Kluwer Academic Publishers..

Grigutsch, S. (1996). "Mathematische Weltbilder" bei Schülern: Struktur, Entwicklung, Einflussfaktoren. Dissertation. Duisburg: Gerhard-Mercator-Universität Duisburg, Fachbereich Mathematik.

Leathman, K. (2006). Viewing mathematics teacher beliefs as sensible systems. Journal of Mathematics Teacher Education, 9(1), 91-102.

Leder G. C., Pehkonen E., Törner G. (Eds.), (2002). Beliefs: A hidden variable in mathematics education? (Vol. 31). Dodrecht: Kluwer Academic Publishers.

Merleau-Ponty, M. (1962). Phenomenology of perception (C. Smith, Trans.). London: Routledge & Kegan Paul.

National Council of Teachers of Mathematics (2000). Principles and standards for school mathematics. Reston, VA: Author.

Padula, J. (2005). Mathematical fiction- It's place in secondary school mathematics learning. The Australian Mathematics Teacher, 61(4), 6-13.

Pajares, M.F. (1992). Teachers' beliefs and educational research: Cleaning up a messy construct. Review of Educational Research, 62 (3), 307-332

Root-Bernstein, R. S. (1989). Discovering. Cambridge, MA: Harvard University Press.

Root-Bernstein, R. S. (1996). The sciences and arts share a common creative aesthetic. In A. I. Tauber (Ed.), The elusive synthesis: Aesthetics and science (pp. 49-82). Netherlands: Kluwer.

Root-Bernstein, R. S. (2000). Art advances science. Nature, 407, 134.

Root-Bernstein, R. S. (2001). Music, science, and creativity. Leonardo, 34, 63-68.

Root-Bernstein, R. S. (2003). The art of innovation: Polymaths and the universality of the creative process. In L. Shavinina (Ed.), International handbook of innovation, (pp. 267-278), Amsterdam: Elsevier.

Sriraman, B (2003a) Can mathematical discovery fill the existential void? The use of Conjecture, Proof and Refutation in a high school classroom, Mathematics in School, 32 (2), 2-6.

Sriraman, B. (2003b). Mathematics and Literature: Synonyms, Antonyms or the Perfect Amalgam. The Australian Mathematics Teacher, 59 (4), 26-31.

Sriraman, B. (2004). Mathematics and Literature (the sequel): Imagination as a pathway to advanced mathematical ideas and philosophy. The Australian Mathematics Teacher. 60 (1), 17-23.

Sriraman,B. (2005). Re-creating the Renaissance. In M. Anaya, C. Michelsen (Editors), Relations between mathematics and others subjects of art and science. Proceedings of the 10th International Congress of Mathematics Education, Copenhagen, Denmark, pp.14- 19.

Sriraman, B., & Dahl, B. (2007). On bringing interdisciplinary ideas to gifted education. In press in L.V. Shavinina (Ed), The International Handbook of Giftedness, Springer Science

Thompson, A. (1992). Teachers' beliefs and conceptions: A synthesis of the research. In D.A.Grouws (Ed). Handbook of research on mathematics teaching and learning (pp. 127-146). Simon & Schuster and Prentice Hall International.

Törner, G. (2002). Mathematical beliefs. In G..C. Leder., E. Pehkonen, E. & G. Törner (Eds). Beliefs: A Hidden Variable in Mathematics Education? (pp. 73--94). Dordrecht: Kluwer Academic Publishers.

Romme, M. A. J., & Escher, A. D. M. A. C. (1993). The new approach: A Dutch experiment. In M. A. J. Romme & A. D. M. A. C. Escher (Eds.), Accepting voices (pp. 11-27). London: MIND publications.

Sternberg, R. J., Grigorenko, Elena L., Singer J.L (Eds) (2004). Creativity: From potential to realization. Washington, DC: American Psychological Association.

Wedege, T., and Skott, J. (2006). Changing Views and Practices: A study of the KappAbel mathematics competition. Research Report: Norwegian Center for Mathematics Education & Norwegian University of Science and Technology, 274 pp. Trondheim

MASS KILLINGS: THE COMMUNICATION OF OUTRAGE

Brian Greer
Portland State University

Preamble

Among poetry about the First World War ("the war to end all wars") one of the most powerful statements of outrage is found in the following lines of a poem by Wilfred Owen about a soldier dying from mustard gas:

> If in some smothering dreams you too could pace
> Behind the wagon that we flung him in,
> And watch the white eyes writhing in his face,
> His hanging face, like a devil's sick of sin;
> If you could hear, at every jolt, the blood
> Come gargling from the froth-corrupted lungs,
> Obscene as cancer, bitter as the cud
> Of vile, incurable sores on innocent tongues
> My friend, you would not tell with such high zest
> To children ardent for some desperate glory,
> The old Lie: *Dulce et decorum est*
> *Pro patria mori.*

(The Latin translates as "It is sweet and proper to die for the fatherland".)

Owen's lines exemplify how it may be possible, through artistic means, to express the horror of violent death. Tatum (2003), in a chapter entitled "The poetry is in the killing" (www.press.uchicago.edu/Misc/Chicago/789934chap6.html) reminds us that some of the most graphic descriptions of killing are to be found in Homer's epics. In this paper, we consider the possibilities of using mathematics, alongside literary and artistic forms, to portray not just individual deaths but the grotesqueness of mass deaths through violence. Though of dubious authenticity, the saying attributed to Stalin that "one death is a tragedy, a million deaths is a statistic" conveys the sense that the scale of the horror of large numbers of deaths is difficult to communicate. Echoing a quotation attributed to George Bernard Shaw that "the mark of a truly educated person is to be deeply moved by statistics", we have proposed (Mukhopadhyay & Greer, 2007) the term "statistical empathy" for the human quality to be so moved.

There follow a number of examples from history and recent times for which attempts have been made, through artistic and mathematical means, to communicate outrageousness. We then consider implications for mathematics education.

B.Sriraman, C.Michelsen, A. Beckmann & V. Freiman (Eds). (2008). *Proceedings of the Second International Symposium on Mathematics and its Connections to the Arts and Sciences* (MACAS2). Information Age Publishing, Charlotte: NC , pp.143-154

Examples of mass death through violence

Historical examples: Hannibal, Napoleon, and Florence Nightingale

> The best graphics are about the useful and important, about life and death, about the universe (Tufte, 1983, p. 177)

Charles Joseph Minard (1781-1870) was a remarkable engineer, educator, and polymath who greatly developed the art/science of visually displaying statistical information. Extensive information on Minard can be found on Edward Tufte's website, www.edwardtufte.com (and see Wainer, 1997). In 1862, Minard wrote a paper on his work from which the following quotations are taken (using the translation by Dawn Finley on the website just referenced):

> The great growth of statistical research in our times has made felt the need to record the results in forms less dry, more useful, and able to be explored more rapidly than numbers alone; thus, diverse representations have been imagined, among others my graphic tables and my figurative maps.
>
> The dominant principle which characterizes my graphic tables and my figurative maps is to make immediately appreciable to the eye, as much as possible, the proportions of numeric results.

In his obituary (Chevallier, 1871; translation by Dawn Finley on Tufte's website), we read that:

> ... in one of his last maps, at the end of 1869, as by a premonition of the appalling catastrophes which were going to shatter France, he emphasized the losses of men which had been caused by two great captains, Hannibal and Napoleon 1st, the one in his expedition across Spain, Gaul and Italy, the other in the fatal Russian campaign. The armies in their march are represented as flows which, broad initially, become successively thinner. The army of Hannibal was reduced in this way from 96,000 men to 26,000, and our great army from 422,000 combatants to only 10,000. The image is gripping; and, especially today, it inspires bitter reflections on the cost to humanity of the madnesses of conquerors and the merciless thirst of military glory.

Both graphs can be seen on Tufte's website and are reproduced in Tufte (1983, p. 176), with the comment (p. 40) on the latter that "it may well be the best statistical graphic ever drawn" and quoting a description of its "seeming to defy the pen of the historian by its brutal eloquence".

Ten years ahead of Minard's depictions of the loss of two armies, Florence Nightingale (1820-1910) published graphics depicting a more topical case, combined with a call to action, namely deaths in the British army in the Crimean War. According to Small (1998), she may have been the first person to use statistical diagrams as a means for persuading people of the need for change. She devised a statistical diagram, that she called the "coxcomb", to depict changes over time in deaths, on the battlefield and in the hospitals, of British soldiers in the Crimean War (1854-56). (The diagram may be viewed at www.florence-nightingale-avenging-angel.co.uk/Coxcomb.htm) With this and other statistical data, she managed to convince the army authorities to make major changes in policy that resulted in substantial decreases in hospital deaths.

Most people know of Florence Nightingale only as someone who made fundamental contributions to establishing the nursing profession. Thanks to a famous depiction in the London Illustrated News in 1855, and a poem by Longfellow in 1857, she is remembered, in one of the most powerful iconic images of popular history, as "The lady with the lamp". It is much less generally known that, as well as her campaigns mentioned above for which she collected masses of data and deployed formidable powers of statistical argumentation, she was a major figure in the application of statistical ideas to social issues, collaborating with such important figures as Quetelet, and also of importance in the development of feminism. The biased nature of her place in the public consciousness is itself an interesting topic. According to Bostridge (2004), she did little actual nursing, being predominantly involved in administration and reform, and:

> ... the major significance of [the work in the Crimea] lies in the way it launched her post-Crimean career of social reform, enabling her to expose "the horrible facts of hospital life, workhouse infirmaries, deaths from famine in India, *all the needless deaths she could see from the evidence of statistical reports*. But she would never have got a hearing on those issues if she had not been the Lady with the Lamp."
> (Emphasis added).

For several decades, hampered by illness, she worked at home, digesting masses of statistics and advising the British government on all kinds of health issues. As reflected in the phrase italicized in the above quotation, this work exemplifies our notion of statistical empathy.

More recent history: Vietnam and Rwanda
These examples are discussed in a powerful chapter by Frankenstein (2007).

The Vietnam Veterans Memorial in Washington, D. C., bearing the names of 57,939 Americans who died in that war, is well known (see, e.g. a description by Tufte (1990, p. 43) of its design effectiveness so that "we focus on the tragic information"). Less well known is *The other Vietnam Memorial* by Chris Burden (www.press.uchicago.edu/Misc/Chicago/789934chap1.html), discussed by Tatum (2003) as follows:

> In the catalogue prepared for a 1992 exhibition of *The Other Vietnam Memorial* at the Museum of Modern Art, Burden is quoted as saying, "I just thought somewhere there should be a memorial to the Vietnamese that were killed in the war. So I wanted to make this book, sort of like Moses' tablet, that would be an official record of all these three million names. I would suspect that we will be lucky if we get twenty-five percent of the names; other ones would be nameless, basically faceless, bodies. . . . I want the size of the sculpture . . . to reflect the enormity of the horror."
>
> ... In order to register three million casualties, Burden took a catalogue that contained four thousand Vietnamese names, transformed them into verbal

> integers, and designed a computer-generated permutation of them. As the exhibition's curator, Robert Storr, observed, "A degree of abstraction necessarily persists. Even so, the war that so many want to consign to the past has never been more actual, with the enormity of the bloodletting at last represented *in toto*. Reckoning the gross facts of history in terms of the fate of individuals, Burden's "Other Vietnam Memorial" thus partially retrieves the Vietnamese dead from statistical purgatory and so from a double disappearance: the 3,000,000 it symbolically lists are the displaced persons of the American conscience.

Artist Alfredo Jaar went to Rwanda in 1994 to try to understand and represent the slaughter of up to a million Tutsis and moderate Hutus. Rockwell (1998) describes Jaar's work:

> Even after 3000 [photographic] images, Jaar considered the tragedy to be unrepresentable. He found it necessary to speak with the people, recording their feelings, words and ideas....In Jaar's Galerie Lelong installation, a table containing a million slides is the repetition of a single image, The Eyes of Gutete Emerita. The text about her reads: "Gutete Emerita, 30 years old, is standing in front of the church. Dressed in modest, worn clothing, her hair is hidden in a faded pink cotton kerchief. She was attending mass in the church when the massacre began. Killed with machetes in front of her eyes were her husband Tito Kahinamura (40), and her two sons Muhoza (10) and Matriigari (7). Somehow, she managed to escape with her daughter Marie-Louise Unumararunga (12), and hid in the swamp for 3 weeks, only coming out at night for food. When she speaks about her lost family, she gestures to corpses on the ground, rotting in the African sun."

Rockwell's review ends with a comment about the numbers: " The statistical remoteness of the number 1,000,000 acquires an objective presence, and through the eyes of Gutete Emerita, we witness the deaths, one by one, as single personal occurrences" (see Rwanda Project 1994-2000 on www.alfredojaar.net)

Gun violence in the US

On the twentieth anniversary of the death of John Lennon, his widow, the artist Yoko Ono, paid for the display of large posters in New York, Los Angeles and Cleveland showing a photograph of blood-stained glasses and the New York skyline with herself in front, bearing the text "Over 676,000 people have been killed by guns in the U.S.A. since John Lennon was shot and killed on December 8, 1980." (See Mukhopadhyay & Greer, 2007, for more details).

How is it possible to make sense of such a large number? One way is to recast it in more graspable form. It corresponds to about 30,000 gun deaths every year. Further computation converts this to about 93 gun deaths per day, roughly equivalent to 4 deaths per hour. In the most concrete terms, this equates to about one death every fifteen minutes.

Of most obvious relevance to students is the fact that many of those killed by guns are children. The most recent published data from the US Centers for Disease Control and Prevention show that 2,827 children and teens died from gunfire in 2003. This figure amounts to "... one child or teen every three hours, nearly eight every day, 54 children and teens every week" (Children's Defence Fund, 2006, 1). By contrast: "The number of children and teens killed by gun violence in 2003 alone exceeds the number of American fighting men and women killed in hostile action in Iraq from 2003 to April 2006" (Children's Defence Fund,

2006, p. 2). Again, a possible communicative resource is restating the figures in a way that connects to students' consciousness:

> • The number of children and teens in America killed by guns in 2003 would fill 113 public school classrooms of 25 students each.
> • The number of children and teens in America killed by guns since 1979 would fill 3,943 public school classrooms of 25 students each.
> • A Black male has a 1 in 72 chance of being killed by a firearm before his 30th birthday.
> • A White male has a 1 in 344 chance of being killed by a firearm before his 30th birthday.
> (Children's Defence Fund, 2006. p. 2).

Kids in the Line of Fire (Violence Policy Center, 2001) provides an in-depth analysis of the link between children, handguns, and homicide. It is based on analysis of homicide data for the five-year period 1995 through 1999. (For a comprehensive timeline of worldwide gun violence in schools, see http://www.infoplease.com/ipa/A0777958.html)

Debate on gun control is polarized and politicized, leading to "furious politics, marginal policy" (Spitzer, 1998, p. 133). A fundamental question is: To what extent can such questions be decided by research? A review sponsored by the National Research Council concluded that: "While there is a large body of empirical research on firearms and violence, there is little consensus on even the basic facts about these important policy issues" (Wellford, Pepper, & Petrie, 2005, p. 1). Nevertheless, the report, in making a number of recommendations for future research, implies that properly done research could settle questions. Thus, at one point it is stated that: "Ultimately, it is an empirical question whether defensive gun use and concealed weapons laws generate net social benefits or net social costs" (p. 6). That is missing a crucial point. No matter how well done the research, it will not settle the question since opinions will differ on how social benefits and costs are to be measured, which is a question of values.

Counting civilian deaths in Iraq
(For a more detailed discussion, see Mukhopadhyay and Greer, 2007).

In October, 2006, the British medical journal, The Lancet, published a paper (Burnham, Lafta, Doocy, & Roberts, 2006) in which the authors reported that:

> We estimate that as of July, 2006, there have been 654,965 (392,979–942,636) excess Iraqi deaths as a consequence of the war, which corresponds to 2.5% of the population in the study area.
> (Retrieved 8/25/07 from http://www.thelancet.com/webfiles/images/journals/lancet/s0140673606694919.pdf)

Note that 654,965 is a point estimate, the number that the analysis identifies as the single most probable; the numbers in brackets are those for a 95% confidence interval. A longer report (Burnham, Doocy, Dzeng, Lafta, & Roberts, 2006) provides more detail and context. In a similar study carried out in 2004 (Roberts, Lafta, Garfield, Khudhairi, & Burnham, 2004) the estimate of excess mortality during the 17.8 months after the 2003 invasion was 98,000, with a 95% confidence interval of 8000–194000 (excluding the data from Fallujah).

In both cases, the estimates were very much higher than others obtained using different methodologies, have been widely contested in the media, and dismissed as not credible by government leaders in the US, UK, Iraq, and Australia. For example, President Bush, questioned by Suzanne Malveaux of CNN at a White House Press Conference said that he did not consider the report credible, that the methodology had been "pretty well discredited" and that he stood by the number 30,000 that he had cited previously. Bush referred to the estimate in the Lancet report as "600,000, or whatever they guessed at". (Retrieved 8/25/07 from www.whitehouse.gov/news/releases/2006/10/20061011-5.html). No further questions were asked on this topic during the press conference. Nevertheless, the President's statement was very widely quoted in the media, often in headlines. Richard Garfield, a public health professor at Columbia University who works closely with a number of the authors of the report commented, as reported in the Christian Science Monitor:

> I loved when President Bush said "their methodology has been pretty well discredited". That's exactly wrong. There is no discrediting of this methodology. I don't think there's anyone who's been involved in mortality research who thinks there's a better way to do it in unsecured areas. (Murphy, 2006)

The appeal to experts by journalists is deserving of analysis. Typically, articles following the publication of the report cite short comments by a number of such experts. The divergence in the opinions cited is typical of what happens when statistical experts give opinions on a complex study. It is the nature of statistical applications of this level of complexity that agreement is not to be expected. There are aspects of the methodology that represent potential weaknesses in the design – indeed, the authors themselves clearly identify and discuss several. To elevate such disagreement to a claim that the methodology has been discredited shows ignorance of the nature of statistical research.

It is no surprise, though lamentable, that the media coverage of the Lancet report has been minimal, underinformed, and undernuanced. Unfortunately, the reaction was not much better from some people who should know better. For example, the group behind the website Media Lens (www.medialens.org) pursued a well-known mathematician, well known for his books, including (ironically) *A Mathematician Reads the Newspaper* (Paulos, 1996). Commenting on the earlier Lancet report, Paulos was quoted in the British newspaper, *The Guardian*, as follows:

> Given the conditions in Iraq, the sample clusters were not only small, but sometimes not random either... So what's the real number? My personal assessment, and it's only that, is that the number is … considerably less than the Lancet figure of 100,000.

After Media Lens commented that they "had not found a single example anywhere in the British or US press of a commentator rejecting estimates of 1.7 million deaths in Congo produced by the same lead researcher (Les Roberts) and offering their own "personal assessment" in this way", Paulos responded:

> I regret making the comment in my Guardian piece that you cite... I still have a few questions about the study (moot now), but mentioning a largely baseless "personal assessment" was cavalier. I should simply have stated my doubts about the study's scientific neutrality ...
> John Allen Paulos Math Dept, Temple Univ" (Email to Media Lens, September 7, 2005, retrieved 8/25/07 from http://www.medialens.org/alerts/05/050906_burying_the_lancet_update.php

It is to be expected that most people will react to the Lancet report in alignment with their political views. This reality seriously questions the naïve notion of sound conclusions being reachable on the basis of objective scientific evidence. Many criticisms of the reports claim that the political views of the authors and of the editor of the Lancet discredit the data. At least in the case of Les Roberts, the authors are anti-war, more specifically opposed to the invasion and war in Iraq, as is the editor of the Lancet, Richard Horton (and as, indeed, is the author of this paper). What are the implications? Are people with such views considered incapable of carrying out studies of this sort and having the findings taken seriously? Such a position rests on the myth of science and mathematics being value-free, ethically neutral, and apolitical. It is worth remembering that *The Lancet* is one of the most highly respected scientific journals. Apparently, however, it should not deal with deaths in war when those deaths are caused by "us". A June 23, 2005 editorial in the Washington Times lamented what it saw as an instance of "egregious politicization of what is supposed to be an objective and scientific journal". Why is it unreasonable that a journal serving a profession whose members take an oath to protect human life should raise issues about the avoidable killing of human beings?

What does this mean for mathematics education?

The tragedies discussed here are clearly illustrative and, sadly, by no means exhaustive. Including what is probably the best-known and documented example, the killing of millions of Jews, homosexuals, gypsies, and others in Nazi Germany, there are too many cases of mass slaughter to mention.

Relevance

I start from the assumption that mathematics can be, and the assertion that mathematics should be, taught as a tool for critically analyzing issues of importance in the lives of students, their communities and society at large, in contrast to the goal of schools being "to keep people from asking questions that matter about important issues that directly affect them and others" (Chomsky in Macedo, 2000, p. 24). In particular, this paper explores the possibility of using mathematics, along with other human communicative tools, to express the outrageousness of mass killings, in particular by promoting statististical empathy. While the teaching of statistics is gaining ground in school curricula in many countries, it is rarely concerned with the analysis of real data about important issues, combined with development of a critical sense of how data and statistical analysis are used and misused in society.

As yet, these ideas have minimally been turned into material for teaching. Some who hear them raise the question whether it would be appropriate. A reviewer of our earlier paper (Mukhopadhyay & Greer, 2007) called the topic "morbid and depressing". The first, obvious, answer is that it is important for students to know. Few, we suspect, would quarrel with the education of young Germans about the Holocaust. By contrast, Americans generally know little about the genocidal acts of Christopher Columbus and the related history of what happened to indigenous peoples of the Americas (Bigelow & Peterson, 1998; Loewen, 1995). A second answer, particularly but by no means uniquely within the United States, is that for many students, especially the older ones, gun violence and military service touch their lives, often tragically.

How do we know what's true?

The paradox of the information age is that we have more information but do not know which of it is, to what extent, true. In researching for this paper using the Internet, I was struck again and again by the fragile authenticity of the material unearthed. For example, while many Western reference books cite Stalin as the source of the statement that "the death of one man is a tragedy, the death of a million is a statistic", the website http://en.wikiquote.org/wiki/Stalin#_note-statistics refers to a statement in the Moscow Times that Russian historians have no record of Stalin saying that. Likewise, I have not been able to find a source for the statement attributed to Shaw that "it is the mark of an educated mind to be deeply moved by statistics".

More importantly, and perhaps increasingly, knowledge and truth are political commodities, making it difficult to evaluate reliability of information and the sources from which it comes. For example, currently a lot of attention is focused on killings in the Darfur region of the Sudan, yet it seems that those in the Congo are much more numerous, yet scarcely reported. If the reader tries, via the Internet, to investigate events in these two regions of Africa, she or he will experience how difficult it is to achieve a balanced view with any confidence.

The example discussed above of the reception accorded the Lancet report puts the problem into sharp focus. If a paper written by highly-qualified scholars of a major university, published in a top scientific journal, is so widely and emphatically discounted, where can we find authoritative information? Accordingly, an increasingly important function of education, including mathematics education, is to provide people with more powerful tools for critically evaluating the information that is available. Or should we just accept the recent suggestion of the social commentator Stephen Colbert that those who pay the most should get to decide what the truth is?

Political and ethical responsibilities of mathematicians and mathematics educators

> This is a great discovery, education is politics! After that, when a teacher discovers that he or she is a politician, too, the teacher has to ask, What kind of politics am I doing in the classroom? (Freire, 1987, p. 46)

Ubi D'Ambrosio (2003) has written passionately about the ethical responsibilities of mathematicians and mathematics educators:

> It is clear that Mathematics is well integrated into the technological, industrial, military, economic and political systems and that Mathematics has been relying on these systems for the material bases of its continuing progress. It is important to look into the role of mathematicians and mathematics educators in the evolution of mankind. … It is appropriate to ask what the *most universal mode of thought* – Mathematics – has to do with the *most universal problem* – survival with dignity.
> I believe that to find the relation between these two universals is an inescapable result of the claim of the universality of Mathematics. Consequently, as mathematicians and mathematics educators, we have to

> reflect upon our personal role in reversing the situation. (Emphasis in original).

D'Ambrosio challenges the stance of political/ethical neutrality adopted by many mathematicians, scientists, and others. As commented by Davis and Hersh (1981, p. 95) this stance became more difficult to justify following the Manhattan Project:

> Individual mathematicians asked themselves in what way they, personally, had unleashed monsters on the world, and if they had, how they could reconcile it with whatever philosophic views of life they held. Mathematics, which had previously been conceived as a remote and Olympian doctrine, emerged suddenly as something capable of doing physical, social, and psychological damage.

As D'Ambrosio asks, do we want to improve mathematics education to make people better able to design weapons of mass destruction? It appears that those working on a new generation of nuclear weapons are highly motivated by the technical challenges and competition (Vartabedian, 2006):

> "I have had people working nights and weekends," said Joseph Martz, head of the Los Alamos design team. "I have to tell them to go home. I can't keep them out of the office. This is *a chance to exercise skills* that we have not had a chance to use for 20 years."
>
> A thousand miles away at Livermore, Bruce Goodwin, associate director for nuclear weapons, described a similar picture: The lab is running supercomputer simulations around the clock, and teams of scientific experts working on all phases of the project "are extremely excited."
> (emphasis added)

Final comment

Each of us has/ all of us have dual existence as an individual and as members of society. As embodied in the other Vietnam memorial, dual to the denial of mass killings is the depersonalization that comes from purging people's names as symbols of their identity. In Alfredo Jaar's work, an image the eyes of one person are replicated a million times towards the same end. Abstraction gives mathematics much of its power, but can be accompanied by loss of humanity, in its most extreme form by reducing people to numbers.

> Counting is the most simple and primitive of narratives -- 1 2 3 4 5 6 7 8 9 10 -- a tale with a beginning, a middle and an end and a sense of progression -- arriving at a finish of two digits -- a goal attained, a dénouement reached. . . The pretence that numbers are not the humble creation of man, but are the exacting language of the Universe and therefore possess the secret of all things, is comforting, terrifying and mesmeric... Counting makes even hideous events bearable as simply more of the same – the counting of wedding-rings, spectacles, teeth and bodies disassociates them from their context – to make the ultimate obscene blasphemy of bureaucratic insensitivity. Engage the mind with numbing recitation to make it empty of reaction.
> Fear of Drowning By Numbers, Peter Greenaway

References

Bigelow, B. & Peterson, B. (Eds.) (1998). *Rethinking Columbus: The next 500 years.* Milwaukee, WI: Rethinking Schools.

Bostridge, M. (2004). *Florence Nightingale's secret war.* The Times, 30 October. Retrieved 8/25/07 from: hnn.us/roundup/entries/8372.html

Burnham, G., Doocy, S., Dzeng, E., Lafta, R., & Roberts, L. (2006). *The human cost of the war in Iraq.* Retrieved 10/11/2006 from http://www.thelancet.com

Burnham, G., Lafta, R., Doocy, S, & Roberts, L. (October 21, 2006). Mortality after the 2003 invasion of Iraq: A cross-sectional cluster sample survey. *The Lancet, 368*(9546), 1421-1428.

Chevallier, V. (1871). Notice nécrologique sur M. Minard, inspecteur général des ponts et chaussées, en retraite, *Annales des Ponts et Chaussées, 2*, 1-22.

Children's Defense Fund (2006). *Protect children not guns.* Retrieved 8/25/07 from http://www.childrensdefense.org/site/DocServer/gunrpt revised 06.pdf?docID=1761

D'Ambrosio, U. (2003). The role of mathematics in building up a democratic society. In Madison, B. L., & Steen, L. A. (Eds.), *Quantitative literacy: Why numeracy matters for schools and colleges. Proceedings of National Forum on Quantitative Literacy, National Academy of Sciences, Washington, DC, December, 2001.* Princeton, NJ: National Council on Education and the Disciplines. (Available at: http://www.maa.org/ql/qltoc.html)

Davis, P. J., & Hersh, R. (1981). *The mathematical experience.* Boston: Birkhäuser.

Frankenstein, M. (2007). Quantitative form in arguments. In D. Gabbard (Ed.), *Knowledge and power in the global economy.* London: Routledge. http://www.media.pdx.edu/Mukhopadhyay/Frankenstein_062206.asx

Freire, P. (with Shor, I.). (1987). *A pedagogy for liberation.* Westport, CT: Bergin & Garvey.

Greenaway, P. (1991). *Fear of drowning by numbers.* Paris: Dis Voir.

Loewen, J. W. (1995). *Lies my teacher told me: Everything your American history book got wrong.* New York: Simon & Schuster.

Macedo, D. (Ed.). (2000). *Chomsky on miseducation.* Lanham, MD: Rowman & Littlefield.

Mukhopadhyay, S. & Greer, B. (2007). How many deaths? Education for statistical empathy. In B. Sriraman (Ed.) *International perspectives on social justice in mathematics education* (pp. 119-136). University of Montana Press.

Murphy, D. (2006). Iraq casualty figures open up new battleground. *Christian Science Monitor*, October 13. Retrieved 8/25/07 from www.csmonitor.com/2006/1013/p01s04-woiq.htm

Paulos, J. A. (1996). *A mathematician reads the newspaper.* New York: Anchor Books.

Roberts, L., Lafta, R., Garfield, R., Khudhairi, J.,& Burnham, G. (2004). Mortality before and after the 2003 invasion of Iraq: cluster sample survey. *The Lancet*, October 20.

Rockwell, S. (1998). Alfredo Jaar at Galerie LeLong in New York. *dArt International, 1(*3), 3;12-13.

Small, H. (1998, March). Florence Nightingale's statistical diagrams. Paper from *Stats & Lamps Research Conference*, Florence Nightingale Museum, London. Retrieved 8/25/07 from: www.york.ac.uk/depts/maths/histstat/small.htm

Spitzer, R. J. (1998). *The politics of gun control* (2nd Ed.). New York: Chatham House Publishers.

Tatum, J. (2003). *The mourner's song.* Chicago: University of Chicago Press.

Tufte, E. R. (1983). *The visual display of quantitative information.* Cheshire, CT: Graphics Press.

Tufte, E. R. (1990). *Envisioning Information.* Cheshire, CT: Graphics Press.

Vartabedian, R. (2006). Rival U.S. labs in arms race to build safer nuclear bomb. *Los Angeles Times*, June 13. Retrieved 8/26/07 from: fairuse.100webcustomers.com/fuj/latimes52.htm

Violence Policy Center (April, 2001). Where'd they get their guns? Retrieved 8/25/07 from: http://www.vpc.org/studies/wguncont.htm

Wainer, H. (1997). *Visual revelations: Graphical tales of fate and deception from Napoleon Bonaparte to Ross Perot.* Mahwah, NJ: Lawrence Erlbaum Associates.

Wellford, Charles F., John V. Pepper, and Carol V. Petrie (Eds.). (2005). *Firearms and Violence: A Critical Review*. Washington, D.C.: The National Academies Press.

"GOOD THEORY" FOR "GOOD PRACTICE" AND VICE VERSA IN THE COMBINATION OF SCIENCE, ART AND TEACHING

Herbert Gerstberger
University of Education Weingarten Germany

Based on several experiences in teaching physics and mathematics in combination with artistic productions, conceptual tools and criteria are sketched which shall support the developement and the evaluation of this approach. After some remarks on metaphor and model, tools from Peircean semiotics are used to investigate more extensive and creative ways of representation. The idea of transversal operations between several aspects of rationality (Welsch) is applied in the raising of an educational hypothesis.

THE STARTING POINT: EXPERIENCES IN AESTHETIC PRODUCTION AND PRESENTATION IN MATHS AND SCIENCE EDUCATION

Until some years ago, two domains of interests and activities existed pretty much separated in my life. As a teacher and scientist I was working in the field of physics and mathematics education with special interests in linguistic and conceptual problems such as *magnitudes*, *terminology* and *metaphor*. In my leisure time on the other hand, I was active in theatre groups and also enjoyed reading and writing poetical texts. In cooperation with a director and teacher of drama, I felt more and more encouraged to try out ways to combine these domains in educational and artistic projects at my university and at schools some of which I shall sketch here shortly.

(1) We offered drama workshops and seminars about *time* with a broad variety of points of view – scientific, philosophical, and artistic. Among the topics we dealt with there were *cosmic and biological evolution, history and lifetime of individuals, moment and duration, velocity, rhythm, metaphors and images of time.* At the end we developed presentations on stage, mostly as a collage of text and elements of body theatre.

(2) In 2005, the Einstein year, we produced a theatre play which was based on the students' genuine questions about relativity and historical and biographical issues. In the questions of physics there, too, *time* was a central question of course, together with space and light, and the students found interesting and entertaining ways of representing history, ideas and mental experiments (*Gedankenexperimente)* using theatrical forms like narration, clowns, songs and sculpture. The development of the production followed a alternating scheme of teaching units and creative work.

(3) Together with a colleague from the arts department, we designed a seminar about *motion.* She presented several relevant approaches in contemporary art like the ones of Tinguely and Fischli-Weiss, but also cartoon, movie, sculpture (stabile - mobile) and action art. Like in the Einstein project mentioned before, I gathered students in mini-seminars in order to expose the basic scientific aspects of motion. The students were then asked to produce their own artistic

B.Sriraman, C.Michelsen, A. Beckmann & V. Freiman (Eds). (2008). *Proceedings of the Second International Symposium on Mathematics and its Connections to the Arts and Sciences* (MACAS2). Information Age Publishing, Charlotte: NC , pp.155-164

works through which they reflected on the central topic of motion, using one of the artistic techniques and styles which had been presented beforehand.

(4) On the occasion of Albrecht Beutelspacher's travelling exhibition "Maths to Touch", at our university a group of students installed a parallel exhibition called "Time to Touch" which essentially consisted of a couple of sculptures. The students gave a short introductory speech and then guided the spectators through the exposition with specific comments about the ideas behind the individual objects and installations.

(5) In an attempt to comment on the concept of *diagrammatic reasoning*, I wrote a piece of fiction accompanied by a didactical essay for math educators. The stories are inspired by Jonathan Swift's Gulliver's Travels, embedded in a fantastic context where ethnomathematical researchers meet several isolated tribes who have kept their specific ways of numerical and relational reasoning (Gerstberger 2006b).

In my present contribution I will not exhibit these activities but I rather take them as a starting point for the explication of some ideas which range from conceptual reflections to the design of a research project. As the central questions I consider those which address the problems of representation and presentation. Even if these words are close to one another, *representation* will lead into the more theoretical realms of semiotics whereas *presentation* will denote a specific step in an educational procedure.

FIRST APPROXIMATION: FROM METAPHOR TO MODEL

How can we legitimate activities as those described above? How can we assign structure and value to them? In order to establish a suitable conceptual frame of reference, as a first step I shall hint at some aspects of *metaphor*. This is because structural features of metaphor, as a rhetoric trope and as a cognitive means, too, might lead to some understanding of more complex semiotic situations like the students' artistic productions. From analogies and differences with linguistic phenomena we might find a first way of understanding more complex structures of topic-guided creativity and performance.

Let's imagine a teacher who, in a spontaneous attempt to give a somehow lively image of geometric features, explains: "The symmetry axis of a straight line *collects* the points which have equal distances to each end. [1]" The verb *to collect* stems from another semantic domain than the rest of the terms which are of a plainly technical register. Implicitly, the symmetry axis might be seen as some person who acts deliberately in a certain ambience which however is in no way specified. The very charm of the metaphorical use of the verb *to collect* lies in the combination of two properties: it has some associative openness to imagine concrete everyday situations, and on the other hand it is apt to transport the definite mathematical concept of *set* as the result of an activity like collecting. If we fix one specific situation as a possible source domain of this metaphor, e.g. a child who collects mushrooms, we can clearly distinguish advantages and disadvantages of this metaphor. The resulting bunch of mushrooms might appear as a picture of the concept of *set.*[2] But in this picture there is no represen-

[1] This metaphor is due to Lietzmann (L. Führer: private communication)

[2] By the transition from the action - and the verb *to collect* - to the result - and the noun *collection* -a *reification* or *hypostatic abstraction* (Peirce) is performed. Halliday (1999) introduced the concept of *grammatical metaphor* for phenomena like this. This concept is further developed in (Gerstberger 2006a).

tation of the geometric properties of the target domain. Even if the spatial position of the mushrooms should have had any significance, it would have been cancelled by the act of collecting them. Therefore, the flower scenario serves as a *model* for the set aspect but not for the geometry. This example hints at the relationship of *metaphor and model* which has been broadly investigated in epistemological studies. I shall use the concept of metaphor in a strictly linguistic sense where, together with metonymy and other forms of figurative speech, it belongs to the category of rhetorical tropes and is effected by a violation of semantic consistency called *catachresis* - which means false use or displacement of a word or a group of words from one context into another. This transfer is due to some similarity or analogy in some relational substructures of the two domains involved. There is a distinction between the traditional poetic metaphor and the more recent concept of cognitive metaphor. Max Black (1962) considers the role of the metaphor in poetical speech as analogous to the role of the model in science. Many instances, however, seem to suggest an interdependence of model and metaphor of the following kind: A model is a more complex and elaborated icon of the primary domain and has to fulfil certain purposes which restrict the freedom of model building. Metaphors can occur in accordance with an already existent model and refer to it in several ways. A newly created metaphor can show up in an occasional speech act; others have become habitual in the scientific community and are used as technical terms (also: "dead" metaphor, lexicalized metaphor). In this respect, a metaphor is *illustrative* rather than *constitutive*. The other way round, in many cases a model is built in a secondary step through and after the systematic elaboration of metaphors. In this case, a metaphor is *constitutive* rather than *illustrative*. In an attempt to distinguish metaphor from other concepts, a definition given by Mary Hesse balances very well sharpness and generality: "We start with two systems, situations, or referents, which will be called respectively the primary and secondary systems. Each is described in literal language. A metaphoric use of language in describing the primary system consists of transferring a word or words usually used in connection with the secondary system." (Hesse 1980: 111f)

In this definition, a broad generality is indicated by the term *system* whereas the confinement to *words* excludes inflationary extensions of the concept of metaphor to sign systems beyond language as e.g. geometry or to physical objects. A metaphor is not understood as belonging to any arbitrary semiotic system and not even as any linguistic entity whatsoever but strictly as a word or a group of words which occurs in a spoken or written context. Usually it is addressed at a language community which is supposed to have the prerequisites to recognize the catachresis as meaningful. A model, in contrast, has to fulfill demands of explicitness and adequacy. The variety of models transcends the domain of language which is evident from Black's list ranging from physical scale models to theoretical models (Black 1969: 219 ff). Therefore a model does not reside in language but in situations and processes – real or imagined.
I hesitate to use Hesse's term *system* because of its several special meanings and I rather go on speaking of *domains* in a more general sense.

SECOND APPROXIMATION: FROM INTERPRETANT TO AESTHETIC PRODUCTION

My thesis now is that artistic productions and presentations which are rooted in scientific topics, have something in common both with metaphor and model, but can be adequately described and understood only within a broader conceptual framework. In the case of the above symmetry axis, the scenario of collecting mushrooms did not work as a model for the geometric situation, but we can certainly find suitable examples: it is easy to produce word problems with hidden treasures or similar narratives which imply a geographic context. As another possibility the domain of the graphical arts is evident. In the examples of this kind, geometry is in a certain way related to another domain which is alien to it. However there are also fields of free and interpretative working with the initial topic of the symmetry axis in planar geometry itself, e.g. if we add a third point and investigate the intersection of the three symmetry axis. The operation which leads from the initial topic to the second domain can eventually be described as *projection*, as *contextualization* or just as *application*. Before this operation can be performed, however, it is necessary to chose or to invent a secondary domain which promises to please our demands.

Summing up all the steps of the procedure described so far, we can identify them as a sequence of sign processes each of which is a *semiosis* in the sense of Morris and Peirce, i.e. the *generation* and *interpretation* of a sign. Let's for the sake of precision look at one of the most frequently quoted definitions of the triadic sign concept by Charles Sanders Peirce:

"A sign, or representamen, is something which stands for somebody for something in some respect or capacity. It addresses somebody, that is, creates in the mind of that person an equivalent sign, or perhaps a more developed sign. That sign which it creates I call the *interpretant* of the first sign. The sign stands for something, its object. It stands for this object, not in all respects, but in reference to a sort of idea. (CP § 2.228)"

In the metaphor example above, the object is the mathematical concept of the symmetry axis and the representamen is the very sentence – written or spoken - which expresses the definition given above. The interpretant is a mental entity which is, according to Peirce, "created" by the representamen. This act of creation seems to be enigmatic here and it could take a psychologist and a neurologist to analyze it thoroughly. In the more philosophical crafts, we could think of consulting authors who dwell in the theories of mental models (Johnson-Laird 1995) and mental spaces (Fauconnier 1994; Brandt 2004). In the above definition this act is not explained at all. We certainly should not understand it naively as a mechanical causation: a mere phrase, be it uttered or written, does not create any determined effect in somebody's mind if it is not embedded in a context of discourse and also in the history of the person whose mind is affected. Moreover, it seems to be more appropriate to concede the act of creation to the addressee of the sign instead of the sign itself. The way Peirce put it would then be understood as something like a metaphorical shift.

The object itself is questionable, too. Is it a Platonic entity which would also exist without being represented by any representamen at all? Note that there are many equivalent definitions of the symmetry axis! Or is it an interpretant in somebody else's mind - the teacher's? This last idea is in better accordance with Peirce' thinking of semiosis as an ensemble of infinitely linked processes where one single triadic unit can be isolated only artificially and which "results in a series of successive interpretants" (CP § 2.303). Thus the above definition must not be isolated either, and its obvious circularity seems to be made on purpose.

So the continuation of semiosis does not only appear at the pole of the object but also at the pole of the interpretant. The mental entity created "by the sign", being a sign, too, "creates" another interpretant. If we strictly follow the definition, this interpretant of second order is again a mental entity. But it can also be taken as an object in its own right and therefore can be expressed by something "external" which can be perceived publicly, and functions as a second representamen. It may be a sigh, a gesture, an utterance, but also a more elaborate product: a painting, the telling of a story, a dance. In metaphor, the interpretant accomplishes a *blend* (Fauconnier 1994) through synopsis of two domains of discourse. In this blend a mutual transfer is expressed by the catachrestic lexeme.

In our concern of aesthetic reflections on scientific topics, I shall treat the interpretant and its representation as a unit without overemphasizing the difference between idea and representation. This is in accordance with Hoffmann (2005:45) who points out that "signs are no transcendental, subjective or mental conditions of possible experience for Peirce, but conditions for which the difference between internal or external plays no role" and especially in mathematics "for any 'concept' or 'mental state' 'external signs answer every question, and there is no need at all of considering what passes in one's mind' (Peirce, NEM I 122)". Specific systems of external signs in mathematics and science are called *diagrams*, and the operation on them *diagrammatic reasoning*. (CP 4.418, 1903) This way of doing has "the *normative* and *compelling* character of the logic of representational systems" but "it seems obvious that we need a certain *freedom* for the genuine *creative* aspects of learning and discovering. Peirce called this 'the highest kind of synthesis', when the mind – 'in the interest of intelligibility' – introduces 'an idea not contained in the data, which gives connections which they would not otherwise have had.' He compares such acts of synthesizing with the creativity of an artist: 'The work of the poet or novelist is not so utterly different from that of the scientific man. The artist introduces a fiction; but it is not an arbitrary one. ... Intuition is the regarding of the abstract in a concrete form, by the realistic hypostatization of relations; that is the sole method of valuable thought. (CP 1.383, 1888)" So Peirce himself seems to envision a continuum of ways of reasoning, i.e. of intentional producing of and operating on signs, which ranges from strict application of conventional rules to artistic creativity, where utmost "more developed signs" arise. Oscillations between thoughts and things, between interpretants and representaments, evolve to a process where the interpretant may have an "emotional" character first, then becomes "dynamical" and finally "logical" This fascinating insight (or claim), however, raises a special tension in epistemic contexts: the tension between essential and accidental, or "truth" and "beauty". But there is no escape: this problem arises already with the very first representation of any idea, because every representamen is a perceivable object or process. As such it cannot be stripped off its purely aesthetic aspects: the colour of the ink, the melody of the utterance, the circumstances where a representamen is produced and presented etc. In the case of linguistic signs, this problem has partly been addressed by speech act theory; in the case of conventional signs (symbols) it is linked to the difference between type and token. On the general semiotic level, however, it seems to become hopelessly complex.

On the practical level and especially in teaching, one special aspect of this problem is the relationship of "correctness" and aesthetic features which range from "clarity" to "beauty". The tension between these aspects might cause conflict as well as synergy. Synergy which exposes the mathematical "core" might please the teacher, whereas it can be the artist's intention to hide a tricky mathematical idea which helped him to produce impressive effects. The other way round, a scientist might tacitly use the tricks he learned from a showman in order to make his talk seductive. Considering these poles, ourselves and our students can consciously make the choice of either exposing a scientific concept explicitly or else using it in a concealed way. We can spread out a variety of artistic means and styles or we can focus on one special kind. Of course, we can learn from the masters of painting or writing[3], from circus and comedy: In one of our seminars, surprisingly a student decided to present the concept of floating in a deliberately mistaken way in order to parody nonsensical science. If one looks

---

[3] Not only from those who use relate to scientific subjects, even though there are many.

around, a surprising richness in research and documentation of "good practice" can be found[4].

Let's look back at our conceptual consideration so far and compare the movements of *transfer* in the three cases: metaphor, modelling and aesthetic production. Using a metaphor, the speaker tacitly and often unawarely exploits a specific similarity of relations in two domains - source and target - which are both represented by language registers. The catachrestical word is an index of this implicit relationship which in itself can be understood as iconicity. In contrast, to build or to use a model is the more effective the more the two domains and the transfer are made explicit. This is sometimes expressed by the use of the very same words or symbols both in source and target domain, e.g. when we use the term "orthogonal" in function spaces. The better understanding of the target domain is the unconcealed aim of building or using models. In our approach to aesthetic production, the scientific topic takes the role of the source which feeds the aesthetic target domain with structural features - and often with genuine beauty, too. The relationship of source and target is not necessarily explicit and the scientific source itself may even remain a secret of the production. However, the scientific topic has to be sufficiently well understood beforehand and one will certainly appreciate a still better understanding by the aesthetic production. We hope that with the emphasis on the aesthetic outcome instead of the scientific roots, the psychological effect of collateral learning and understanding can be stimulated. This advantage should be still more powerful by social effects when the production is followed by a presentation in front of an audience.

TRANSVERSAL OPERATIONS BETWEEN COGNITIVE AND AESTHETIC ASPECTS

The Janus-faced representamen can be seen as a starting point for dwelling deeper into the relationship between epistemic or cognitive rationality on the one hand and aesthetic rationality on the other. In philosophical tradition until now we can observe a long and multifaceted debate about aspects of *rationality*, where epistemic and aesthetic rationality are completed by ethics as a third aspect. The distinction between these three aspects is linguistically represented in all categories: syntax, semantics, and pragmatics. Accordingly in logics, we have three fundamental categories of judgments: epistemic, ethical and aesthetic. (1) The epistemic judgment, of course, is essential in science; it is the propositional way in which facts are expressed. (2) The category of ethical judgment is about allowing and forbidding, commitment and entitlement (Brandom). (3) The category of aesthetic judgments traditionally comprises the realms of beauty and therefore seems to apply to music and the fine arts mostly. However, etymology tells us that "aesthetic" derives from the Greek "aisthesis" which approximately means sensual awareness and cognition in general. In this broad view Gernot Böhme has proposed a theory of "aisthesis" which re-examines former philosophies of aesthetics. In everyday life we are always using these three categories more or less consciously. But what are the specific characteristics of aesthetic rationality? Let's in a first approach understand aesthetic production of whatever kind as the creation and/or presentation of something (an object or a process) with awareness of and emphasis on its perceivability. In science, these products – texts, diagrams, apparatus, and experimental ensembles – are semiotic means which mediate something else, usually an epistemic or didactical question within a well defined context. Features like shape, colour, style are unavoidable but neglected in the scientific focus. Being

---

[4] Especially Kubli (2005); Negrete (2005), Schlichting (2006); conferences like MACAS1 and mathematica e cultura (Emmer et al. 1996-2006), and initiatives like "Science on Stage".

present all the same, however, these features and their possible collateral effects can be envisaged in their own right. This is what Böhme calls "Inszenierungswert", i.e. the surplus value of putting on stage, which contributes to the quality of any product in addition to the classical economical categories of utility value and exchange value. Epistemic and aesthetic aspects of rationality correspond to two kinds of *experience*, as Andrea Kern points out in her interpretation of Kant (Kern 2000): On the one side, aesthetic experience is autonomous and independent of "ordinary" experience because of its lacking an outer referent. Perceived aesthetically, the relevant aspects of an object are its immanent relationships and the way it affects our senses and taste. Moreover, aesthetic experience is even able to bring the act of perception itself to the fore. This self-referential structure might yield a special kind of *pleasure* – as the counterpart of the pleasure which accompanies intellectual understanding. On the other side, experience cannot be its only object and therefore aesthetic experience is always linked to a basic ordinary one. According to a still more complex consideration, the "ability to present something and at the same time to thematize the presentation itself as a presentation seems to us to be an indication of the greatness of the works of art. Such a presentation is always also the destruction of any unmediated perception of presentations – the destruction of presentation. Such presentations do not represent reality, the world, but, rather, reflect our activity in the world. Representation does not consist of the objects it designates (Fichtner 2005:187)". Well, in the case of non-representing art this thesis should be modified, but this case is not relevant to our project. So we might hope to gain a little bit of this greatness of art, the more so as we employ art not primarily in a contemplative way but rather in productive and even theatrical activities.

The ability to distinguish between several aspects of rationality, to recognize each of them in its special quality, to use and also to connect them in an even creative way reminds of what the philosopher Wolfgang Welsch calls *transversal reason*. It appears as a competence of second order because it is based on sensibility in each single aspect. Considering this, in our educational approach we do not only appreciate and foster aesthetics in its own right, but we emphasize and use it also in contexts where cognitive and epistemic tasks have to be accomplished. In the last paragraph I will shortly sketch this approach.

TEACHING, TEACHER TRAINING AND RESEARCH

It is in good agreement with educational philosophy that all the three aspects of rationality are to be considered essential and to be combined in educational practice. It is valuable to see ethical aspects in matters of plain knowledge and to be scientifically honest e.g. in creative writing as it is good in general to be flexible enough to look at one particular focus from different points of view. Especially the combination of intellectual work with didactic presentation belongs to a teacher's everyday job and it combines cognitive and epistemic subjects with means of teaching which have indispensable aesthetic aspects. But students, too, can and should be inspired in this way which will foster not only intellectual growth but also personal development. Each utterance, every corporal action, of course every experimental demonstration which is made in class happens "on stage". So, even if it primarily communicates some scientific content, we must not neglect the Böhmean "Inszenierungswert". In this respect, for both teachers and students we can modify one of Paul Watzlawick's axioms of communication in stating: "You cannot *not* present" and we can identify one aspect of the "transversal" competence mentioned above as *presentational competence*. It is certainly related to *modelling competence* and *representational competence* (Michelsen 2005), which are more fundamental and less dispensable for academic tasks. But when we go beyond the hard core of maths and science education, presentational competence will not range as luxury.

In an interdisciplinary research project called APPLE[5] , in cooperation with a colleague, Elisabeth Schlemmer from the department of educational studies we have designed a way of educational practice in four steps: (1) teaching and learning some subject matter, (2) aesthetic production as a creative reflection thereof, (3) presentation of the latter in front of an audience (the class) and (4) reflection. We aim at the definition of presentational competence and we also explore ways to develop this competence, and to verify it in well defined concretisations and facets. In our teacher training seminars, students of maths and science education are equipped with the background in educational studies as well as in aesthetic theory and drama. They are also inspired to dwell deeper into specific arts according to their individual preferences. At the same time they develop their own sub-projects which are realised in schools. At the moment we are working on filtering out precise questions and hypothesis' which can then be tackled in a proper empirical way.

REFERENCES

Böhme, G. (2001). *Aisthesis. Vorlesungen über Ästhetik als allgemeine Wahrnehmungslehre.* München: Wilhelm Fink

Black, M. (1962). *Models and Metaphors.* Ithaca and London: Cornell UP

Brandom, R. (1994). *Making it Explicit. Reasoning, Representing and Discoursive Commitment.* Cambridge, Ma. etc.: Harvard UP

Brandt, P.A. (2004). *Spaces, Domains and Meanings. Essays in Cognitive Semiotics.* Bern etc: Peter Lang

Emmer, M. et al.(ed.). (1996-2006). *matematica e cultura.* Milano: Springer

Fauconnier, G. (1994). *Mental Spaces: Aspects of Meaning Construction in Natural Language.* Cambridge etc.: Cambridge UP

Fichtner, B. (2005). *Reflective Learning – Problems and Questions Concerning a Current Contextualisation of the Vygotskijan Approach.* In Hoffmann et. al. (Eds.) *Activity and Sign. Grounding Mathematics Education* (pp. 179-190). New York: Springer

Gerstberger, H. (2006a). *Formen der Metaphorik in Physik und Mathematik.* In Girwidz, R. et al. (eds.) *Lernen im Physikunterricht. Festschrift für Prof. Dr. Christoph von Rhöneck* (pp. 137-146). Hamburg: Verlag Dr. Kovač

Gerstberger, H. (2006b). *Ein narrativer Zugang zum semiotischen Blick auf mathematische Themen.* Journal für Mathematik-Didaktik 27 (2006) H. 3/4 (285-299)

Halliday, M.A.K., Mackensen, Ch. M.I.M. (1999). *Construing Experience Through Meaning. A Language Based Approach to Cognition.* London & New York: Cassell

Hesse, M. (1980). *Revolutions and Reconstructions in the Philosophy of Science.* Brighton: The Harvester

Hoffmann, M.H.G. (2005). *Signs as a Means for Discoveries. Peirce and His Concepts of "Diagrammatic Reasoning", "Theorematic Deduction", "Hypostatic Abstraction", and "Theoric Transformtion".* In Hoffmann et. al. (eds.) *Activity and Sign. Grounding Mathematics Education* (pp. 45-56). New York: Springer

[5] Aesthetic Production and Presentation in Learning and Education

Johnson-Laird, P.N. (1995). *Mental Models. Towards a Cognitive Science of Language, Inference, and Consciousness*. Cambridge, Ma.: Harvard UP

Kern, A. (2000). *Schöne Lust. Eine Theorie der ästhetischen Erfahrung nach Kant.* Suhrkamp: Frankfurt/M.

Kubli, F. (2005). *Mit Geschichten und Erzählungen motivieren. Beispiele für den mathematisch-naturwissenschaftlichen Unterricht.* Köln: Aulis Verlag Deubner

Michelsen, C. (2005). *Expanding the domain: Variables and functions in an interdisciplinary context between mathematics and physics.* In Beckmann, Michelsen, Sriraman (eds.) *Proceedings of the First International Symposium of Mathematics and its Connections to the Arts and Sciences* (pp. 215-226). Hildesheim: Franzbecker

Negrete, A. (2005). *Fact via Fiction. Stories that Communicate Science.* In Sannit (ed.) Motivating Science (pp. 91 – 94). Luton: The Pantaneto Press.

Peirce, C.P. (CP). *Collected Papers of Charles Sanders Peirce.* 1931-1958. Cambridge, Ma.: Harvard UP

Peirce, C.P. (NEM). *The new elements of mathematics by Charles Sanders Peirce* (Vol. I-IV). The Hague-Paris/Atlantic Highlands, N.J., 1976: Mouton/Humanities Press

Schlichting, H.J. (2007). *Kann die Auseinandersetzung mit der Kunst beim Physiklernen helfen?.* In Höttecke (ed.) Naturwissenschaftlicher Unterricht im internationalen Vergleich Reihe: *Gesellschaft für Didaktik der Chemie und Physik* Bd. 27, 2007

Welsch, W. (1996). *Vernunft. Die zeitgenössische Vernunftkritik und das Konzept der transversalen Vernunft.* Frankfurt/M.: Suhrkamp

MODEL ELICITING ACTIVITIES REVISITED – A REPORT FROM A CASE STUDY OF THE PENALTY THROW PROBLEM

Steffen M. Iversen
Roskilde Gymnasium (Denmark)

In this paper, the results of a case study are presented. The study was the outcome of continued collaboration of researchers meeting at MACAS 1, and deals with the possible connections between students' performances when working with traditional math tests and open-ended real-world tasks respectively. About 200 students enrolled in a calculus course at the University of Southern Denmark worked in groups on a Model Eliciting Activity (Lesh & Doerr, 2003), and data from this along with data from both traditional standardised pre- and post-tests were collected and analysed. It was concluded that little correspondence can be detected between the students' performances in the two types of mathematical domains, and some possible explanations for this are offered in the paper.

INTRODUCTION

The case study reported in this paper found its inspiration at MACAS 1,[1] particularly in the plenary presentation of the topic *Model Eliciting Activities* by Richard Lesh (see Lesh & Sriraman, 2005). After the symposium a continued collaboration between the author, Lesh and his colleague Christy J. Larson was initiated, and this enabled the planning, carrying through and evaluation of the case study presented below.[2]

The purpose of the study was to explore the possible relations between students' performances when working on traditional standardised tests and "messy" real-world problems involving mathematics respectively. Taking Lesh & Sriraman (2005) as a starting point two linked research questions was formulated:

1. What are the relations between students' performances in standardised tests and their performances when working with open-ended problems involving mathematical modelling - especially Model Eliciting Activities?
2. If students who perform poorly in the standardised tests but well in the modelling problem (or vice versa) can be identified – how can we explain these shifts in performance?

THEORETICAL FRAMEWORK

The last century in the history of mathematics is characterized by the increasing influence of applied mathematics. In such different fields as engineering, economics, biology, and medicine applied mathematics has played, and still plays a more and more important role in

[1] The 1st International Symposium of Mathematics and its Connections to the Arts and Sciences. May 18-21, 2005, University of Schwaebisch Gmuend: Germany.
[2] A more detailed description of the study can be found in Iversen & Larson (2006).

B.Sriraman, C.Michelsen, A. Beckmann & V. Freiman (Eds). (2008). *Proceedings of the Second International Symposium on Mathematics and its Connections to the Arts and Sciences* (MACAS2). Information Age Publishing, Charlotte: NC pp.165-172

new development and breakthroughs (e.g. Burkhardt, 2006 and Steen, 2005). One of the great challenges in the contemporary work of mathematics education researchers and teachers of mathematics is how this should be reflected in the classrooms. A reasonable, yet not exhaustive, answer to this is an increased focus on mathematical modeling in the daily teaching practice of mathematics. There is therefore a general agreement between teachers and researchers of math education alike that modeling should make up an important part of both present-day and future educations in mathematics (Gravemeijer & Doorman, 1999; Lesh & Doerr, 2003; Burkhardt, 2006).

Modes of assessment

Revealing what kind of mathematical knowledge and which mathematical competences students possess is not an easy task. Most of the currently applied standard assessment modes do *not* provide an adequate picture of students' abilities in mathematics (Niss, 1999). Especially the ability of the students to work with meaningful contextualized problems using higher order mental skills is not considered in standardized tests (Watt, 2005). Such tests often focus on a small subset of the mathematical ideas and understandings that are valued and useful when working with open-ended real-life problems (Resnick & Resnick, 1993), and consequently standardized tests *"tend to emphasize what is not important, while neglecting what is important."* (Lesh & Clarke, 2000, p.117).Testing and illuminating a non-representative subset of goals and objectives traditional tests results therefore can not be extrapolated to represent a satisfying description of the whole spectre of students' mathematical abilities (Stephens, 1988; Watt, 2005). As a consequence of this Lesh and Clarke (2000) suggest that testing the students should involve letting them engage in open-ended task-based problems that involves relevant mathematical ways of thinking, thereby revealing a broader picture of the students mathematical capabilities. In the study presented in this paper this is what will be attempted. The concrete open-ended problem that was presented to the students in the case study can be characterized as a *Model Eliciting Activity* (MEA) in the sense of Lesh & Doer (2003), and was constructed in accordance with the six concrete principles of instructional design that operationally define such problem solving activities.

THE PENALTY THROW PROBLEM

The case study involved about 200 students enrolled in a calculus course at the University of Southern Denmark's *Faculty of Science and Engineering.*[3] At the first session of the course the students all did a standardized pre-test individually. Subsequently, they worked on a MEA problem named *The Penalty Throw Problem* (PTP) in small groups. It was made compulsory that every group later had to hand in their solution to the PTP in writing. During this session the students were all observed by both one of the researchers and a set of teaching assistants, providing valuable notes for later interpretation. At the end of the seven-week calculus course the students did a written standardized examination, and their solutions were made available to the researchers, and could therefore serve as a pre-test with respect to the case study.

[3] Off the whole group of students 178 and 199 did the pre- and the post-test respectively, and 201 did the MEA problem.

The PTP is a MEA that was designed of the author especially for this case study, and was inspired by the well-known *Volleyball Problem*[4] (Lesh & Doerr, 2003; Lesh & Sriraman, 2005). In the PTP the students' job is to describe a procedure to pick out the three handball players (of a given squad) that are most suitable for throwing the penalty throws in the upcoming world cup tournament.[5] For both the PTP and the Volleyball Problem the data available to the students are heterogeneous in nature, consisting of both quantitative and qualitative information. In order to develop powerful and re-usable solutions to the problems the students has to be able to (a) compare and estimate a variety of qualitative and quantitative information (b) quantify qualitative data (c) accomplish the preceding goals using iterative processes that do not depend on a single formula and (d) generate visual representations (e.g. graphs or charts) to compare profiles of individual players or teams (Lesh & Sriraman, 2005, p. 20). All of these being demands that separates the modelling problems from traditional standardised tests in mathematics.

ANALYSIS OF STUDENTS' DATA

In order to compare the students' solutions to the PTP and the pre- and post-test respectively we developed an *Assessment Tool* rooted in the so-called *Quality Assurance Guide* designed especially for the assessment of MEAs (Clarke & Lesh, 2000). The Assessment Tool provided us with the possibility of quantifying the students' solutions to the PTP thereby allowing a statistical comparison to the students' scores on the standardized pre- and post-tests to be made. Using this conceptual tool we were also able to identify key characteristics of the students' solutions – and thereby spot when interesting conceptual systems (or models) had been developed by the students.

Key characteristics of student solutions

The students' solutions to the PTP were in general diversified and covered a wide spectrum.[6] Still, some general features can be deduced. Firstly, most of the conceptual tools or models used by the students were quite simple in a mathematical sense. Even some of the best solutions used only simple mathematical operations as for instance linear functions, ranking or scaling of numbers. In general we identified three main characteristics that separated the groups that did well on PTP (here referred to as High MEA Achievers) from the groups that performed poorly on the PTP (Low MEA Achievers).

1. *Data relations* - The High MEA Achievers generally considered relations between the different types of given data systematically. This could include drawing lines of tendency or the like. The Low MEA Achievers only considered data relations intuitively.
2. *Representational and presentational media* – The group of High MEA Achievers often used different kinds of media (graphs, charts, tables etc.) to develop and present their models whereas the Low MEA Achievers either did not use any media at all or only used medias with little explanatory power.

---

[4] Also described in detail in Lesh & Sriraman (2005).

[5] At that particular time this was the women handball world cup played in St. Petersburg, Russia, December 2005, in which the Danish national team was participating.

[6] There were also over 70 groups.

3. *Notation* – All of the groups that were characterized as High MEA Achievers used some sort of mathematical notation (e.g. standard x and y-notation or Camel Case), whereas the Low MEA Achievers used almost none.

The quality and usefulness of the students' solutions were not alone characterized by which explicit mathematical tools they used, but moreover by how the they were able to *use* these (often simple) tools to describe and conceptualize the problem when developing their solution.

Quantitative Analysis

The pre- and the post-tests were done individually and the PTP was completed in groups. We therefore analysed the collected data quantitatively by looking for correlation between both individual PTP-performances and individual pre-test performances and group PTP-performances and the sum of groups' pre-test performances respectively. This should provide us with a more satisfying picture of possible relations between the students' performances in the two kinds of mathematical activities.

In both cases the collected data showed very little (if any) statistically correlation between the students' pre-test performances and their PTP-performances (fig. 1 & 2).[7] This was also the case when comparing the students' post-test performances with their PTP-performances. Only when comparing the pre-test scores with the post-test scores some correspondence could be recognized (see Iversen & Larson, 2006). So both of the standardized tests gave different pictures of the students' mathematical abilities than the PTP. This supports the claims of Lesh and others that (a) There exists a large group of students that are highly capable of using mathematics in a powerful way when engaged in open-ended task-based situations – and some of these students are characterised as low achievers by traditional assessment modes, and (b) there are also a large group of students who perform well on standardised tests, but a lot of these students have a hard time applying their mathematical competences to real-world problems in productive and useful ways. As a minimum the data collected suggests that standardised tests do not necessarily do a good job identifying able modellers in the classrooms.

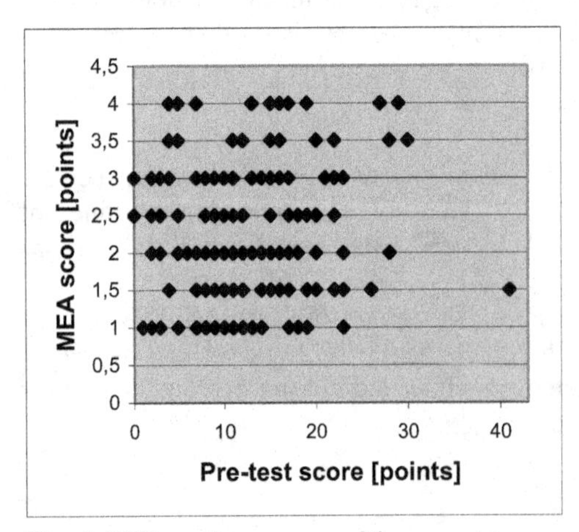

Fig. 1: PTP-performances with respect to Pre-test performance [by individual]

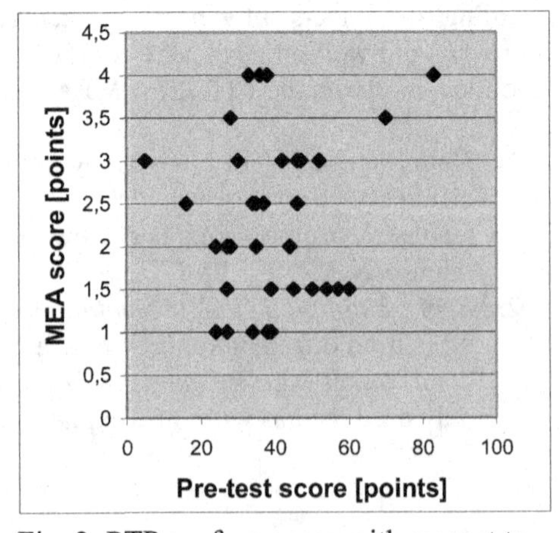

Fig. 2: PTP-performances with respect to Pre-test performance [groups of 3 students]

[7] Also displayed in Iversen & Larson (2006, p. 286-287).

Qualitative analysis

After the search for statistical correspondence we subsequently used the quantitative results to identify two interesting subsets of the students. The first subset consists of five groups who performed poorly on the pre-test, but did well on the PTP. In contrast to this the second subset consists of four groups who performed well in the pre-test, but failed to construct a useful model in the PTP.[8] These two subsets of groups were further analyzed hoping that this could help us explain the lack of correlation between the some of the students' performances in the standardized tests and the PTP respectively.

The groups in the first subset with low pre-test performances and high PTP-performances all used quite simple mathematical tools in sophisticated ways thereby constructing models that took into account most of the given data in relevant ways. This could be by using multi-phase ranking systems or linear weighted scoring schemes supplemented by qualitative procedures to ensure the inclusion of all the data. Some of the groups also used lines of tendency to explore relations in the data provided, thereby being able to judge the importance of the different kinds of data in their final models. These groups demonstrated clearly that simple mathematics can be a powerful tool when used properly to engage in real-world problems.

The other subset of groups had a tendency to ignore large amounts of data without justification, and seldom advanced their models to more than initial levels. They focused on small areas of investigation at a time, often trying to fit bulky pieces of information into neat prefabricated mathematical constructs. They basically had their focus on the mathematical components of the problem and therefore failed to develop useful models for making sense of the complex situation. Almost no mathematization was detectable in their solutions and their final models ended up using primarily simple ranking or picking.

THE IMPORTANCE OF STUDENTS' ATTITUDES TOWARDS MATHEMATICAL MODELLING

The analyses indicate that little connection exists between students' traditional mathematical abilities (computation and deduction) and their modelling competencies. This is of course a wrong, and far too radical, conclusion to make based solely on our case study. Another study by Maass (2006), indicate that their *do* exist some relations. One possible explanation of why we failed to locate any correspondence could perhaps be that our study did not particularly consider factors as students' attitudes and beliefs towards modelling in mathematics and their following motivation for engaging in open-ended modelling problems. Maass (2006, p. 138) concludes that especially students' negative attitudes towards modelling basically will hinder their ability to develop usable models with relevant mathematical content. Burkhardt (2006, p. 191) supplements this by concluding that some of the students who are usually low-achieving shows substantial improvement in mathematics, when introduced to the modelling aspect of mathematics. It seems that the traditional teaching in mathematics has "turned these students off", and that modelling holds the potential of displaying the relevance of mathematics, and thereby wake an interest for the subject in them.

---

[8] These two subsets of groups can be identified in figure 2 as belonging to the upper left quadrant and the lower right quadrant respectively.

APPRECIATING THE VALUE OF USING SIMPLE MATHEMATICS IN COMPLEX SETTINGS

Although the pre-test results clearly reveal that the groups that did well on the pre-test, but poorly on the PTP, are in fact capable of handling advanced mathematical objects in standardized forms, these group members are not able to mathematize the PTP in a reasonable way. The opposite is the case for the groups that ended up with the reverse results. These groups are in spite of their trouble to solve traditional math problems able to use simple mathematical objects in powerful and complex ways when developing their models in their work with the PTP.

In Iversen & Larson (2006) we therefore propose adding another explaining factor to the discrepancy between students' performance in the standardized tests and the modelling task respectively. MEAs, and many other types of modelling, often involve solving tasks in complex real-world settings using mostly conceptual tools that can be characterised as simple in a purely mathematical context.[9] On the other hand, the conceptual tools needed to do well in the standardized tests are quite complex mathematically, but is only applied in simple contexts. The often heard catchphrase that learning traditional mathematics makes you able to survey and solve complex problems may have to be adjusted. There may be an important discrepancy also between simple thinking using complex mathematics and complex thinking using simple mathematics.

The groups that managed to develop shareable and re-usable models when working with the PTP, succeeded in focusing their mental processes on interpreting, organizing, and conceptualizing instead of focusing entirely on the mathematical objects involved in the process. Acknowledging the importance of attitude and beliefs pointed out above we could say that these students were able to balance their reasoning with their emotions in a way that allowed them to construct powerful conceptual systems.

"Using creative, critical problem-solving, decision-making and innovative thinking processes, and being able to evaluate, judge and predict consequences this group of students shows characteristics that can best be described as holistic complex thinking."
Iversen & Larson, (2006, p. 290).

Still among this group of students were in fact some who performed very poorly in the standardized tests, and so our study indicates that it is truly possible for students that are identified as low achieving by traditional mathematics assessment to engage meaningfully in complex real-world tasks whose solutions necessitate the ability to construct powerful and complex conceptual systems (or models) by using mostly simple mathematical tools available at many different educational levels.

CONCLUSIONS

In the study little correspondence was detected between the students' performance on the open-ended modelling task called the Penalty Throw Problem and the standardized pre- and post-test respectively. This, of course, is not equivalent to the fact that no connection exists at all. Instead we call attention to the fact that no straightforward one-to-one coherence between students' modelling abilities and their achievements in traditional assessment modes in mathematics are clearly demonstrated. It is suggested that part of this can be explained by the fact that much modelling involves simple mathematics applied to complex settings whereas

[9] See also Gainsburg (2006).

much traditional testing in mathematics involves handling advanced mathematical objects in very simple contexts. Standardized assessment does therefore not necessarily provide us with a qualified and usable picture of the students' mathematical competences and perhaps talented modellers are ignored and never identified in the classrooms. This constitutes a serious problem in an age where future-oriented professions hunger for people who are not necessarily able to score well on standardized tests but instead are able to make sense of complex systems, and develop powerful and re-usable conceptual tools that draw on a variety of disciplines (Lesh, Hamilton & Kaput, 2006). It is another well known fact that topics not being tested in the end of a course often disappear out of the classrooms. This is also very much the case in the teaching of mathematics. According to Niss (1999) the distance between the goals of the mathematics education community and the assessment modes most often used is increasing. A failure to test students' modelling abilities therefore may mean a failure teaching the students to develop powerful mathematical models in the daily teaching practice and this will just increase the demand for qualified modellers in the future.

We do not advocate that modeling should replace all of the traditional teaching and testing in mathematics, but keeping the results of the presented study in mind a shift of paradigm may be necessary. In this we agree with Lesh & English (2005) that claim attention should be shifted from asking what kind of algorithms and computations students can memorize correctly to asking what kind of complex open-ended problems they can describe effectively. Only in this way – by respecting mathematics' many important connections to the arts and sciences – can the future of mathematics education reflect the society that surrounds us in a reasonable and productive way.

References

Burkhardt, H. (2006). Modelling in Mathematics Classrooms: reflections on past developments and the future. *Zentralblatt für Didaktik der Mathematik*, 38 (2), 178-195.

Clarke, D. & Lesh, R. (2000). Formulating Operational definitions of Desired Outcomes of Instruction in Mathematics and Science Education. In Kelly, A. E. & Lesh, R. (Eds.). *Handbook of Research Design in Mathematics and Science Education* (pp. 113-?). New Jersey: Lawrence Erlbaum & Associates.

Gainsburg, J. (2006). The Mathematical modelling of Structural Engineers. *Mathematical Thinking & Learning,* 8 (1), 3-36.

Gravemeijer, K., & Doorman, M. (1999). Context problems in realistic mathematics education: A calculus course as an example. *Educational Studies in Mathematics,* 39, 111–129.

Iversen, S. M. & Larson. C. J. (2006). Simple Thinking using Complex Math vs. Complex Thinking using Simple Math. *Zentralblatt für Didaktik der Mathematik*, 38 (3), 281-292.

Lesh, R. & Doerr, H. M. (Eds.) (2003). *Beyond Constructivism: Models and modeling perspectives on mathematics problem solving, learning, and teaching*. Mahwah, NJ: Lawrence Erlbaum Associates.

Lesh, R. & Sriraman, B. (2005). John Dewey Revisited – Pragmatism and the models-modelling perspective on mathematical learning. In A. Beckmann, C. Michelsen, & B. Sriraman (Eds). *Proceedings of the 1st International Symposium of Mathematics and its Connections to the Arts and Sciences*. May 18-21, 2005, University of Schwaebisch Gmuend: Germany. Hildesheim, Berlin: Verlag Franzbecker, pp. 7-31.

Lesh, R. & English, L. D. (2005). Trends in the evolution of models & modeling perspectives on mathematical learning and problem solving. *Zentralblatt für Didaktik der Mathematik*, 37 (6), 487-489.

Lesh, R., Kaput, J. & Hamilton, E. (Eds.) (2006). *Foundations for the Future: The Need for*

New Mathematical Understandings & Abilities in the 21st Century. Hilsdale, NJ: Lawrence Erlbaum Associates.

Maas, K. (2006). What are modelling competencies? *Zentralblatt für Didaktik der Mathematik*, 38(2), 113-142.

Niss, M (1999). Aspects of the nature and state of research in Mathematics Education, *Educational Studies in Mathematics,* 40, 1–24.

Steen, L. A. (2005). *Math & Bio 2010: Linking Undergraduate Disciplines*. Mathematical Association of America Press.

Watt, H. M. G. (2005). Attitudes to the Use of Alternative Assessment Methods in Mathematics: A Study with Secondary Mathematics Teachers in Sydney, Australia. *Educational Studies in Mathematics,* 58, 21-44.

CHALLENGES FOR INTERDISCIPLINARY TEACHING OF MATHEMATICS AND THE SCIENCES IN UPPER SECONDARY SCHOOL

Stinne Hørup Hansen
University of Southern Denmark

Teachers' and students' handling of an interdisciplinary project with physics, chemistry, biology and mathematics in upper secondary school is the focus of the paper. An interdisciplinary project on radiation is described and the challenges are presented through cases regarding the teachers interdisciplinary tolerance, multi-disciplinary teaching by a single-disciplinary teacher, and the students handling of the interdisciplinarity. Mathematics and chemistry were the subjects which teachers and students found hardest to integrate with the other subjects. The research found that the teachers were not sufficiently trained in interdisciplinary teaching, therefore the integration of all four subjects was; to a large extend, left to the students.

INTRODUCTION

In august 2005 a new reform of the Danish educational system, which increased the demands for interdisciplinary teaching, was implemented. A successful increase of the cooperation between the subjects requires that both students and teachers possess the didactical qualifications necessary to handle interdisciplinarity.

The present article is a description of an interdisciplinary course for physics, biology, mathematics and chemistry in a first year class in a Danish urban Upper Secondary School. I was fortunate to get the opportunity to participate in the design, planning and implementation of the course in the early stages of my PhD. The reform demands that the schools establish teacher teams to handle three interdisciplinary courses: General Language Understanding, Elementary Science, and General Study Preparations. As a preparation for the reform, the teachers of mathematics, physics, biology and chemistry were assigned to conduct an interdisciplinary course in the spring of 2005. The demands and challenges within the course are comparable to the problems the teachers are currently challenged with in the implementation of Elementary Science.

I will discuss the challenges I observed when four subjects and particularly four teachers collaborate. The exemplary problems will be presented through a set of cases. Finally I will suggest how one could strengthen the quality of interdisciplinary courses and teacher collaboration.

PROJECT DESCRIPTION

Topic

The overall topic of the project was chosen to be radiation, since this topic is a part of the curriculum in physics and applies the mathematical notions of exponential functions and half-time constants. A decisive factor in the choice of topic was whether the teachers found it relevant to approach the topic from their primary discipline.

B.Sriraman, C.Michelsen, A. Beckmann & V. Freiman (Eds). (2008). *Proceedings of the Second International Symposium on Mathematics and its Connections to the Arts and Sciences* (MACAS2). Information Age Publishing, Charlotte: NC , pp.173-180

Concept

The objective of the project was to let students write a documentary report on a topic related to radiation, which they were to present orally. The report was produced in groups selected by the teachers.

The concept of the project can be related to the criteria which, according to Mitchell (1993), must be satisfied in order to facilitate the stimulation and maintenance of students' interests in a topic. My focus in this paper is how students and teachers handled the fact that the project is interdisciplinary. Consequently I will not address the notion of interest and which effect the project had on students' interest in natural science.

Professional criteria

Within all four disciplines it is mandatory for the students to make and hand in a range of assignments. These assignments include, among other things, solving mathematics assignments and producing documentary reports of experimental work in biology, physics and chemistry courses. This means that the present project was bound to be concluded with a documentary report in which the individual disciplines' criteria as to the scope of such a report were satisfied. Only the mathematics teacher provided the students with a leaflet on which he described the criteria in mathematics. Among other things, the leaflet informed that:

> The report must include the following:
>
> 1) Give an account of the meaning of exponential development, and how this connects to radioactivity.
> 2) Give an account of the meaning of half-time.
> 3) A description of what single logarithmic paper is and how it is used in connection with exponential growth.
> 4) Give an account of how we perform exponential regression on TI-89.
>
> Find examples that illustrate the individual items.

The practical course process

As preparatory homework for the project, the students were to read a textual material on radiation. The first day of the project the students had a physics lesson. The physics teacher chose to use that lesson to demonstrate various types of radiation so as to stimulate students' interest in the overall topic. The students experienced that a burned out light bulb could, light if it was elevated in a glass of water in a microwave oven. Further, the students observed how light is spread when sent through a grating. This, in turn, led to a brief treatment of waves and the electromagnetic spectrum. For the remainder of the lesson the group chose their topic.

Three groups chose to write on cosmic radiation, two groups chose to write on radiation and cancer while the last group chose to write on the world of colors.

The project was held over a three week period.

ETHNOGRAPHIC METHODS

The objective of the project was to designing an interdisciplinary course for mathematics and the sciences where the students experienced a weakening or even elimination of the boundaries between the four subjects. This required close collaboration with the teachers during the design of the project for the elimination of the boundaries to be a central part of the course structure. I advocated for the project to be concluded with a report presenting material from all four subjects and an oral presentation. The benefit of this active collaboration with teachers in the planning of such courses is that one, as a researcher, gain insight into the planning process and that one is able to influence the planning process. For an elaboration of the benefits and drawbacks of such a collaboration effort between researches and teachers refer to Paul Cobb (2000).

During the course I used a technique called "participating observer" (Zevenbergen 1998) as the ethnographic method for gathering empirical data. When you as a researcher position yourself as participating observer you are considered to be an "additional teacher" and not a silent and anonymous observer.

I was present during the entire course and observed the work of the students as well as their contact to the teachers. I concluded the project with questionnaires and 1½ year after the project I returned to the class to interview students from each group. In this paper I have chosen to apply the results of my field notes from observations of students' conversations internally in the group as well as of their contact with the teachers and on the interviews of the students. I will put special emphasis on mathematics role in the project.

RESULTS/OBSERVATIONS

The results are presented by two exemplary cases.

Case 1: The teachers' way of tolerating interdisciplinarity

The biggest challenge for the teachers was to contribute constructively to the interplay between the subjects. The mathematics teacher in particular found it very difficult to see possibilities in the interdisciplinary work. As shown in the following conversation between the mathematics teacher and one of the groups, the teacher insisted that the students should hand in a separate mathematics report when he discovered that the students were not working on the mathematics part during a mathematics lesson.

The following is an extract of the conversation between a group and the mathematics teacher (MT):

Mathew: "we haven't done any math yet. We figured that we would bump in to some math if we just continued with the project."

Karen: "We just need some compounds and their half life's, then we can calculate on that. We want something that fits in with our project."

MT: "You must make a separate report, which only deals with mathematics."

Many teachers, and therefore also students, regard mathematics as a subject that is isolated from the sciences, and do not possess the interdisciplinary competences that are required to integrate mathematics in subjects such as physics (Michelsen, 2006). Michelsen (2001) studies how students can model exponential growth by simulating radioactive decay. Modelling provides the possibility for mathematics to interact with the sciences, and it would have been advantageous to make use of this method to integrate mathematics in this specific project. In connection to this I refer to a report from the ministry of education regarding competencies and learning mathematics from 2002 which contains a list of suggestion on how to renew mathematics teaching.

When asked the question:" How do you think the teachers like interdisciplinary teaching?" one of the boys replied:

> "I think they find it troublesome, definitely. But maybe it can give a better understanding and interest the students more, which is also sort of motivating for the teachers (…) at the same time it is also a lot of work and a bit complicated to engage in (…) maybe it differs how much the teachers like that you work on something that has nothing to do with their subject"

When you work on an interdisciplinary project based on a specific subject, but without detailed descriptions of the demands from the subjects involved, it is likely that more emphasis will be on one subject over the others. In most cases it can be difficult to make a homogenous report if all subjects must be represented equally. Hence, a softening of the borders between the subjects requires that the subjects are willing to be compromised (and can be compromised in accordance with the mandatory curriculum) and let the students decide the weightings of the subjects involved.

The teachers' lack of interdisciplinary tolerance resulted from two aspects. First of all there were certain curriculum requirements that must be met, in order to spend a relatively large amount of lessons on the project. Second, the teachers were very reluctant to cross the borders to the other subjects. As a result of this I found the teachers trying to pull the students back into their own "subject-home ground" regardless of the topic the students had decided to cover. Following the implementation of the Reform in 2005 several evaluative reports of the reform have been published. In one report from the Danish Ministry of Education from 2006 (Frederiksen *et al.* 2006) there is an evaluation of the subject cooperation . According to the report, mathematics has become more application-oriented in order to participate in interdisciplinary projects. This accommodates the students' expectations during the current project when they said: "We figured that we would bump in to some math if we just continued with the project". The report also describes that the mathematics teachers feel that the change has a negative effect on the students' mathematics skills and that the change degrades mathematics to a mere tool for solving problems within other subjects.

Case 2: Students handling of interdisciplinarity

During the project process, the students had an alternating single-subject and multi-subject approach to the project. During meetings with their teachers they were forced to focus on the subject represented by the teacher, whereas their group work allowed them to work interdisciplinarily. The students spent a considerable amount of time taking due account of all subjects and the different demands.

Boy:

> "I think it is one of the most difficult assignments we have had (referring to the mathematics part). It is the first time ever that we have something really upper secondary school-like. I was expecting it to be like this all the time. It is also hard to think about all the subjects at the same time."

One of the groups had a meeting with the chemistry teacher (CT):

Tina: "We are going to make an experiment where we measure radiation indoors and outdoors this afternoon."

CT: "How will you get some chemistry into this?"

Tina: "Can we make an experiment with free radicals? Just a small one?..."

Petra: "We only have some chemistry regarding DNA."

CT: "what is your comprehension of DNA? A DNA molecule has a chemical structure. You could write about that."

Petra: "How about the catastrophe in Chernobyl. Can evolution occur in connection to that? Could we combine this with mutation?"

Astrid: "Hasn't northern lights something to do with chemistry?"

CT: "Yes."

Tom: "Hasn't γ-radiation something to do with chemistry?"

CT: "Yes, it is also very much physics. I think you are going to do experiments with γ-radiation in physics."

Astrid: "Can you make experiments with northern lights?"

CT: "Northern lights is something with excitated electrons."

Astrid: "That they decay?"

CT: "No, that they jump from one energy level to another."

Astrid: "Isn't that enough chemistry?" [the student hopes the chemistry teacher is satisfied if they write about northern lights.]

In the above extract of a conversation between the chemistry teacher and one of the groups it is clear that the students are very confused about how to incorporate all the subjects. In addition it is unclear to them which subject they are dealing with when they talk about radiation and northern lights. They are asking the chemistry teacher for help but are not receiving clear answers. In the end it is left to the students to incorporate all four subjects in the combined report.

The report (Frederiksen *et al.*, 2006) documents that the biggest problems regarding disciplinary interplay are within the sciences. The most difficult subject to integrate is chemistry while biology was integrated with other subjects most successfully. This is in agreement with the findings of the current project, where the students are asking the chemistry teacher for help, without receiving must, whereas the students hardly asked any questions to the biology teacher.

In an interview I asked a girl how the cooperation between the subjects could be improved. She replied:

> "It could make the lessons more interesting, but I think it is a problem with the teachers. I don't think they want to bother."

This is in accordance with the report (Frederiksen *et al.* 2006), where the students find that the teachers are rather sceptical towards the subject cooperation within the sciences and often too many subjects are involved.

The reform, however, demands that the teachers engage in teamwork and another evaluation of the elementary courses in Upper Secondary School shows that both teachers and students are satisfied with the collaboration (Dolin *et al.* 2006). However, the majority mention that Elementary Science has been conducted more as parallel subject-teaching rather than genuine interdisciplinary subject-collaboration (the authors understanding of parallel subject-teaching is: teaching of a main topic in two or more disciplines in the same period of time but with no obvious interaction).

WHAT DOES IT TAKE TO IMPROVE INTERDISCIPLINARY TEACHING IN PRACTICE?

In May 2006 the ministry of education published a report that presents the results of an investigation of how the organization of teacher teams and the execution of interdisciplinary projects proceeded in the first semester after the reform was implemented.

The Report states that the execution of the interdisciplinary sequences in organized teams is a revolutionary way of preparing lessons in upper secondary school. As official interdisciplinary text-books were not available in the beginning of the reform one of the major challenges for the teams was to design their own teaching material. According to the report, the sequence of Elementary Science has been the most difficult for the teachers to handle. The report offers two main explanations; namely the lack of official teaching material and the fundamental differences between the subjects. According to the report a solution to this problem is to develop the required teaching material. In addition I suggest more interdisciplinary training of the teachers in order to improve their elementary science competencies. This is necessary if the cooperation between the subjects must become truly interdisciplinary and not merely parallel subject-teaching. According to the report by Frederiksen *et al.* (2006) the implementation of subject cooperation must be looked upon as an ongoing process. The teachers are undergoing a process meanwhile they must teach according to these new interdisciplinary demands.

The interplay between the natural sciences has been prioritized with the introduction of the reform of the Danish Upper Secondary School in 2005. Now the challenge is for the management, educational researchers, textbook-developers, and the teachers themselves to facilitate the process, the teachers are undergoing to acquire the competences required to implement true and meaningful interdisciplinarity.

References

Dolin, J., Hjemsted, K., Jensen, A, Kaspersen, P., Kristensen, J. (2006). Evaluering af grundforløbet på stx, Institut for Filosofi, Pædagogik og Religionsstudier, Syddansk Universitet, 1-53.

Frederiksen, L.F., Kaspersen, P., Wiese, L.B. (2006). Evaluering af arbejdsformer og fagligt samspil i stx, hhx og htx efter gymnasiereformen, Institut for Filosofi, Pædagogik og Religionsstudier, Syddansk Universitet, 1-44.

Michelsen, C. (2001). Begrebsdannelse ved domæneudvidelse. Elevers tilegnelse af funktionsbegrebet i et integreret undervisningsforløb mellem matematik og fysik. Syddansk Universitet, Syddansk Universitets Trykkeri.

Michelsen,C. (2006). Functions: A modelling tool in mathematics and science. *Zentralblatt für Didaktik der Mathematik*, 38:269-280.

Mitchell,M. (1993). Situational Interest - Its Multifaceted Structure in the Secondary-School Mathematics Classroom. *Journal of Educational Psychology*, 85:424-436.

Niss,M., Jensen,T.H., Andersen,T.B., Andersen,R.W., Christoffersen,T., Damgaard,S., Gustavsen,T., Jess,K., Lange,J., Lindenskov,L., Meyer,M.B., and Nissen,K. (2002). *Kompetencer og matematiklæring. Ideer og inspiration til udvikling af matematikundervisning i Danmark.* Undervisningsministeriet, Uddannelsesstyrelsen, Roskilde.

Undervisningsministeriet. (2002). Bedre Uddannelser. Undervisningsministeriet, Uddannelsesstyrelsen.

Zevenbergen R . (1998). Ethnography in the Classroom. In: Malone JA, Atweh B and Northfield JR (eds) Research and supervision in mathematics and science education. Lawrence Erlbaum Associates, Mahwah, New Jersey, London, 19-38.

FERMAT MEETS PYTHAGORAS

Thilo Hoefer
Univ. of Education Schwäbisch Gmünd / Staufer-Gymn. Waiblingen

Ever since the TV series 'Baywatch', the job of a lifeguard has been perceived as a very exciting one by teenagers. Therefore it constitutes a motivating starting point for a sequence of lessons connecting the physics and mathematics. To be more precise, a connection between Fermat's principle on the one hand and Pythagoras's theorem on the other will be established, including problem solving with functional extremas, assisted by a graphing calculator.

Introduction

Exercises dealing with physical problems have been a component of mathematics lessons for a long time. To solve such problems, mathematical methods are used to connect physical dimensions with each other. The interpretation of the obtained mathematical results is often purely theoretical and hard to understand for many pupils. In only few cases will the (mathematics) teacher prove these results by showing a physics experiment, so that pupils may truly experience the problem and its solution.

The following example will show a sequence of lessons which uses a different, interdisciplinary approach to teaching this type of problem solving. It includes the understanding of the problem and of its solutions in real (physical) surroundings. The physical content is central to optics, but also permits the gaining of mathematical experience in a special manner. The application of graphing calculators (GC) or computer algebra systems (CAS) enables teachers to deal with many topics between the two subjects, mathematics and physics, in the classroom, although the mathematical skills of their pupils would often not allow to do this (without the media mentioned above).

The initial situation in the physics class

The model of the beam of light is often taught at the beginning of optics lessons in physics. The course of beams is followed and predicted, the latter being done in many cases by using the phenomena of reflexion and refraction of light. Predicting the course of a beam of light which is reflected at some surface will not cause a major problem after pupils have become aware of the fact that the angle of incidence equals the angle of reflexion. However, there will be a problem in understanding the whole concept of refraction without knowing trigonometric operations like the sinus: To predict the angle of refraction one needs to solve the equation $\sin(\alpha)/\sin(\beta) = n_\alpha/n_\beta$ (using the predetermined refractive indices n_α and n_β and the given angle of incidence α). Even if pupils are able to cope with this equation, there will be the problem of having used two different rules for two phenomena, which both stem from one physical principle, Fermat's principle. With the knowledge that pupils have achieved at this point, there is no chance to connect them.

B.Sriraman, C.Michelsen, A. Beckmann & V. Freiman (Eds). (2008). *Proceedings of the Second International Symposium on Mathematics and its Connections to the Arts and Sciences* (MACAS2). Information Age Publishing, Charlotte: NC , pp.181-186

Why Fermat's principle is not contained in the curriculum[1] of primary physics

Fermat's principle can be summarised in one sentence, which is easy to understand: Between two given points, the beam of light follows the course which requires the shortest period of time (e.g. Vogel 1995[2], p. 174). This principle remains valid even if the beam of light is forced to make a detour, as a result of a mirror being used. The reflected beam of light follows the fastest course from A – via a mirror – to B (Vogel 1995, p. 173).

One reason why using Fermat's principle with a mirror does not cause any major problem is the fact that the course of the beam of light is completely inside one medium (e.g. air). There is an increase in complexity when the beam of light travels across (two) different media, as the velocity of light varies in different media. The shortest, i.e. fastest course is now different from the geometrically shortest way between the two points A and B. When trying to calculate the fastest course using different velocities, a rather complex target function has to be minimized. As Pythagoras's theorem is used twice (once for each medium), the target function[3] is the sum of two root terms. This being taught in a beginners' (physics) class, pupils will certainly be unable to use the tools of analysis (e.g. derivatives) which are required to solve this type of minimizing problem. They will be able to use Pythagoras's theorem, to thereby create a target function and state a reason why this target function has to be minimized to satisfy Fermat's principle. But then, they will not be able to find a solution. That is the reason why so far, Fermat's principle has not been included in the curriculum of beginners' physics classes when using the model of the beam of light.

New aids: GC and CAS

With the latest version of the curricula in Baden-Württemberg (MKJS 2004, p.97), the GC has become a prescribed aid for the (about 14-year-old) pupils of the 7th/8th grades at the Gymnasium. All school forms (e.g. Realschule) or schools in other parts of Germany have at least central computing laboratories, most of them with the opportunity of using CAS. Especially in classes equipped with GC, pupils get a tool which helps them to sketch even curves they would not be able to sketch on their own because of the skills in analysis which are needed, but still missing. Thus it becomes possible to minimize the above-mentioned target function, aided by a GC or CAS.

How Pythagoras and Fermat can help a lifeguard

The job of a lifeguard is to rescue drowning people. Therefore the lifeguard has to reach the drowning person as quickly as possible. So the problem is how to get from his current position (usually on the beach) to that of the person in the water, who is in need of aid. The velocity of the lifeguard when he is running on the beach will be higher than when he is swimming in the sea. So the fastest way to reach the person in most cases is not the same as the linear distance. The lifeguard should cover a larger distance running on the beach, so that the distance which has to be covered swimming becomes shorter.

Once this background information has been understood, Fermat's principle can be formulated in a new way: Light acts like a perfect lifeguard. Because of its clarity, the problem of the lifeguard and the ensuing formulation of Fermat's principle are particularly suitable for the

[1] To be more precise: It is not contained in the physics curriculum for the 8th grade (up to 15-year-old pupils) in Baden-Württemberg, Germany.

[2] Own translation

[3] For further information see the worksheet "The lifeguard's problem"

beginning of the sequence of lessons. Supporting the motivational factor, one could even start with a short scene from "Baywatch", showing a lifeguard on duty.

The whole sequence of lessons could be constructed in the following way:

First, the target function is developed by using Pythagoras's theorem and the uniform motion equation (s=v*t). One should be aware of the fact that pupils will not have solved many extremal problems. As a consequence, it is expected that there will not only be difficulties at the end of such a problem solving process, but even at the beginning, when the target function itself has to be developed. Therefore pupils should be guided tightly at the beginning, for example with the step-by-step calculation of the time needed for a concrete course[4] (cp. Worksheet 1).

After this "careful" beginning, the focus will be on the variable point where the lifeguard enters the water. With this focus, the target function for the required time, which is subject to the choice of this variable point, will be developed. Aided by GC or CAS, the curve of the target function will be considered, directly followed by an interpretation of its meaning to the problem. Hence the need for the low point in this curve will (hopefully) be created.

Having solved the lifeguard's problem, Fermat's principle will be formulated by the teacher in analogy to the "perfect lifeguard" problem. Thus, the new content will be new only in the context of light, not in that of the mathematical skills needed. Furthermore this is the chance to strengthen those skills by solving another problem of this kind, this time in combination with optical physics and (if not again needed) without the steps presented by the teacher to facilitate understanding. Pupils will receive specifications for the speed of light in air and water as well as the description of an experiment with a laser pointer placed outside a tank of water and trying to hit an object under water with its beam. They will then have to calculate the point where the beam of the laser pointer should make contact with the surface of the water, so that the object under water will be hit (cp. Worksheet 2). Of course, this problem should not only be solved theoretically, but also proven in a live experiment.

After these steps, related mathematical and physical contents should be dealt with separately. In mathematics, the structure of the solution could be discussed and generalized for upcoming problems with extreme values. This could result in a kind of "instruction manual" describing the steps needed to solve this type of problem, along with the order in which these steps should be executed. In physics, this experimental example should be generalized, with the aim of describing optical refraction. At the same time, everyday phenomena like the (refracted) glance into water should be explained. The content dealt with will be the same in both cases, though the physical treatment of the problem starts from the opposite side (The course of the beam of light is observed in an experiment, so it is assumed and not sought. The question is now: Why is it like that?). And of course now the law of reflection should also be concluded from the application of Fermat's principle.

Does it make sense to deal with functions that pupils can only handle with the aid of CAS or GC?

To answer this question one should bear in mind that functions can not only be represented by an algebraic term, but also, for example, by verbal description, tables and graphs (cp. e.g. Beckmann 1999, Leuders / Prediger 2005, …). And pupils are able to solve problems with

[4] The course could be a guess by pupils on the question which is the fastest way of covering the distance. After every pupil has fixed and calculated his own guess, all proposals can be compared in class and the pupil with the best result can be declared "best lifeguard".

functions represented, for example, by graphs, even if they cannot handle the terms. So why shouldn't it be acceptable to use CAS or GC to change the representation of the function into one that is appropriate for solving the problem with the pupils' own skills? Especially in the problem introduced above, pupils are unable to calculate the function's minimum, but could change the function's representation into a table by calculating one pair of variates after the other. Then they could draw the function's graph by drawing the points and interpolating them. Thus they would create the graph with the help of which they could solve the minimizing problem (graphically). So pupils could at least approximately solve the problem. If pupils know how to do this, there is no reason why CAS or GC should not be used to solve the problem more efficiently and more precisely. It can even make sense to sometimes use functions with algebraic representations that pupils will not be able to deal with: Various didactic research has shown that the skills of pupils are often one-sided (cp. Beckmann 1999, PZ 1990; Vinner / Dreyfus 1989, ...). Their understanding of the concept of functions is often limited or based on mathematical operations only. Useful experiences with different representations in non-mathematical surroundings are hard to find. The proposed way of teaching a unit is one way of crossing the boundaries of mathematics in order to develop an extensive idea of the concept of functions.

Future prospects

Fermat's principle is often taught in advanced classes (17-year-old pupils). The way of teaching it which has been outlined above could also be interesting for those classes, however, it should be taken to a higher level. In this case, the point where the beam of light hits the water should not be predicted, but should be measured, followed by the calculation of the velocity of light in the liquid medium. This leads to a non-trivial solution that needs the first derivative of the function of time depending on the mentioned contact point. After this solution has been obtained, points should be predicted as mentioned above, the prediction of experimental outcomes being a fundamental aim in physics. This idea of a sequence of lessons is to be taught in the Comenius programme ScienceMath in the future.

Acknowledgement

The author would like to thank Verena Moeller (Staufer-Gymnasium Waiblingen) for being a big support in the translation of this article.

References:

Beckmann, A. (1999). Der Funktionsbegriff als Unterrichtsgegenstand zu Beginn des Mathematikunterrichts in der zweijährigen Höheren Berufsfachschule. *Journal für Mathematikdidaktik,*20(4), 274-299.

Beckmann, A. (2000). Bereitet der Mathematikunterricht auf die Mathematik im Physikunterricht vor?. *Mathematica didactica ,23*(1), 3-23.

Leuders, T. / Prediger, S. (2005). Funktioniert's? – Denken in Funktionen. *Praxis der Mathematik,* 47(2), 1-7.

PZ (Päd. Zentrum Rheinland-Pfalz) (1990). Funktionen und Graphen. *PZ-Information Mathematik,* Bad Kreuznach.

Vinner, S. / Dreyfus, T. (1989). Images and Definitions for the Concept of Function. *Journal for research in Mathematics Education,* 20(4), 256-366.

Vogel, H. (1995). Gerthsen Physik (18). Springer, Berlin.

MKJS Baden-Württemberg (2004). Bildungsplan Gymnasium 2004. Stuttgart.

Worksheet 1: The lifeguard's problem

Lifeguard Mitch is standing in front of his tower when he discovers that a person in the water is in trouble. The direct way from his tower to where the water hits the beach is 50 m. From there, it is 50 m straight on through the water and then another 50 m orthogonally to the first line, until Mitch reaches the person in trouble (fig. 1). He knows his speed on the beach (7 m/s) and in the water (2 m/s). To reach the person as quickly as possible he runs to a point Q first, where he enters the water and swims directly to the person. The distance at the beach is u and in the water is w (fig. 1).

Fig. 1 The situation of the lifeguard

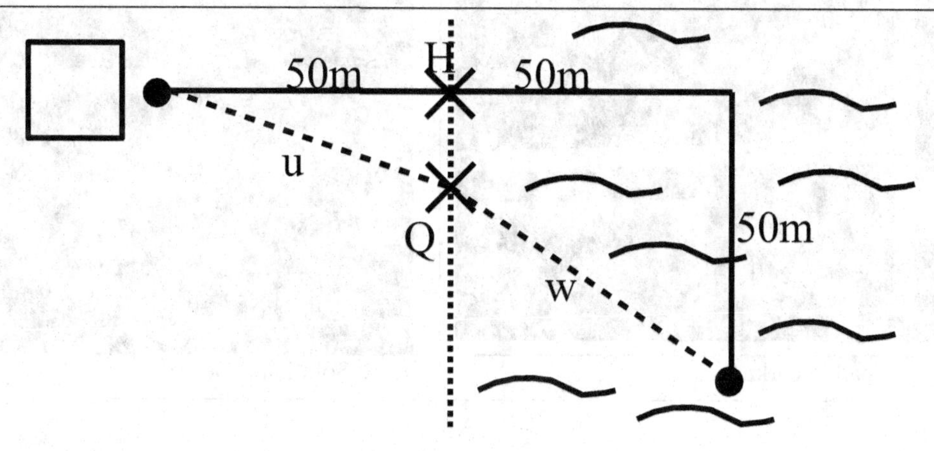

Exercises:

a.) How much time would Mitch need if he ran directly to the water (to H) and then swam directly to the person in trouble?

b.) He could also run to the point on the beach where the water line is orthogonal to the line between Mitch and the person in trouble. He could then swim the direct 50 m to the person. How long would it take him then?

c.) Imagine you are the lifeguard (with the same speeds). Choose a point on the water line very quickly, to which you would run directly and from which you would swim to the person. How long would it take you?

d.) Compare your solution to those of your fellow students. Who is the best lifeguard?

e.) Mitch chooses a point Q at the distance of x to H. Calculate the total time in subject to x.

f.) The result of e.) is a function t(x). Take a graphing calculator and look at the curve of the function. Describe everything you can see from this curve.

g.) Point out the distance x that will be best for Mitch.

Worksheet 2: Fermat's principle

The speed of light is not the same in every medium. For instance, it is 300,000 km/s in the air, about 200,000 km/s in glass, and in water it is approximately 225,000 km/s. Besides, one characteristic of light is that it 'acts' like a perfect lifeguard. That means that the course of a beam of light between two points A and B is always the fastest way from A to B. This is called Fermat's principle, named after its discoverer Pierre de Fermat (1601-1665).

Exercises:

A little water snail wants to have some light in its shell. So the beam of a laser pointer outside the water has to hit the shell (cp. fig. 1a).

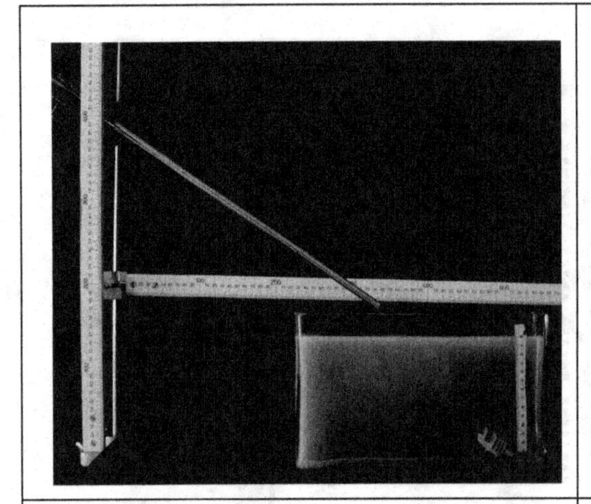

Fig. 1: Snail in darkness

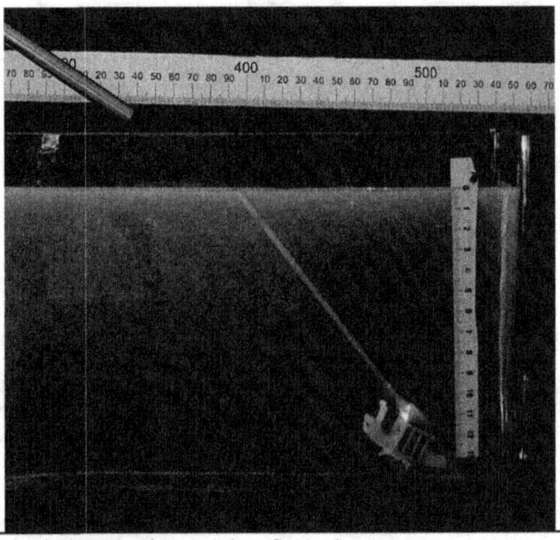

Fig. 2: Solution to be found

- **a.)** If one measures the interesting points and lengths in figure 1 and copies them into a cartesian coordinate system, one will get the laser pointer in L (0/40), the water line along a line with y=14 and the shell of the snail in S (50/3) (all units are cm). Draw this coordinate system with L, g and S (use a suitable scale).
- **b.)** The laser pointer can be adjusted, so that its beam will hit the water line in Q (x/14). Choose any Q and add it to the coordinate system you have drawn in a.).
- **c.)** Where does Q have to be placed, such that the total way from L via Q to S will be as fast as possible (use the speeds of light given in the text above).
- **d.)** With the help of Fermat's principle, discuss the following points: What will happen if one points the beam of light at Q as calculated in c.)? And what will happen if one points it more to the right/left?

MATHEMATICAL LITERACY
- THROUGH SCIENTIFIC THEMES AND METHODS

Astrid Beckmann
University of Education, Schwäbisch Gmünd Germany

In a project of the University of Schwäbisch Gmünd scientific themes and methods are integrated into mathematical lessons of German secondary schools. In the paper the aspects "phenomenon" and "experiment" are investigated in regard to an advancement of mathematical literacy.

1 MATHEMATICAL LITERACY

Mathematical literacy is the ability to apply mathematical knowledge on manifold and context related problems in a functional, flexible and sensible way.

Mathematical literacy contains both

- "formal knowledge", that is the ability to use (simple) structures and methods/calculations
- "applicable knowledge", that is a competent use and the linking-up of mathematical concepts and structures as well as the ability of mathematizing and modelling in unknown situations.

International studies like PISA investigate the mathematical literacy of students. In the description of the tasks and the interpretation of the results 6 competence-levels are distinguished. Mathematical literacy is understood to require higher competence-levels. For example competence-level IV includes the ability to argue in not very well known contexts and to handle linear models of real life situations. Level V additionally requires the ability to handle complex algebraic and functional models and the ability to interprete such models, to solve problems gradually and to explain relations between algebraic formulas and reality. Level VI includes the ability to build complex algebraic models of unknown real life situations, the ability to find complex solutions and to handle algebraic expressions and to generalize. In Germany only every sixth student reached level V[1]. Especially at secondary schools like junior high school (German Realschule) not even 10 % reached this level, and at secondary schools of the lower qualification level (German Hauptschule) not even 1 % were successful (PISA 2003). The missing of "applicable knowledge" is obvious, whereas "formal knowledge" exists in many cases.

Over-formalisation, however is not only a problem of a restricted understanding of mathematics or of missing motivation, but it leads to a wrong and unattractive image of mathematics in society. This calls for mathematical literacy as an important educational aim as well. In general it contains the ability of a person to recognize and understand the role of mathematics

[1] The situation is similar in many other countries.

B.Sriraman, C.Michelsen, A. Beckmann & V. Freiman (Eds). (2008). *Proceedings of the Second International Symposium on Mathematics and its Connections to the Arts and Sciences* (MACAS2). Information Age Publishing, Charlotte: NC , pp.187-196

in the world, the ability for sensible mathematical judgements, and the ability to use mathematics in a way which meets the person as a constructive, active and reflecting citizen (PISA 2003).

2 LEARNING ENVIRONMENTS AND MATHEMATICAL LITERACY

The advancement of mathematical literacy requires a learning environment which does not only consider "formal knowledge", but also "applicable knowledge". It has come to be recognized that the learning of concepts and reasoning should be less formal. A clear differentiation between the idea/matter of a concept and the calculus concerning it is important (Danckwerts & Vogel 2006). Additionally there is a consensus that multidimensional connections to reality are needed in mathematical lessons (e.g. Houston et al. 1997, Henn 1997) to prepare students for exploring new contexts and handling linear and complex models of real life situations. Students should be confronted with many different traits of reality and get the possibility to act in them and thus to tackle even new situations. This includes authentic experiences, verbal and argumentative examination, modelling activities and interpretation of the model (see e.g. also Kaput 1994, Niss et al. 1991).

This leads to the idea of integrating scientific themes and methods into mathematical education. Scientific work is – simply said – the observation and modelling of reality. Progress in science is based on mathematizing and the verbal examination of the model. Scientific discoveries need mathematics. All new technologies are based on mathematical ideas. Scientific applications can be developed and described through mathematics. Apart from this, scientific themes and methods contain more aspects than simple mathematical considerations (see also e.g. Golez 2005, Michelsen 2005, Höfer 2007).

3 MULTIDISCIPLINARY APPROACHES AS A CHANCE FOR MATHEMATICAL LITERACY

An analysis of the mathematical and scientific curriculum shows that they have topics in common. Examples are *circle*, *mapping*, *symmetry*, *dimension units*, *inequations* and *variables*. There are also common methods, such as *acquiring theorems*, *mathematizing*, *calculating*, *finding of functional relations* etc. (Beckmann 2003). The parallel, multi- or interdisciplinary approach can lead to application-oriented and network learning. Students get the chance to learn concepts in unknown contextual settings, in order to become aware of multidimensional and interrelated aspects - important conditions for mathematical and scientific literacy.

Example *Circle*

Usually 7th or 8th grade students (12 to 14 years old) are confronted with the theme *circle* in both Mathematics and Physics. In Mathematics they are aquainted with propedeutic ideas, a systematic course follows in 9th or 10th grade. The mathematical interest focuses on the circle as a set of points, whereas in physics the circle is viewed as a circular path followed by a specific mass. In Physics the theme is motivated by realistic situations like merry-go-rounds on fairs. Observations of the world around us and special phenomena stimulate experiments. Additionally the dynamic aspect of circle-points can be visualised by constructions with compasses or by dynamic geometry systems. These methods are known from mathematical lessons. Physics uses mathematical concepts. To describe the movement of bodies on a circular path it uses for example *radius*, *circumference*, *tangent* and π. In a joint approach these different perspectives are connected to a multidimensional concept:

Mathematics	Physics
Definition: A circle is the set of all points with the same distance r from a given centre M.	Circular motion: Every circular motion is based on a force. Without a force a body would not move or would move rectilinearily with constant velocity. This can be shown by the following experiment: - Hand takes a string with a body at its end. - Hand stimulates rotation. - Hand lets the string go (hammer throwing)
Circle and line: The special lines to a circle are characterized by drawings of possible positions relating to the circle. Tangent: a line which touches the circle in one point.	Investigation of the circular motion: - direction of the motion: tangent - direction and dimension of force: right-angled to the tangent - central force: always acting force which is directed to the centre of the circle and conserves the circular motion
Theorem: A tangent to a circle is at right angles to the radius of the circle at its point of contact.	Observation/Theorem: The force acts right-angled to the direction of motion. In the above experiment: If the hand lets the string go, it flies tangentially away.

4 SCIENTIFIC MATTERS AND METHODS IN MATHEMATICAL LESSONS

Scientific matters and methods do not automatically guarantee successful interdisciplinary mathematical lessons. Rather it can be surmised that the way they are used is decisive for mathematical learning. This will be explained by the following examples of scientific aspects:

Physical theorem

Physical theorems in the form of algebraic terms like $v = \frac{s}{t}$ are the most widely used scientific themes in mathematical lessons. Often they are used like formulas or as material for activities related to equations. The risk is an over-formalisation. On the other hand they are a chance for an application-oriented learning: theorems are the result of the investigation of reality and modelling activities. Their discussion can lead to new conclusions and to predictions. The algebraic term admits an estimation of dependencies and changes between quantities. The way algebraic terms are developed and used in the lessons forms the basis for the advancement of mathematical literacy. For example: interdependence between pressure and height of a liquid:

You could give your students the formula about the dependence of pressure and height of a liquid, that is $p = \rho hg$ (g = constant of gravity), and the task to calculate the pressure p in the deep h of a liquid with a given specific gravity ρ.
Or You could give your students some experimental equipment (e.g. a water jar, a manometer and a ruler), tell them the following story about the barrel of Pascal and let them find a solution:

> The mathematician and physicist Blaise Pascal (1623 – 1662) claimed that he could burst a barrel of wine with only a few glasses of wine[2].

Phenomena

Physics deals with inanimate nature. Starting point for scientific findings is the observation of phenomena in their given or natural surrounding. Aim is their description through theorems and their explanation (answers to how and why). Phenomena are a chance for application-oriented learning and for interdisciplinarity in mathematical lessons: phenomena appear in many different real life situations. Phenomena triggers active discussion and they can lead to various simple or complex models. The risk is not to find the right balance between qualitative exploration and the deepness of mathematizing. Depending on the learning group it can be enough to cognize the functional relation concerning the phenomenon in a merely qualitative way.

Experiment

A central scientific method is the experiment. It serves to check on, verify or falsify a hypothesis and to create new hypotheses. Experiments do not need to be quantitative; they can simply show phenomena, which lead to new explanations and hypotheses. Experimental activities are a big chance for mathematical learning as they make it more action-oriented and add new perspectives to the exploration process. Experiments have a special meaning for the learning of the concept of function. The experimental steps correspond to its different aspects or stages of learning (Beckmann & Litz 2005). A comprehensive understanding of the concept of function can be seen as a major precondition for mathematical literacy (see also Michelsen & Beckmann 2007).

5 SCIENCE AND MATH – CLASSROOM RESEARCH ACTIVITIES IN GERMAN SECONDARY SCHOOLS

5.1 Description of the project

In a project of the University of Schwäbisch Gmünd scientific themes and methods are investigated at school. The project started in 2003 and is mainly concentrated on the learning of the concept of function (Beckmann 2004, Beckmann & Litz 2005, Beckmann 2006 and 2007a).

In this paper the aspect of the learning environment is focussed while it centres on the scientific aspects *phenomena* and *experiment* in mathematical lessons in regard to the development of mathematical literacy. Questions are:

---

[2] Pascal used the interdependence between pressure and height and applied it to real life. (The formula could be found through an experiment before). The pressure increases proportionally with the height of the liquid. A ten times higher height leads to a ten times higher pressure. Pascal took a small long pipe, put it into the barrel, climbed up on the balcony of a house and filled wine into the pipe.

Are phenomena and experiments in mathematical lessons a contribution to "applicable knowledge"?

Do they support

- learning in complex situations?
- an applicable discussion?
- modelling activities?

The students in the project are 14 to 16 years old and attend German secondary schools of lower, middle and higher qualification level: German Hauptschule, German Realschule (like junior high school) and German Gymnasium (that is : Grammar school).

In two, three or four lessons (each 1.5 hours) the students are confronted with worksheets, materials and experiments. They have to tackle the topic on their own and answer the question about the dependence of two interrelated quantities. The students´ activities were documented by video-, audiotape, minutes and worksheets.

The observation of **phenomena** in a given surrounding is the starting point for scientific discoveries and investigations. In the project the students have to discuss a phenomenon depicted by a text and a picture. In some cases the students can use laid out material. On the worksheets the students were asked to work intensively at the phenomenon. Only later they should use the equipment for an experimental investigation. Examples for phenomena given on the worksheets are the following:

Phenomenon I: a rolling ball on a slope: dependence between time and distance in an accelerated motion.
Kind of impulse: text and picture, pupils´own experiences/ imagination

If you could choose to play soccer on a slope or on a plain, surely you would choose the plain ground. Obviously the ball acts differently on a slope and on a plain surface.

Exchange experiences in the group.

Suppose there is only a slope. The ball runs down the hill. You run behind the ball and want to catch it. How quickly do you have to run in comparison to the movement of the ball?

Describe the situation until you catch the ball.

Discuss the situation in the group.

(Simon 2005, p. 134)

Phenomenon II: A closed air-pump: dependence between air volume and pressure.
Kind of impulse: Text and picture, material, authentic experience

On the table you can see an air pump. Extract the lope (pipe).

What is striking?

Talk about it in the group.

Now look at the sketch of the (closed) pump.

Describe more exactly: What happens during the shifting in? What/ Which quantities change?

Discuss it in the group.

air

Lope/ pipe

Phenomenon III: moving away from a source of light: dependence between intensity of light and distance.
Kind of impulse: Text and picture, pupils´ own experiences/ imagination, material, authentic experiences with material (idealising the real life situation), models

Imagine you walk into a very long tunnel. You cannot see the end. How does the intensity of light (brightness) change (if there are no cars around/ no car´s lighting)?

Talk about it in the group.

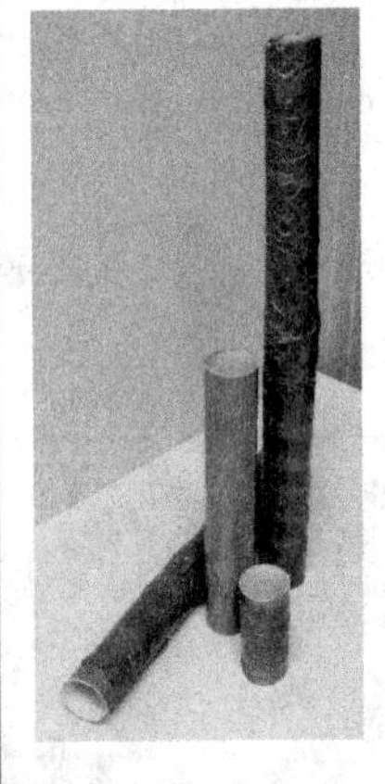

Now take a cardboard tube. Hold it at the window (the window forms the source of light – sunlight). Observe the brightness.

Now take another pipe with another length. Hold it at the window. Observe the brightness. Compare.

Talk about it in the group.

Which diagram describes the situation best? Assume and tick.

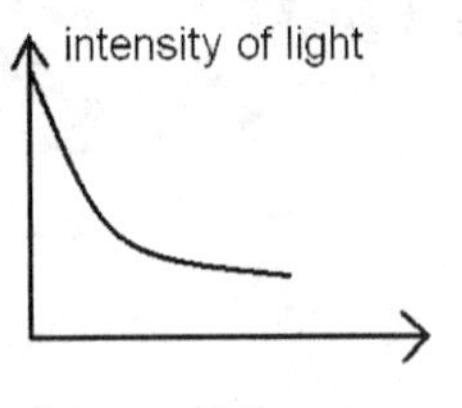

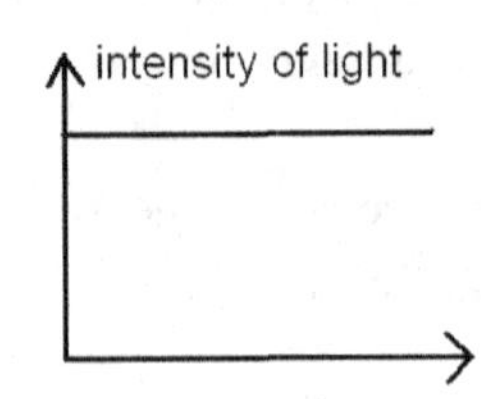

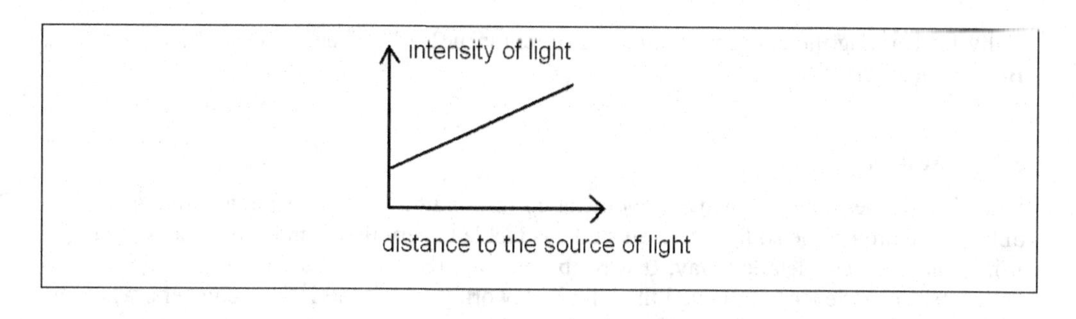

Phenomenon IV: A moving car: dependence between time and distance in a steady motion. Kind of impulse: Text and picture, pupils′ own experiences/ imagination, material, authentic experiences with material (idealising the real life situation), specialisation, question about interdependence/ request for experimental activities

Imagine you sit in this car and the car 1. starts at a traffic light, 2. drives around a corner, 3. moves on a straight highway for a long time.

Describe the different motions of the car.

Talk about it in the group.

Equipment: On the table you can see a car, a tape measure, stopwatches.

Take some minutes to acquaint yourself with the car. Let it go. Which of the above mentioned motions (1st, 2nd, 3rd?) is most similar to the car′s motion?

Discuss it in the group.

Let the car go 20 cm, 40 cm etc. Could you say something about the times the car needs? Which correlation do you assume?

Talk about it in the group.

General task:

Describe the relation between distance and time needed. Check: Does the relation correspond to your above assumption? Describe the special characteristics of the relation.

Phenomena typically stimulate further investigation, mostly through experimental activities. **Experiments** idealise the phenomenon and reduce it to a few flexible quantities. In the project the students have to observe, measure and describe the relation between two quantities given in an experimental setting. Impulses are hints at real-life situations (phenomena like described above). The modelling activities mainly deploy tables and graphs and their verbal interpretation and description. High achievers are asked to look for a term which describes the situation best. The groups are confronted with different kinds of experiments concerning proportional, quadratic, cubic, inversely proportional, and other functional relations. The experiments were developed for the use in mathematical lessons (simple assembly) and especially for learning the concept of function (Beckmann 2004). They are described in detail in Beckmann 2006.

5.2 Results

Scientific themes concern many phenomena corresponding to very different situations of reality. In the project the so far involved students tackled every theme intensively and discussed it in a more or less detailed way. It was obvious that the concern about reality and the students′ own experiences motivated the talk. Noteworthy is that the lower achievers, who normally are rather taciturn, lively talked about the phenomena.

Example (see phenomenon IV above: Describe the different motions of the car, here: starts at a traffic light):

Student S1 moves his hand upwards.

> S1: That rises like this.
> S2: Here it moves.
> S1 starts the experimental car (it moves with constant velocity)
> S2 (protests) It should *start* at a traffic light.
> S1: Here it moves.
> S2: It is getting quicker.
> S2: Or try it on the floor.
> S1 simulates the starting move with the experimental car on the floor. He repeats the movement. The students watch carefully.
> S2: It accelerates.
> S1: Yes.
> S2: Yes, that′s it!

An important result of the project is that the concern to reality leaded to a reasonable discussion: The discussion started on a general level and mostly ended in a concrete talk about the interdependence of the two quantities. The students were stimulated to build a hypothesis and to investigate it in an experiment. Mathematical modelling especially in a graph was used to investigate the hypothesis. It is interesting that the connection between the phenomenon and the model (constructed in the experiment) remained present during the whole work.

Example (extract from minutes[3]):

> The students discuss the relation between the intensity of light and the distance from the source of light. In a first measurement they have obtained two values.
> S1: Normally it is not proportional.
> S2: Eh, then let′s do it; then we′ll know.
> S1 draws a system of co-ordinates. Together the students discuss the division of the axis and plot it. Finally the measurement is filled in.
> S2: That is not proportional.
> S1: Sure?
> S3: Except that the line runs like this.
> S2: That is not proportional, as I said before.
> The students take more tubes and take many measurements, before they verify their assumption.

Example (from a final report of a students′ group):

---

[3] see phenomenon III above

> "The second project – we called 'light and tunnel'. The further the car drives into the tunnel, the darker it becomes. The first tube had a length of 9.7 cm. When we held it at the window the intensity of light was some 36 lux. When we held the tube with 30 cm length at the window, we could only measure 0.1 lux. In an extra graph we could read exactly, how much intensity there is at the beginning of the tunnel and in how far it decreases."

The results of a short final test following the experimental activities confirm that there is an increasing capacity of reasoning ("applicable knowledge"): 8th graders of a class of the German Hauptschule were confronted with diagrams of functional relations between velocity and time:

> The picture shows the movement of a car.
> In which traffic situation is the car?

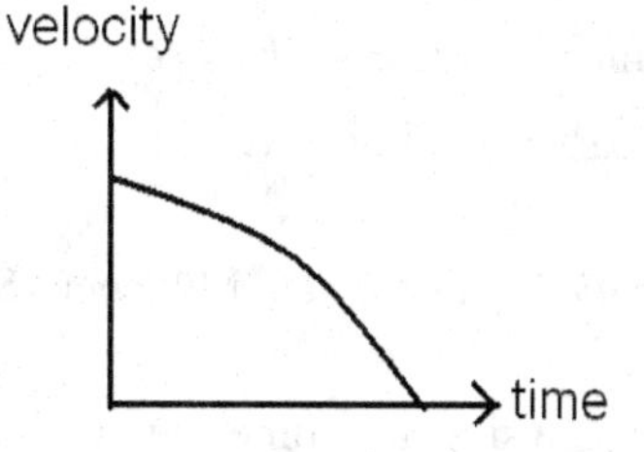

A typical answer of students without experiences with phenomena and experiments is: "The car is rolling downhill." It is typically that the answer describes a static view at the diagram and not a functional relation (Schlöglhofer 2000). However the answers of the final test were: "The car slows down.", "It stops at the crossroads.", "It has to brake because of a barrier." "The car gets into a traffic jam." Here the students regard the functional relation. Remarkable is that they connect the simple mathematical model with an imaginative story of reality and own experiences.

6 FINAL REMARK

In the project special scientific aspects were integrated into mathematical lessons of secondary schools in Germany. The scientific matters and methods, especially *phenomena* and *experiments*, trigger important activities in regard to the development of mathematical literacy. The students were motivated to discuss various real-life situations. The vivid talk of the students and the ability they showed by connecting real situations with mathematical models rais the hope that they are better prepared for mathematizing unknown situations than the average student who took part in the international studies cited above.

Additionally social skills and self-management were furthered. The students learned to work on their own and to co-operate und communicate in a team. A student wrote in his final comment:

> "I liked the experiment "Free fall", because here you do not need to work on your own. You have to solve the task in a team, because one has to drop the ball and the other has to stop the time. This is most difficult here: to work in a group."

Finally the students experienced that mathematical knowledge is more than "formal knowledge". On the question "What did you learn?" a student answered "... that maths was not only calculating."

References

Beckmann, A. (2004). Funktionsbegriffserwerb und Kompetenzerwerb durch Experimente. *Beiträge zum Mathematikunterricht*, Augsburg 2004, Hildesheim, Berlin (Franzbecker) p. 77-80

Beckmann, A., Litz. A. (2005). Learning the concept of function through experimental activities. In: Beckmann, A. Michelsen, C., Sriraman, B. (Hg.): *Proceedings of the first International Symposium of Mathematics and its Connections to the Arts and Sciences.* Hildesheim, Berlin (Franzbecker), p. 215-226

Beckmann, A. (2006). Experiments for learning the concept of function. *Experimente zum Funktionsbegriffserwerb – Ausgearbeitete und erprobte Vorschläge für den Mathematikunterricht der Sekundarstufe I und der frühen Sekundarstufe II.* Köln (Aulis Verlag Deubner)

Beckmann, A. (2007): Funktionsbegriff durch Experimente. *mathematik lehren*

Beckmann, A. (2007a). Non-linear functions in secondary school of lower qualification level, The Montana Mathematics Enthuisiast 14/2

Danckwerts, R. & Vogel, D. (2006). *Analysis verständlich unterrichten*, München (Spektrum Akademischer Verlag/ Elsevier GmbH)

Golez, T. (2005). Calculus between mathematics and physics: real-time measurements – A great opportunity for high-school teachers. In: Beckmann, A. Michelsen, C., Sriraman, B. (Hg.): *Proceedings of the first International Symposium of Mathematics and its Connections to the Arts and Sciences.* Hildesheim, Berlin (Franzbecker), p. 201-214

Henn, H.-W. (1997). Mathematik als Orientierung in einer komplexen Welt. *MU* 43/5, p. 6-13

Höfer, Th. (2007). Mathematik und Physik im Dialog. *MU* 1/2, p. 37-45

Houston, S. K. & Blum, W. & Huntley, L.D. & Neill, N. T. (Hg.) (1997). *Teaching and Learning Mathematical Modelling,* Chichester (Albion)

Kaput, J. (1994). The Representational Roles of Technology in Connecting Mathematics with Authentic Experience. In: Biehler, P. et al. (Hg.). *Didactics of Mathematics as a Scientific Diszipline,* Dordrecht (Kluwer, Academic Publishers)

Michelsen, C. (2005). Expanding the domain. Variables and functions in an interdisciplinary context between mathematics and physics. In: Beckmann, A. Michelsen, C., Sriraman, B. (Hg.): *Proceedings of the first International Symposium of Mathematics and its Connections to the Arts and Sciences.* Hildesheim, Berlin (Franzbecker), p. 201-214

Michelsen, C. & Beckmann, A. (2007). Förderung des Begriffsverständnisses durch Bereichserweiterung – Funktionsbegriffserwerb und Modellbildungsprozesse durch Integration von Mathematik, Physik und Biologie. In MU 1/2 , p. 46-57

Niss, M. & Blum, W. & Huntley, I. (Hg.) (1991). *Teaching of Mathematical Modelling and Applications,* New York u.a. (Horwood)

Schlöglhofer, F. From photo graph to function graph. Vom Foto-Graph zum Funktions-Graph. *Mathematik Lehren.* (Dec 2000) (no.103) p. 16-17.

Simon, S. (2005). *Fußball-Europameisterschaft der Frauen*, England 2005, FF-Magazin,

INTERACTIVE MODELING IN VIRTUAL SPACE: OBJECTS OF CONCRETE ART

Heinz Schumann
University of Education Weingarten/Germany

1. Introduction

The pluralism of styles in 20th century art (see, e.g. Walther 2000) is also reflected in the design of three-dimensional objects and sculptures (see, e.g. Rowell 1986). One of the many styles is the school of Concrete Art, which developed rapidly after World War II, especially in Europe (see, e.g. Lauter 2002) and which is best represented by geometric three-dimensional objects. Concrete Art is characterized by a puristic „geometrism“ (Rotzler 1977). – In short, „concrete“ artists use the universal language of geometry to make their non-explicit conceptions concrete while „abstract“ artists tend to abstract from natural shapes until they have extracted their inner geometric substance. – The three-dimensional objects of Concrete Art are based on fundamental elements of spatial geometry, i.e. prism, pyramid, cylinder, cone, sphere and platonic solids (see, e.g. Guderian 1990), which are also elements of general education in geometry.

In 1949, Max Bill (1908-1994), Swiss sculptor, painter, designer, architect and art theorist and one of the leading protagonists of Concrete Art, published his programmatic thesis “The Mathematical Approach in Contemporary Art”, in which he stated his opinions concerning the relationship between art and mathematics resp. geometry:

“By a mathematical approach to art it is hardly necessary to say I do not mean any fanciful ideas for turning out art by some ingenious system of ready reckoning with the aid of mathematical formulas. So far as composition is concerned every former school of art can be said to have had a more or less mathematical basis. There are also many trends in modern art which rely on the same sort of empirical calculations.... I am convinced it is possible to evolve a new form of art in which the artist's work could be founded to quite a substantial degree on a mathematical line of approach to its content. This proposal has, of course, aroused the most vehement opposition. It is objected that art has nothing to do with mathematics; that mathematics, besides being by its very nature as dry as dust and as unemotional, is a branch of speculative thought and as such in direct antithesis to those emotive values inherent in an aesthetics; and finally that anything approaching ratiocination is repugnant, indeed positively injurious to art, which is purely a matter of feeling.... Now in every picture the basis of its composition is geometry or in other words the means of determining the mutual relationship of its component parts either on plane or in space... It must not be supposed that an art based upon the principles of mathematics, such as I have just adumbrated, is in any sense the same thing as a plastic or pictorial interpretation of the latter. Indeed, it employs virtually none of the resources implicit in the term "Pure Mathematics".

The art in question can, perhaps, best be defined as the building up of significant patterns from the everchanging relations, rhythms and ratios of abstract forms, each one of which, having its own causalty, is tantamount to a law in itself.... Thus the more succinctly a train of thought was expounded and the more comprehensive the unity of its basic idea, the closer it would approximate to the prerequisites of the Mathematical Approach to Art. So the nearer

B.Sriraman, C.Michelsen, A. Beckmann & V. Freiman (Eds). (2008). *Proceedings of the Second International Symposium on Mathematics and its Connections to the Arts and Sciences* (MACAS2). Information Age Publishing, Charlotte: NC , pp.197-208

we can attain to the first cause or primal core of things by these means, the more universal will the scope of art become - more universal, that is, by being free to express itself directly and without ambivalence; and likewise forthright and immediate in its impact on our sensibility. …the orbit of human vision has widened and art has annexed fresh territories which were formerly denied to it. In one of these recently conquered domains the artist is now free to exploit the untapped resources of that vast new field of inspiration I have described with the means our age vouchsafes him and in a spirit proper to its genius. And despite the fact the basis of this Mathematical Approach to Art is in reason, its dynamic content is able to launch us on astral flights which soar into unknown and still uncharted regions of the imagination." (This essay has been reprinted in several countries and languages. (This English version of Max Bill's essay first appeared in 'Arts and Architecture', Los Angles, No. 8, 1954.)

Computer tools for interactive geometric construction in virtual space in the context of synthetic geometry (Schumann 2005a) enable us to model objects created by artists of the Concrete Art school, thus making them accessible in new ways (Schumann 2005b). – In addition the virtual space can be used as media for designing physical arts objects and for exhibit virtual arts objects. Interactive modelling is not the same as modelling by computer-graphic programming, which is more efficient but also more difficult and which will not be discussed here. Modelling always is accompanied by an individual interpretation of the modeled object of art!

On the other hand, interactive tools for interactive construction in virtual space enable us to construct objects according to our own imagination and according to individual design concepts – or else to modifiy generated computer-graphic models in virtual space. A special modification that should be mentioned is the dynamic modelling of static objects of art by animation.

Objects modeled and modified in virtual space are not subject to gravity; further, the tool options give us possibilities that hardly exist in physical space. In short, the results of modelling and/or modification have only a virtual existence. It is hardly possible to achieve the same spatial effect as in physical reality and modelling in virtual space also enables no distinction between small and large object formats. The effect created by the artist's choice of material is neglected when modelling in virtual space, and also the material-dependent technique of creation employed by the artist differs widely from modelling using a computer-graphic tool. Physical replicas made from the original materials used by the artist have quite a different quality.

In our opinion, both aspects of modelling in virtual space are important for a concept of instruction that bridges the gap between spatial geometry and art, far beyond the current concepts of „visual communication" and media education (e.g. Walch 2003, see also Section 3.3 of this contribution).

In the following, computer-graphic modelling and modification will be illustrated by selected examples of some protagonists of Concrete Art in Middle Europe, i.e. objects by Max Bill, Alf Lechner, Gerard Caris und Anton Stankowski.

The tool we employed was Cabri 3D, which was originally developed for interactive construction in virtual space in the classroom. The objects selected all have a polyhedral structure or may be approximated by such a structure as the computer tool is well suited for generating polyhedral structures. Greater plasticity by special surface texturing and shading

was not intended. The modelling and modification procedure both requires and improves spatial perception, knowledge of spatial geometry and skill in handling the computer tool.
The modelling examples are based on pictures (e.g. photographs; in principle, a body scan of the whole object provides more detailed information for modelling which may be lost in a single photograph). In contrast to photographs, models in virtual space can be viewed from all sides and even give us views that are impossible in reality. The loss of tactile perception is negligible – after all, touching is prohibited in exhibitions and art galleries -, but there is of course a loss of aura as compared to the original objects in their original environment.
The nets as polyhedral 2D models generated in virtual space and their modifications can be folded and made available as physical objects by printing and folding.

2. Modelling and Model Modifications (Examples)
2.1 Max Bill (1908-1994)
Max Bill developed the beginnings of an an aesthetic theory of shapes, which he defined as follows: "Shape is that which we meet in three-dimensional reality. Shape is all we can see. But upon hearing or thinking the word, it means more than just something that exists by chance. We always attribute to the concept of shape a certain quality...We know that not every shape is beautiful per se. Beauty always means relative beauty.“ (Bill 1952, S.6). - Max Bill uses sections of the torus, sphere and cube to arrive at aesthetically pleasing motives of halved solids. Of the infinetely many possibilities of cutting a cube in halves by plane sections, he selected those that meet certain conditions (Figs. 1/2): The cutting planes must (necessarily) go through the center point of the cube and should also go only through corners or midpoints of the cube edges (see e.g. Schumann 1995). These conditions are met only by the following sections: Section parallel to a side face (cutting plane: square); section along the diagonal of a side face (cutting plane of ratio $1:\sqrt{2}$); section through two diametrally opposed cube corners and a midpoint of one cube edge (cutting plane: rhombus of side length $\sqrt{5}:2$ and height $\sqrt{30}:5$); section through the midpoints of a union of edges of non-parallel edges (cutting plane: regular hexagon of side length $\sqrt{2}:2$).

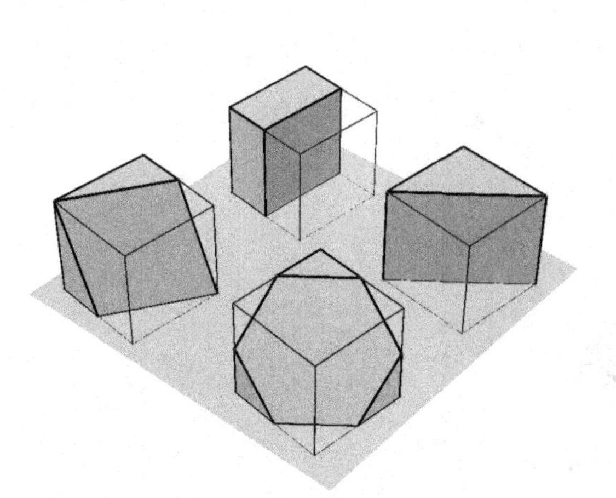

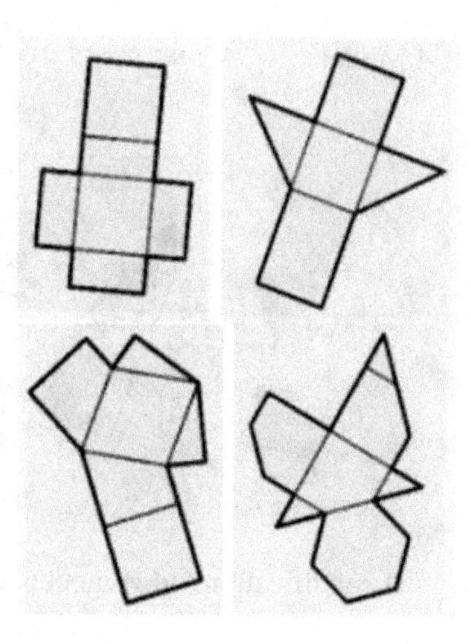

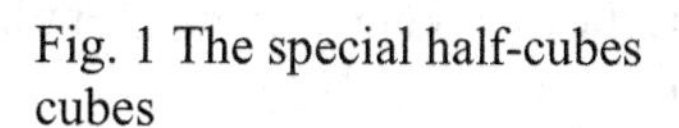

Fig. 1 The special half-cubes

Fig. 2.1-4 Nets of the special half-cubes

For an urban park in the city of Jerusalem, Max Bill assembled the four pairs of half-cubes generated by his special sections in a particular arrangement (Fig. 3, design photo)

Fig. 3 „schtatt e schtadt e schtatt", installation of four half-cubes, 1973-1985

Let us reconstruct a model from this image. Fig. 4 shows the finished model.

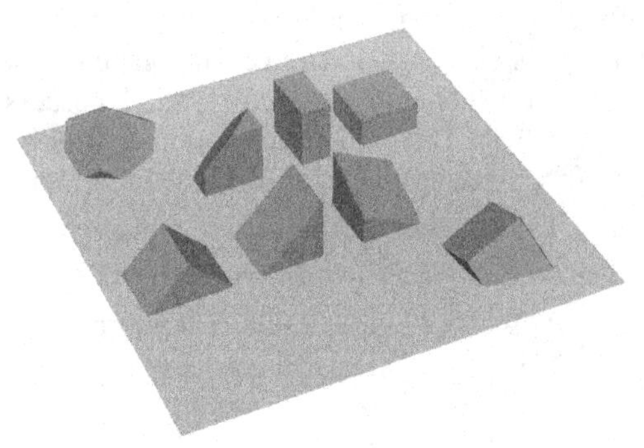

Fig. 4 Model

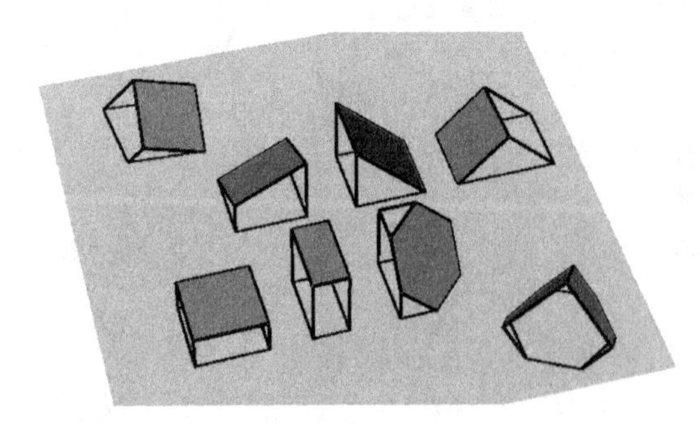

Fig. 5 A modification (edge models with "roofs")

Figure 5 shows an example of models in which some of the faces become "roofs". Further modifications can be achieved by changing the relative positions of the half-cubes.

2.2 Alf Lechner (*1925)

In his own words, Alf Lechner characterizes his design philosophy as follows: "There is so much complexity in simplicity that one cannot be simple enough. Real discoveries can be made only in the very simplest of shapes, as is well known. The more complex a shape, the less of its essence is visible." (cited in Schreiber 2005). Lechner's work focused on simple and usually monumental steel objects, of which he also documented the design and construction process. As an example of his objects derived from an original shape, we are going to model the object presented in Fig. 6. It is interesting to note that Alf Lechner was another artist who focused on cube sections for his steel structures.

Fig. 6 Tilted cube, 1975 (PH Freiburg)

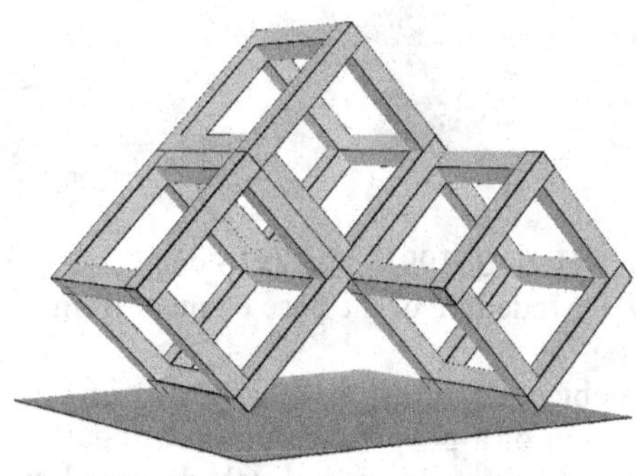

Fig. 7 Model

The model (Fig. 7) can be viewed from all sides. – By a similar modelling process, the user can generate his own cube structures à la Lechner.

2.3 Gerard Caris (*1925)

Apart from his work on two-dimensional pentagonal structures, Dutch artist Gerard Caris also worked on spatial structures made up of regular dodecahedrons (see Guderian and Volkwein 2000). Guderian discussed the regular dodecahedron, the biggest of all platonic bodies, in the history of culture and art (Guderian and Volkwein 2000, S. 12-15). Of the many dodecahedron sculptures by Gerard Caris, we select two which are typical of his work.

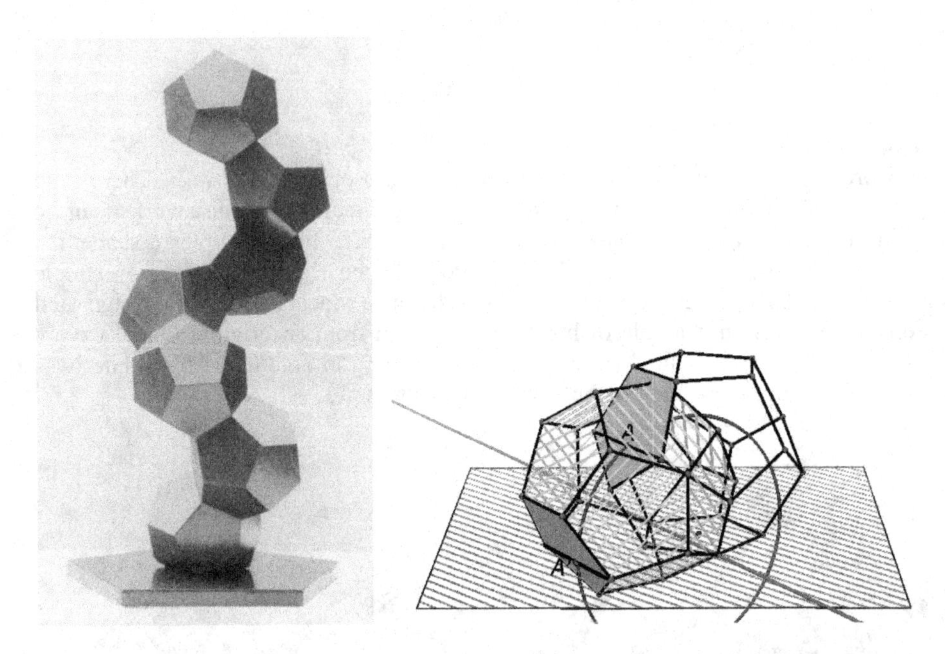

Fig. 8 Helix 3, 1993, 36x13x13 cm
Fig. 9 Modelling of the base element from the regular dodecahedron

The object Helix 3 (Fig. 8) consists of 6 regular dodecahedra spiraling upward from a base in the form of a prism standing on a regular pentagon. First, we are going to model the base element by rotating a dodecahedron and cutting off the projecting part (Fig. 9/10). We are then from below, going to attach the 5-sided prism as base plate (Fig. 11). The printed nets (Fig. 12.1/2) can be used for physical modelling. Finally, we are going to place the dodecahedrons on the base, one by one, as in the picture (Fig. 13); the stability of the model is illustrated by the axis vertical to the base plate through the center of the base prism. In the final step, we are going to make the model look like the original (Fig. 14). We can now view our model from all sides.

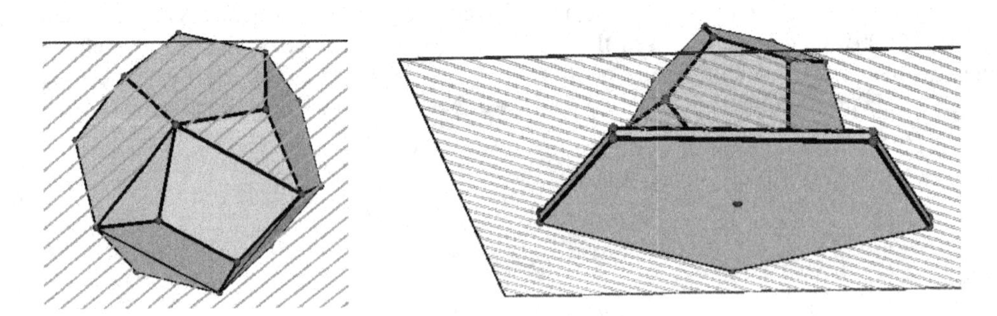

Fig. 10-11 Modelling the base plate

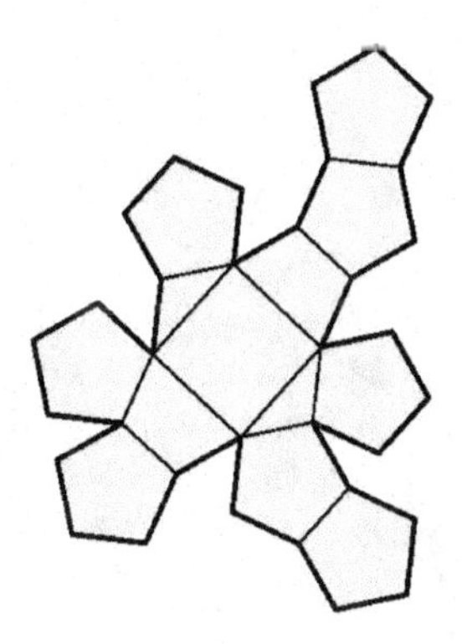

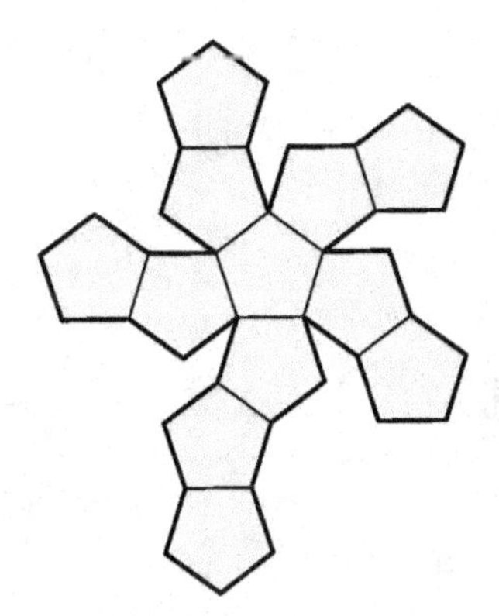

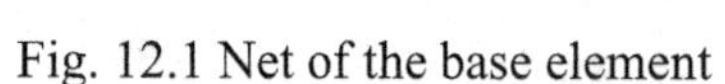

Fig. 12.1 Net of the base element

Fig. 12.2 Net of the regular dodecahedron

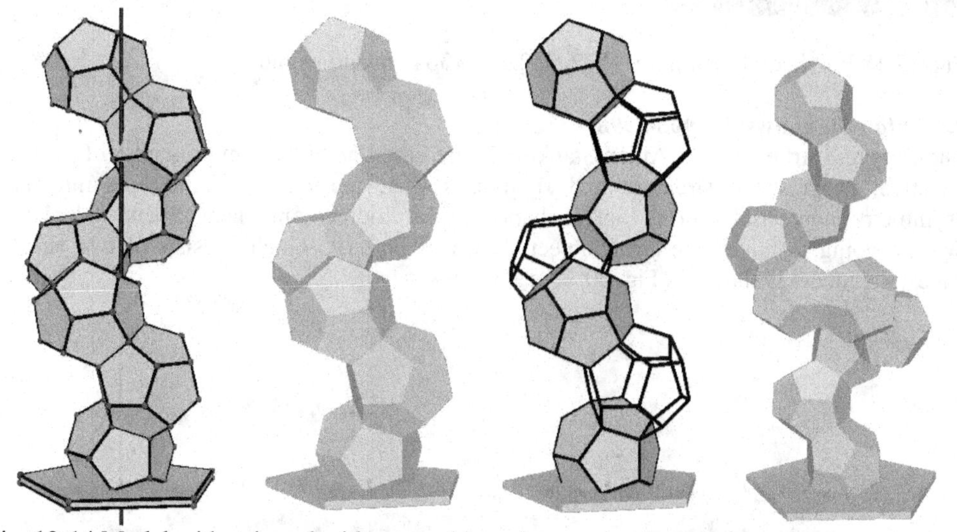

Fig. 13-14 Model with axis and without marking of corners and edges
Fig. 15 Modification 1; Fig. 16 Modification 2 (of 9 dodecahedrons)

A modification may consist of alternating elements with surfaces or edges represented (Fig. 15, example) or of more than six dodecahedra, e.g. nine (Fig. 16), ensuring that the resulting structure is stable. The infinite countable number of possible combinatorial variations of such helical structures gives us much room for free design, and this goes also for the even more varied structures consisting of branched helixes. The designs of helixes consisting of dodecahedra can be applied to other Platonic solids as well, e.g. the icosahedron, or to combinations e.g. of octahedra and icosahedra.

An impressive monumental sculpture by Caris is shown in Fig. 17; it consists of six intermeshing regular dodecahedrons and has a core of congruent cubes each of which is inscribed into a dodecahedron (Model in Fig. 18).

Fig. 17/18 Polyhedral Sculpture, 1977, 332x364x364 cm and its model

2.4 Anton Stankowski (1906-1998)

The Concrete Art objects of Anton Stankowski are based on his design principles of positive-negative, progression, structure and rhythm, scatter, tension and balance, symmetry - asymmetry, non-serial - serial, logical shaping and irrational component (Stankowski 1979). We are going to model the small object shown in Fig. 19, which consists of two sheared cubes; the model is shown in Fig. 20.

Fig. 19 Small object Figur Nr.14, 1998-2005, 13 x 12 x 12 cm, stainless steel - left
Fig. 20 Model

This model can be generalized into a double parallelepipedon of congruent rhombuses; individual bodies are obtained by folding the net in a corner of the body, followed by point mirroring. By parallel displacement of the basic body, one obtains the superposed body which can be moved along the diagonal through the basic body. Of course the modified model may not always be stable in its position.

Finally, we are going to model Stankowski's metamorphosis „Vom Kreuz zum Quadrat" (From cross to square), 1975 (Stankowski 1979, p. 140), which consists of twelve phase images (Fig. 21) from a spatial metamorphosis in a continuous animation, which can't be represented by a print media.

Fig. 21 Phases of spatial animation of the metamorphosis

3. Aspects of Generalization and Classification

3.1 Computer-graphic Modelling of Objects of Constructive Art

Interactive computer-graphic modelling and modification of objects of Constructive Art can be applied also to objects of other concepts of Constructive Art (Rotzler 1977), e.g. minimal art (Schumann 2006b). We are going to show just one example by SoLeWitt (*1928): Fig. 22 Photograph of the original, Fig. 23 model)

Fig. 22 Variation on three different kinds of cubes, 1967
Fig. 23 Model

3.2 Classification of Sculptures with a Mathematical Structure

The three-dimensional objects of Constructive Art, including those of Concrete Art, represent only a very narrow section of the many sculptures with a mathematical structure which according Zalaya & Barrallo 2003 can be classified as

Sculptures with a *geometric shape structure*:

Polyhedra
Curved surfaces (quadrics, rotational surfaces, ruled surfaces, minimal surfaces, non-orientable surfaces etc.)
Topological sculptures
...

or

Sculptures with a geometric „*Operating structure*":

Symmetric structure
Boolean structure
Transformation structure
Modular structure
...

A three-dimensional object of Concrete Art can be defined as belonging to both classification aspects. The objects generated with Cabri 3D and described in this contribution may thus be classified as polyhedral, symmetric, transformative and modular.

3.3 Interdisciplinary Project

In the context of school geometry, the subject of interactive modelling and modification of 3D objects of Concrete Art can be introduced in interdisciplinary projects of art and mathematics/ geometry. The virtual space as a medium for art design and reconstruction, art representation and reception and art action is not yet being aware in practising interdisciplinary projects for geometry and art at school.

4. Literature

Bainville, E., Laborde, J.-M. (2004): Cabri 3D 1.2 und 2.0. (Software). Grenoble: Cabrilog. www.cabri.com

Bill, M.: form – eine bilanz über die formentwicklung um die mitte des XX. jahrhunderts. Basel: Karl Werner, 1952

Guderian, D.; Volkwein, P.: Gerard Caris. Gestalten und Forschen mit dem Pentagon. Ingolstadt: Museum für Konkrete Kunst, 2000

Guderian, D: Mathematik in der Kunst der letzten dreißig Jahre. Ebringen i. Br.: Bannstein-Verlag 1990

Hüttinger, E.: Max Bill. Zürich: ABC Verlag, 1977

Lechner, A.: Maß und Masse. Katalog zur Ausstellung 11.12.1981-7.2.1282. Städtische Galerie Regensburg

Lauter, M. (Hg.): Konkrete Kunst in Europa nach 1945. Ostfildern-Ruit : Hatje Cantz, 2002

Rotzler, W.: Konstruktive Konzepte. Zürich: ABC Verlag, 1977

Rowell, M. (Hg.): Skulptur im 20. Jahrhundert. München: Prestel, 1986

Schreiber, A.: Komplexe Einfachheit. www.artnet.de/Magazine/news/schreiber/ schreiber04-20-05.asp, 2005

Schumann, H.: Körperschnitte. Raumgeometrie interaktiv mit dem Computer. Bonn: Dümmler, 1995

Schumann, H.: Dynamische Raumgeometrie. In: Beiträge zum Mathematikunterricht 2005. Hildesheim: Franzbecker 2005, S. 533-536

Schumann, H.: Interaktives Modellieren im virtuellen Raum. In: LOG IN Heft Nr. 133 (2005), S. 55-61

Schumann, H.: Interaktives Rekonstruieren und Modifiziern von Objekten der konstruktiver Kunst im virtuellen Raum. In: Beiträge zum Computereinsatz in der Schule, 2006 (20. Jg.) Heft 1 Dynamische Raumgeometrie IV, März, S. 77-126

Stankowski, A.: Bildpläne – mit Skizzen, Texten und Bildern zur konkreten Malerei. Stuttgart: Edition Crantz, 1979

Volkwein, P.(Hg.): Museum für Konkrete Kunst Ingolstadt. Heidelberg: Edition Braus,1993

Walch, J: Festum. Medien im fachlichen und überfachlichen Unterricht. Kurseinheit 8.5. Medienverwendung im Fach Kunst. Hagen: FernUniversität, 2003

Walther, I. F. (Hg.): Kunst des 20. Jahrhunderts. Köln: Taschen, 2000

Zalaya, R. & Barrallo, J.: Classification of Mathematical Sculpture. www.mi.sanu.ac.yu/

BUILDING VIRTUAL LEARNING COMMUNITY OF PROBLEM SOLVERS: EXAMPLE OF CASMI COMMUNITY

Viktor Freiman , Nicole Lirette-Pitre, Dominic Manuel
Université de Moncton

1. Introduction

In 2005 during the *MACAS1* conference, we have reported first results of the study of building interdisciplinary online communities of learning using WIKI in mathematics and science didactic courses (Freiman, Lirette-Pitre, 2005). New 2006 results have been discussed during the ICMI colloquium *Espace Mathématique Francophone* in Sherbrooke, Canada (Freiman & Lirette-Pitre, 2006) and at the *Atlantic Educators Conference* in Fredericton, Canada (Lirette-Pitre & Freiman, 2006). Our exploratory study indicates that collaborative virtual spaces of sharing and discussing subject specific and interdisciplinary issues as well as collaborative problem solving may contribute to the creation of new learning opportunities for all groups of participants. In this paper, we are going to discuss our new project **CASMI** (Communauté d'Apprentissages Scientifiques et Mathématiques, www.umoncton.ca/casmi) whose goal is to bring together schoolchildren, their teachers and university students enrolled in initial teacher training in order to promote mathematical and scientific literacy based on solving contextual challenging problems, interactions and discussions and use of technology.

2. Virtual learning communities in mathematics

An unprecedented growth of Web based educational resources allows Klotz (2003) to affirm that in mathematics, as in other disciplines, the Web is expanding our concept of the classroom itself, changing what is learned and how it is learned. This affects student-teacher relationship, and provides access to new types of mathematical activities and resources that can be used by teachers in order to differentiate mathematical challenges so they can meet educational needs of all groups of learners, by learners themselves to they can have access to the learning tools unavailable in the classroom, and also, by other persons who just want to have pleasure doing some mathematics.

Several recent studies report a positive effect of virtual problem-based environments on pupils' motivation toward mathematics. The study of the NRICH project shows that 1) on-line resources are not suitable solely for the most able but have something to offer pupils of nearly all abilities, 2) enrichment is not only an issue of content but a teaching approach that offers opportunities for exploration, discovery and communication, and 3) effective mediation offers a key to unlocking the barriers to engagement and learning (Piggot, 2004):.

Another example of a pedagogically powerful virtual environment has been created within the project Math Forum (mathforum.org). It is built on the idea of interaction between members of a virtual community that interact around the services and resources participants generate together. These interactions provide a basis for participant knowledge building about mathematics, pedagogy, and/or technology. The interactions also contribute to what can be described as a Math Forum culture that encourages collaboration on problem posing and problem solving (Renninger & Shumar, 2002).

B.Sriraman, C.Michelsen, A. Beckmann & V. Freiman (Eds). (2008). *Proceedings of the Second International Symposium on Mathematics and its Connections to the Arts and Sciences* (MACAS2). Information Age Publishing, Charlotte: NC , pp.209-222

According to Pallascio (2003), in a virtual research community such as *Agora de Pythagore* (*http://euler.cyberscol.qc.ca/pythagore/*) the students face learning situations in which they will be actors and also creators of their own knowledge via argumentative discourse. Use of technology as knowledge building and communication tool does affect the traditional classroom setting, in which learning happens in the 'didactical triangle' of teacher-knowledge, student-knowledge, and teacher-student relationships.

Taurisson (2003) describes this change using a metaphor of 'explosion of didactical triangle' which is being replaced by a dynamic and complex structure with numerous interactions that develop new properties, evaluate constantly and auto-regulates the functioning of teaching-learning process. One of the important elements of the named above virtual communities is the active participation of pre-service teachers in problem solving activities and discussions, analyzing student's reasoning and giving a personalized formative feed-back. Upon Bednarz (2004), this phenomenon brings also new option for teacher training in the community of practice.

The Internet site CAMI created in 2000 at the Faculté des sciences de l'éducation at the Université de Moncton, Canada, is another example of this process in which K-12 schoolchildren from the French Canadian New Brunswick community could solve challenging mathematical problems posted online and send their solutions using e-mail and soon after receive a personal feed-back from pre-service teachers.

Our own experience with the CAMI project indicates that combining non routine challenging problems (Sheffield, 1998) and technology supported communication between schoolchildren and university students enrolled in initial teacher training (Renninger & Shumar, 2002) create new opportunities for New Brunswick (Canada) francophone schoolchildren to develop their problem solving and mathematical communication abilities (Vézina & Langlais, 2002, Freiman, Vézina, & Langlais, 2005), and hopefully attract more children to science and mathematics.

3. Some results form our previous studies

Survey and interview data from schoolchildren, teachers and university students collected over the 6 years long CAMI project (Vézina & Langlais, 2002, Freiman, Vézina, & Gandaho, 2005, Freiman & Manuel, 2007) show that many schoolchildren like problems. They say the CAMI problems are good, interesting, amusing and enjoyable. To work with our problems helps them to learn how to solve problems and to improve their math skills. Children also say they like the CAMI computer environment saying that it is a good site, with nice colors; it allows them to use computers and Internet. For some of them, working with CAMI makes math class more interesting. Our findings suggest also that the university students find CAMI beneficial for the development of mathematical communication in young children. The majority of university students see the CAMI project as valuable tool to penetrate into a genuine child's thinking. They seem to be united with the opinion that the project helps them to learn how to understand a child's reasoning, to be open to a variety of strategies and ways of communication that children use and provide a child with a meaningful formative assessment. Therefore, they want the project to remain a part of their math education training.

According to the interviews conducted with eight teachers involved in the New Brunswick Laptop in Schools Initiative (Fournier, and al., 2006), teachers find that the CAMI project motivates schoolchildren in solving complex mathematical problems of varying difficulty levels that meet learning needs of schoolchildren of different ages and abilities. It allows schoolchildren to communicate mathematically, discuss their different strategies applied with their peers and become more autonomous problem solvers. The possibility of getting a personal feed-back from university students is also seen as a positive element because it helps them to improve their communication and reasoning skills in order to make it understandable to another person. The problems allow children to discover new mathematical concepts going beyond the prescribed curriculum, and develop their own strategies (Freiman, Manuel, 2007).

While our data indicate positive impact of the CAMI project on all categories of participants: schoolchildren, university students and teachers, many questions remain unanswered and several pedagogical and technological issues need to be addressed. In the next sections, we will analyse some technological and pedagogical solutions proposed in the process of transforming a mathematical problem solving on-line activity into a multidisciplinary problem-based collaborative virtual community now known as CASMI (Communauté d'Apprentissages Scientifiques et Mathématiques Interactifs, www.umoncton.ca/casmi) which has been established since October 2006.

4. CASMI model: learning at large

Fig 1: Index page of the CASMI community. By selecting from the menu located on the left side, each member can solved the problems posted, consult the analysis of the last problems posted, propose a problem, go in the discussion forum, consult the bank of problems already posted, search for specific problems and consult the site tour. A surprise box is also present where special contest or other things can be posted during the school year.

Using the conception of learning at large as a long term perpetual process with feed-back that emerges from and beyond school contexts, where learning objects are personally defined or adapted to personal needs of learner. This being compared to the "learning at short" which defined by Jonnaert, & Vander Borght (2004)as a classroom based, short term, limited in time, initializing at large learning school curricula driven process. We see CASMI virtual community as learning and teaching resource to construct bridges between the traditional classroom learning and the learning at large.

We extended this vision by including in our learning community not only schoolchildren and teachers but also school administrators and parents as well as curriculum developers, university students and professors, software developers, programmers, management team members, all of them as learners working collaboratively to the well being of the community.

Our model of CASMI community presented in Fig. 2 contains six sections which are all interconnected and interrelated: problem space, communication space, initial teacher training space, software development space, management space and research space.

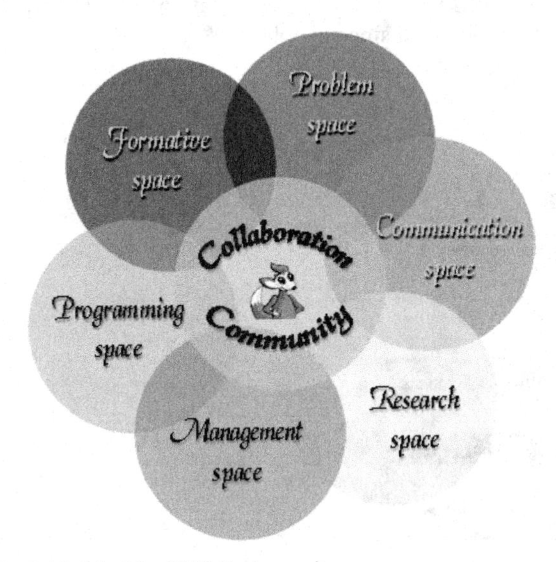

Fig 2: Model of the CASMI community

4.1 Problem space

The problem space represents the key element of CASMI community as community of problem solvers.

We consider problem solving as a key element of life long learning as defined by a model implemented in PISA 2000 and 2003 studies of scientific, mathematical and reading literacy (OCDE, 2000). This dynamic model challenges traditional classroom settings in which students learn to recall and to apply subject specific facts and procedures. It requires the ability of each educated individual to successfully adapt to changing circumstances in complex real world situations. New kinds of knowledge and skills of a cognitive, metacognitive and a motivational nature have to be developed in order to be able to organise and regulate their own learning, to learn independently and in groups, and to overcome difficulties in the learning process, to be aware of their own thinking processes and learning strategies and methods.

Using contextual and open-ended problem situations to asses these abilities, PISA is following modern theories in problem solving such as moving from formal to informal problem context.

Zimmerman, & Campillo (2003) point at shortages of formal and problem contexts which involve well-defined tasks with an exact solution that can be interesting and challenging but would not require such important skills as problem sensitivity and self-definition and thus are less dependent to the development of perceptions of efficacy, outcome expectations, and goal orientation. Following these ideas, we see in CASMI a community of practice which provides each member with an opportunity to solve science, mathematics, chess and, in the near future, reading problems in informal context thorough problem anticipation, regulating open-ended task context, seeking necessary information, development of high levels of behavioural competence, and multiple sources of motivation that sustain long-term recursive solution efforts.

With CASMI problems we aim to propose an adequate challenge to each member of the community disregarding the age, education level or status. We try also to cover variety of topics and contexts. All problems are split into four categories named after animals (manchot – penguin, girafe – giraffe, dauphin – dolphin, hibou – owl. Although each group represent different levels of difficulty (manchot – lower; hibou – highest), we do not put emphasis on it leaving to each member to choose a problem of its interest and encouraging to try all of them.

Mathematical problems are related to four domains of study established by the local New Brunswick French K-12 mathematics curriculum: numbers and operations, algebra, space and shapes, statistics and probability with the respect of four didactical principles: problem situation solving, mathematical communication, mathematical reasoning and making links. Science problems have a more general character and propose everyday life situations that may solicit scientific inquiry and would nurture members' curiosity. Many questions are also related to major curricula topics such as life sciences, chemistry, physics, Earth matters, and space. The chess problems are mostly enigmas that feature particular strategies such as checkmating, winning an important chess material, looking for multiple moves combinations and developing critical judgements about particular position and strategically driven reflective and analytical thinking.

The following shows an example of a mathematic problem that was posted on the CASMI website in March 2007.

Fig 3: Example of a penguin level mathematic problem on the website. By clicking on the left button under the problem, schoolchildren have access to a space where he can answer the problem.

4.2 Communication space

To construct our bank of authentic problems in mathematics, we need to address several issues raised by Nason, Woodruff (2004). They claim that the inability of "textbook" math problems to elicit ongoing discourse during and after problem solving process as well as representational tools' limitations create important obstacles in establishing and maintaining computer supported collaborative learning environments. The authors propose the use of authentic mathematical problems that involve students in the production of mathematical models that can be discussed, critiqued and improved as well as comprehension modelling tools enabling students to represent problems adequately facilitate online student-student and teacher-student hypermedia meditated discourse.

That is why we pay a particular attention to the development of a variety of communication tools that include not only the possibility for each member to send a solution, but also to propose a problem and to participate in the discussion forum. These tools help to build a communication space. Along with a membership, each member has a personal password protected space (portfolio) that keeps all tracks of personal communication (solutions and formative analysis of each solution, as well as proposed problems, personal micro-community (peers, parents, teachers), and personal information that can be modified at all times. Fig 4 is an example of a schoolchildren's portfolio page.

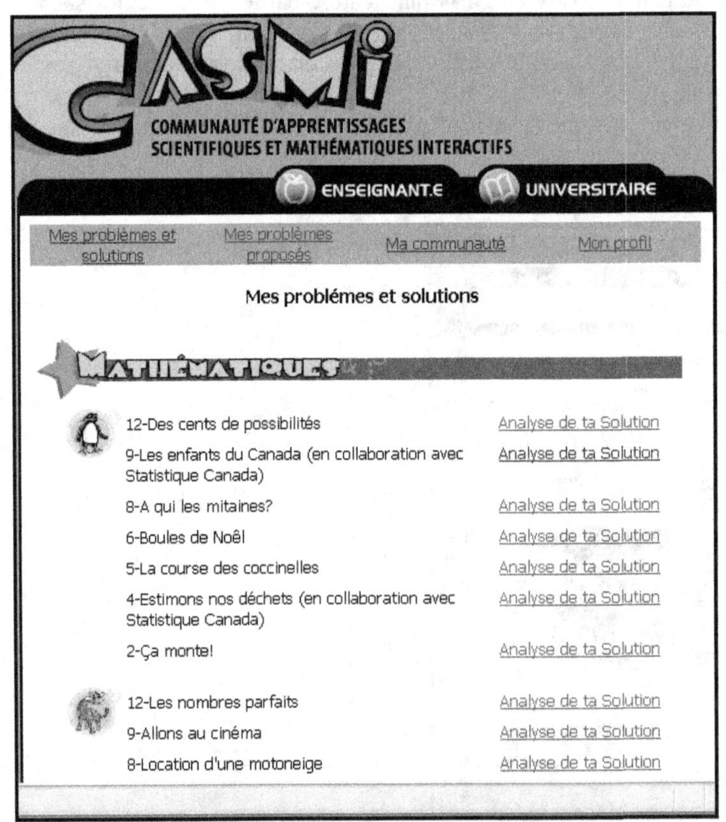

Fig 4: By clicking on the link next to the arrow, all members of the CASMI community have access to their portfolio, which you can see in this capture. Each member can check the problems he or she solved, problems he or she proposed, share his portfolio with other members and change his personal information.

At the same time, the common space available to all members keeps track of collective knowledge: community news, new problems of the week, bank of all problems with the link to the CAMI archive, analysis of the last problems posted with some interesting solutions, discussion forum, links, and surprise box.

Each solution sent via the CASMI electronic response form is analysed by university students enrolled in the mathematics didactic courses and students-experts who are members of our management team. We use common evaluation criteria: problem interpretation and definition, strategy, execution as well as ability to communicate and reflect on the solving process and obtained results. A personal formative feed-back is then put in each member's portfolio. Via micro-community communication tool located in the personal portfolio page link (see Fig 4), teachers can have access to their students' portfolios.

While retroacting on schoolchildren responses, we aim to give a friendly, encouraging and at same time constructive evaluation in order to guide them in their learning process, to improve their problem solving and communication skills and finally, to encourage them to keep up doing science and mathematics challenges.

4.3 Training space

Taking place in all activities within CASMI, university students involved into initial teacher training face many challenges. Among them is the ability to solve mathematical problem which goes beyond direct application of formal rules and operations that characterize application problems. Coming into the mathematics didactic class with different mathematical backgrounds, they find difficult open-ended problems that require interpretation of the problem context and defining it in terms of mathematical models and relationships, building original cognitive and metacogntive tools of problem representations and process of solution. They need also to be able to explain mathematically their strategies and process of solution and making critical judgements about validity of their solutions and applied strategies. However, being able to solve problem meeting our criteria is only the first step in their learning process.

The next step requires an ability to analyze childrens' solutions. Students need to learn to extract from these solutions underlying epistemological, conceptual, didactical and linguistic concepts, anticipate possible interpretation, strategies, difficulties, misconceptions and to think how to apply the general criteria into the concrete problem. After that, they must understand concrete solution and produce a personalized feed-back not only telling children about their success or failure, but rather guide them into the process of learning. In the case of the virtual context, additional obstacles may be related to the fact that there is no direct contact with the author of the solution.

As a course project, the university students are asked to write a reflective analysis of their experience featuring children's work, their own implications, the impact of the project on their further practice in schools, and the utility of such projects in the didactic courses.

In our further plans is the enlarging of the training space by working with in-service teachers helping them to integrate the project into their teaching and also by working directly with children developing auto-regulation skills and helping them to become better problem solvers. A discussion forum is seen as potential rich place for such activities.

4.4 Technological space

The process of transformation of CAMI project into CASMI community required search for innovative technological solutions conforming our pedagogical vision. A new dynamic database structure has been created in order to make CASMI virtual space more viable, interactive and attractive. It contains two interfaces: a user interface that allows schoolchildren, teachers and university students to join the community by filling in an electronic form and choosing a username and a password (see Fig 2), and an administrative interface which allows the management team to coordinate all activities of the website (see Fig 5). The user interface was created by professional web designer using pedagogically approved principals of combinations of menus, pictures, colors, texts and backgrounds (Agosto, 2004; Loranger & Nielson, 2005; Lynch et Horton, 2001; Nielson et Loranger, 2006; Shneiderman & Plaisant, 2005).

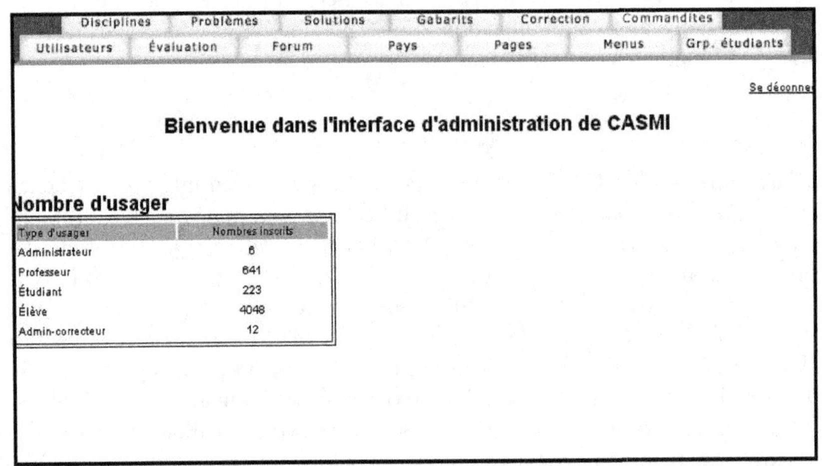

Fig 5: View of the administrative interface of the CASMI community.

In the CAMI environment, the solutions could be sent once without any possible return to the text of solutions to make changes. In the CASMI database, every member can return to work on the problem at any time while the problem is posted on the web site. The last version is always saved. The new solution form gives also several new important features like inserting clipart and drawings, making tables, inserting links, emoticons, and a variety of formatting tools and thus offers more flexibility of personal choice to represent solutions and to express emotions. The following is a view of the form.

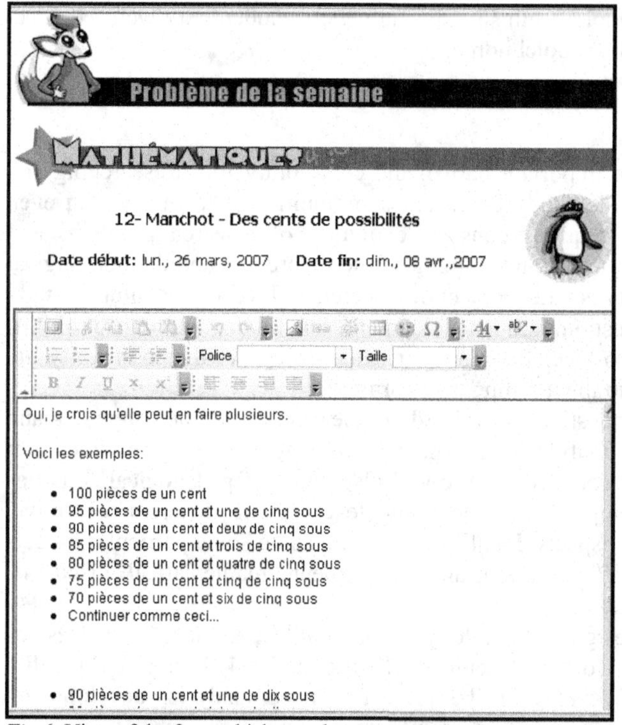

Fig 6: View of the form which members use to solve the posted problems.

The bank of previous problems is now being accompanied by a more powerful and sophisticated search engine allowing search by keywords, subjects, themes, and study domains (like numbers, shapes, etc.).

Unlike with the CAMI project where an important part of communication was organized via e-mail exchange, all communication in the CASMI community goes through database related features that make it more efficient and dynamic. For example, our CAMI survey data showed that only 30% of schoolchildren always received a personal feed-back on their solutions which was mostly due to the erroneous e-mail addresses, e-mail blocking software, unstable network and all other possible reasons (Freiman, & Gandaho, 2005). Now, using each member's personal portfolio, we can guarantee that each feed-back will be received. This portfolio keeps also tracks of all activities. CASMI also offers new options for communication using discussion forum and proposing own problem.

On the administrator's side of the virtual space, CASMI offers a multitude of flexible and dynamic tools for organizing and managing educational activities with options to add, modify, and remove subjects, themes, problems, solutions, analyses, questionnaires, comments, information topics, and membership privileges.

<u>4.5 Management space</u>

The management team is responsible for developing and maintaining all virtual and classroom related pedagogical activities of the CASMI community. In the virtual part, we ensure the preparation and posting of new problems, coordination and realisation of all work related to the evaluation and analysis of the solutions, monitoring and mentoring discussions at the CASMI forum, communication with members (collective and individual), as well as coaching and tutoring on how to use different options. The on-site activities include seminars and workshops with teachers, school administrators, university students as well as direct presentations in the classrooms with schoolchildren.

4.6 Research space

Setting up a research agenda is an important part of the community life. Researching any education setting is not an easy task. Study of virtual communities of learning is an even more complicated enterprise that requires constant evolution of research problems and innovative approaches to the construction and refining of the theoretical and methodological framework. The CASMI community is a new object of research and we are planning to study it on both levels – macro-level: questioning the development, functioning and impact of it as the whole system as well as a micro-level: zooming particular aspects of the community life and featuring special cases related to the teaching and learning.

Our first set of research questions is related to the process of development and implementation of all six spaces that form our model ensuring at the same time their compatibility, effectiveness, and synchronism. We will also study of pedagogical activities provoked by CASMI putting emphasis on learning teaching in different cognitive, metacognitive, social and affective aspects. Finally, we will focus on some particular aspects of CASMI activities such as differentiation and enrichment, mathematical discourse, scientific inquiry setting up case studies.

Using a Research-Based Design methodology (Bell, Hoadley, & Linn, 2004; Design-Based Research Collective, 2003; Collins, Joseph, & Bielaczyc, 2004; Wang et Hannafin, 2005). We'll go through several cycles of the project – need analysis, development, evaluation, and new need analysis and new research questions. Research data will be collected using different quantitative and qualitative tools: questionnaires, interviews with all groups of participants, story telling describing in details what happens and why. We will also set up a statistical monitoring of all activities and participants' online behaviour. A special study will be conducted on the interest related factors and process of learning. Patterns of interactions will be collected from our database, systematized and analysed.

5. Some preliminary observations

To conclude our paper, we will share some preliminary observations made during first six month of project activities. First off all, the CASMI community seems to attract many schoolchildren and teachers since the numbers of inscriptions is constantly increasing approaching 5000 to the beginning of the April, 2007. Among them, more than 4000 are schoolchildren from Grades K-12, more than 450 are teachers, more than 200 are university students and 20 are members of the management team.

Since October, 2006, we have posted 144 new problems of the week, 48 in each subject: math, science, and chess. We also received an important number of solutions which vary from 800 to 1000 by-weekly. Math is the most popular subject which is probably due to the success of the CAMI project but the increasing number of science and chess solutions may indicate growing interest of the community in these subjects.

Each solution is being analysed and commented by a university student – pre-service teacher by means of a CASMI electronic form that has been created and validated by the management team members. This formative feed-back emphasises positive elements regarding problem solving strategies, communication skills and reflective thinking. Also, it aims to give schoolchildren some challenges thus helping them to improve their work.

Our first impression from our students' reports made during didactic courses shows many positive aspects of this kind of communication. Comparatively to the CAMI site, all retroactive work with solutions has become easier. However, pre-service teachers find that the direct contact with schoolchildren is sill missing and electronic solutions are sometimes not clear enough to understand children's reasoning. Some are afraid of not being able to give right comment.

Regarding new sections (comparatively to CAMI) of proposing problems and participations in the discussion forum we are still missing reliable research data. However, we can see that some teachers started to propose their own problems to be posted as a problem of the week and others ask their students to compose a problem via the CASMI electronic form. The number of problems received to date surpasses the capacity of the problem of the week routine and we are searching for innovative solutions of sharing these problems with the whole community. For now, we see our discussion forum as a possible place of this collaborative work.

During the first weeks of CASMI, many schoolchildren actively started to post their messages some of which reflected their emotions, impressions and feelings regarding the new site. But, for the smaller number of member it became the tool of chatting online without any reference to the problem solving activities. In order to maintain our community as a learning place, we decided to close temporarily the forum and to re-open to open it with a moderator who decides what message can be posted and provokes some desirable discussions.

Overall, our first year experience can be seen positively in terms of community building but many pedagogical and technological issues are still to be resolved.

6. References

Agosto, D. E. (2004). Design vs. Content: A study of adolescent girls' website design preferences. International journal of technology and design education, 14, 245-260.

Bednarz, N. (2004). Former les futurs enseignants à la didactique ou par la didactique? Maurice Sachot et Yves Lenoir (dir.) Les enseignants du primaire entre disciplinarité et interdisciplinarité : quelle formation didactique ? Sainte-Foy : Les Presses de l'Université Laval

Bell, P., Hoadley, C., M., & Linn, M. C. (2004). Design-based research in education. In M. C. Linn, E. A. Davis & P. Bell (Eds.), Internet environments for science education. Mahwah, NJ: Lawrence Erlbaum Associates.

Collins, A., Joseph, D., & Bielaczyc, K. (2004). Design research: Theoretical and methodological issues. The Journal of the Learning Sciences, 13(1), 15-42.

Design-Based Research Collective. (2003). Design-based research: An emerging paradigm for educational inquiry. Educational Researcher, 32(1), 5-8.

Freiman, V., Lirette-Pitre, N. (2006). WIKI MATH-SCIENCE : un outil de débats interdidactiques pour la formation initiale des enseignantes et des enseignants au Nouveau-Brunswick. Présentation au Colloque Espace Mathématique Francophone, Sherbrooke, 27.05 – 31.05.2006.

Freiman, V.& Lirette-Pitre, N. (2005). Innovative Approach of Building Connections Between Science and Math Didactics in Pre-Service Teacher Education Using Wiki-Technology. In: Proceedings of The First International Symposium of Mathematics and its

Connections to the Arts and Sciences,19th – 21st May 2005, Schwäbisch Gmünd, Germany, Eds A. Beckmann, C. Michelsen & B. Sriraman, Pädagogische Hochschule, Schwäbisch Gmünd University of Education, Verlag Franzbecker, Hildesheim, Berlin, pp. 162-173.

Freiman, V., & Manuel, D. (2007). Apprentissage des mathématiques. Dans S. Blain, et al. Les effets de l'utilisation de l'ordinateur portatif individuel sur l'apprentissage et l'enseignement. Rapport final. Présenté au MENB par le CRDE et l'équipe de recherche ADOP, Université de Moncton, Mars, 2007.

Freiman, V., Vézina, N.& Gandaho, I. (2005). New Brunswick pre-service teachers communicate with schoolchildren about mathematical problems: CAMI project. Zentralblatt fuer Didaktik der Mathematik, Vol. 37, No.3, pp.178-190.

Freiman, V., Vézina, N.& Langlais, M. (2005). Le Chantier d'Apprentissages Mathématiques Interactifs (CAMI) accompagne la réforme au Nouveau-Brunswick : Mathématique virtuelle à l'attention du primaire, disponible en ligne : http://spip.cslaval.qc.ca/mathvip/rubrique.php3?id_rubrique=18.

Freiman, V., & Gandaho, I. (2005). New Curriculum Reform in Action: New Brunswick's pre-service teachers communicate with schoolchildren on mathematical problems via Internet site CAMI. In Proceedings of World Conference on Educational Multimedia, Hypermedia and Telecommunications 2005 (pp. 2816-2820). Norfolk, VA: AACE.

Fournier, H., Blain, S., Essiembre, C., Freiman, V., Villeneuve, D., Cormier, M., Clavet, P.(2006) Project ADOP: A Conceptual and Methodological Framework for Assessing the Effects of Direct Access to Notebook Computers. Brief paper presented at ED-MEDIA 2006, Orlando, Fl. 26.06 – 30.06.2006

Jonnaert, Ph. et Vander Borght, C. (1999) Créer les conditions d'apprentissage. Un cadre de référence pour la formation didactique des enseignants, Bruxelles : De Boeck-Université

Loranger, H. & Nielson, J. (2005). Teenagers on the web: 60 usability guidelines for creating compelling websites for teens. Fremont, CA: Nielsen Norman group.

Lynch, P. J. & Horton, S. (2001). Web Style Guide (2 ed.). New Haven: Yale University Press.

Nason, R.A . & Woodruff, E. (2004). Online collaborative learning in mathematics: Some necessary innovations. In T. Roberts, (Ed.), Online collaborative learning: Theory and Practice (pps. 103-131). London : Infosci.

Nielson, J. & Loranger, H. (2006). Prioritizing Web Usability. Upper Saddle River, NJ: New Riders Press.

Pallascio, R. (2003). L'agora de Pythagore : une communauté virtuelle philosophique sur les mathématiques. Dans : Taurisson, A., Éd. Pédagogie.net : l'essor de communautés virtuelles d'apprentissage. PUQ.

OECD (2000). Measuring Student Knowledge and Skills : A New Framework for Assessment. http://www.pisa.oecd.org/dataoecd/45/32/33693997.pdf

Renninger, K. Ann, and Wesley Shumar (Eds) (2002). Building Virtual Communities: Learning and Change in Cyberspace. Cambridge, England: Cambridge University Press,.

Sheffield, L., Éd. (1998). Development of mathematically promising students. NCTM.

Shneiderman, B. & Plaisant, C. (2005). Designing the user interface: Strategies for effective human-computer interaction (4 ed.). Boston: Pearson Addison Wesley.

Taurisson, A., Éd. (2003). Pédagogie.net : l'essor de communautés virtuelles d'apprentissage. PUQ, 2003.

Vézina, N. et Langlais, M. (2002). Résolution de problèmes, communication mathématique et TIC : l'expérience du projet CAMI. Nouvelles de l'AEFNB, 33(5), 9-12.

Wang, F. & Hannafin, M. (2005). Design-based research and technology-enhanced learning systems. Educational Technology Research & Development, 53(4), 1042-1629.

Zimmermann, B., Campillo, M. (2003). Motivating Self-regulated Problem Solvers. In: Davidson, J., Sternberg, R. The Psychology of Problem Solving. Cambridge University Press.

VISUALIZATION AND UNDERSTANDING IN MATHEMATICS

Jessica Carter
Department of Mathematics and Computer Science,
University of Southern Denmark

We address understanding in mathematics in relation to visualization. We will use a description of understanding suggested by Ajdukiewicz (as is also explored in Sierpinska,1994), where understanding is described in the following way: ``a person P understands an expression if on hearing it he directs his thoughts to an object other than the word in question" (Ajdukiewicz, quote from Sierpinska 1994, pp. 28-29). We illustrate this description by examples where we take what is understood to be some expression (e.g., a definition or a proof) and the object that the thoughts are directed to consists of (possibly mental) diagrams or pictures. The major example comes from a part of analysis, more precisely from free probability theory, that was introduced by Voiculescu in the 1980's. The example is intended to illustrate that mathematicians also think in terms of pictures and diagrams, but that these often are removed in the final representation of the results. If we grant that the description of understanding given above is accurate, then I suggest that understanding for research mathematicians is not necessarily different from understanding for students. Finally, I will relate the description of understanding to previous work on the nature of mathematical objects.

Introduction

The topics of visualization and explanation in mathematics are currently discussed in the philosophy of mathematics. Although the topic of understanding can be seen as connected both to visualization and explanation, not much has been written about it.[1] This paper considers the topic of understanding in relation to visualization.

I will explore a description of understanding formulated by Ajdukiewicz. This description is also explored in Sierpinska, (1994). Understanding is described in the following way: ``a person P understands an expression if on hearing it he directs his thoughts to an object other than the word in question" (Ajdukiewicz, quote from Sierpinska 1994, pp. 28-29).
Although there may be many ways to understand this statement, we shall here only be interested in when the object that the thoughts are directed to is a (possibly mental) diagram or picture. I will present a number of examples illustrating how I understand this statement. The major example is taken from actual research mathematics and is intended to show that mathematicians also think in terms of pictures. One consequence is that understanding for a mathematician can be described in the same way as understanding for a student.

---

[1] A few authors have addressed these topics, for example, J. Tappenden (2005) and J. Avigad (Understanding proofs. To appear in The Philosophy of Mathematical Practice forthcoming for Oxford University Press). Avigad explains understanding in terms of certain abilities, i.e. describing what you should be able to do in order to claim that you have understood a proof. Tappenden's approach is more in line with this paper, as he makes a case that visualization plays a role in explaining fruitfulness of certain frameworks.

B.Sriraman, C.Michelsen, A. Beckmann & V. Freiman (Eds). (2008). *Proceedings of the Second International Symposium on Mathematics and its Connections to the Arts and Sciences* (MACAS2). Information Age Publishing, Charlotte: NC , pp.223-232

In the last part of the paper, I will connect the description of understanding to a philosophical description of mathematical objects presented in (Carter, 2004).

Visualization, explanation and understanding

Visualization, in terms of using pictures and diagrams in proofs, has played an immense role in mathematics. When talking about proofs relying on diagrams one can not but mention Euclid's Elements. However, during the 19th century, the use of pictures in proofs became discredited. Mathematicians of the time worked on the increase of rigour in mathematical reasoning which meant abandoning the use of intuition and geometric reasoning. One example of this thought at play is Cauchy's Cours d'Analyse from1821. Mancosu (2005) explains how the use of diagrams has again won acceptance because of developments in, for example mathematics, mathematics education and computer science.

With respect to understanding proofs, pictures seem to be very useful. There are even examples of proofs using only pictures. One example is the proof of the equality $1+2+...n = n(n+1)/2$ by drawing a picture consisting of dots representing the numbers 1, 2, 3,... . The sum can be obtained from the area of this figure:

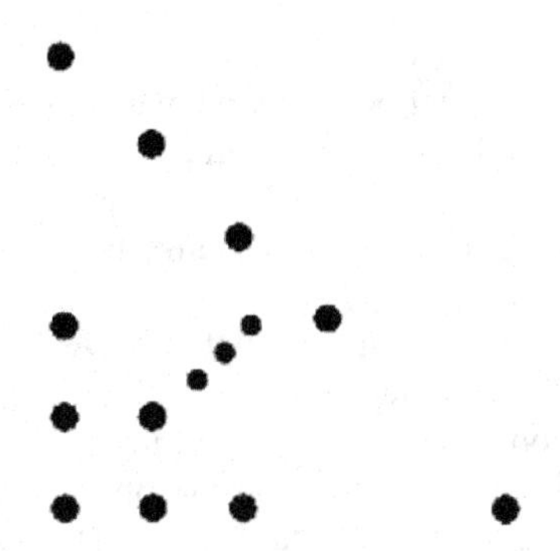

Explanation is, of course, connected to understanding, as explanations are given in order to provide understanding. When discussing explanation in mathematics, it is often referred to a classic paper by Steiner (1978). Steiner also discusses proofs and draws a distinction between proofs that demonstrate and proofs that explain. Steiner's aim is to describe what is meant by an explanatory proof. One of the descriptions that Steiner discusses is that a proof is explanatory if it can be visualized. But he finds that this criterion is too subjective. Instead he settles for the following description: A proof is explanatory if it makes ``reference to a characterizing property of an entity or structure mentioned in the theorem, such that from the proof it is evident that the result depends on this property" (p. 143). Note that it may also be possible to visualize in a picture or diagram how a result depends on properties of certain objects. Consider the diagram of the dots representing the sum $1+2+...n = n(n+1)/2$ above. This diagram shows that the numbers can be arranged so that they represent part of a square, and this gives the desired result.

There are also other descriptions of explanation in mathematics. Sometimes when mathematicians talk about explanations, they refer to facts that are more general than what is to be explained. As an example of this could be the fact that complex analysis explains the behavior of certain things in real analysis. I doubt, however, that this description is very different from Steiner's characterization. By this characterization, a proof of a fact, for example in real analysis, would have to refer to certain properties of the objects embedded in complex analysis.

Turning to the question, whether there is a difference between understanding for mathematicians and students, referring to more general or underlying facts could hardly make things easier to understand for students. Hersh (1993) claims that proofs play different roles for mathematicians and for students. He writes that the role of proofs for a mathematician is convincing, whereas for students it is understanding. This claim seems to be questioned by other mathematicians: ``There is no doubt about it, many results are, in the first place, proved by sheer brute force. ... Subsequently other people, impressed with the result, look at it, try to understand it and finally dress it up in a manner which makes it look appealing, makes it look elegant. ...If you want other people to understand the essential ingredient of an argument, it ought in principle to be simple and elegant" (Atiyah (1974), p. 234).

I wish to insist that understanding can be described in the same way for students and for mathematicians. We now investigate the description of understanding referred to above. We will exclusively consider this description in the case where what is understood is some expression (e.g., a definition or a proof) and the object that the thoughts are directed to consists of (possibly mental) diagrams or pictures. In the next section, we shall give a number of examples illustrating this description.

Note that several authors have claimed that there are different levels of understanding. Connes and Changeux (1995) distinguish 3 levels. The first level consists of mechanical operations, operations that follows a certain procedure and could also be performed by a computer.[2] The second level of understanding involves being able to adapt and criticize a method to a particular problem. This kind of activity results in an understanding of both the aim and the mechanism of the procedure (Connes and Changeux, p. 87). Finally the third level of understanding refers to the creation of new methods or results. I do not find it appropriate to characterize the first level as understanding. For a person to understand something, it requires that the person understands the underlying mechanisms.

Understanding using pictures

A simple example illustrating the act of understanding an expression by directing ones thoughts towards a picture consists of an understanding of the fact that $7 \cdot 8 = 56$. A picture that could illustrate this is a rectangle, where 7 and 8 corresponds to the (length of the) sides and 56 is the area. It could also be pictured as 7 rows of 8 dots. Such a picture could also help understand (visualize) other properties of the product, for example, that $8 \cdot 7 = 7 \cdot 8$. Note that the pictures may provide more information than what is contained in the expression. In the case of the product, we can visualize that multiplication is commutative.

The last example concerns more recent mathematics. My aim in giving this example is to

[2] Interestingely, Connes also places much of what goes on in high-school mathematics, ``tracing graphs of curves and doing kinematic calculations" in this level.

show that mathematicians often think in terms of pictures, but that these pictures are removed when the results are published. The pictures or diagrams are then replaced with formal expressions and algebraic reasoning. Thus it may still be the case that pictures are not accepted as proofs. But the pictures are closer to what the mathematicians think about.

The example we will consider concerns part of analysis, more precisely free probability theory. This part of mathematics was introduced by Voiculescu during the 1980'ies. The aim of this theory is to translate notions from classical probability theory into notions from operator algebra, hoping to be able to solve problems in operator algebra using techniques from probability theory. So far a number of techniques have been developed proving different results.

The paper that we will discuss is titled `Random Matrices and K-theory for exact C^*-algebras'. It is written by U. Haagerup and S. Thorbjørnsen and published in 1999.

The main part of the paper concerns finding a lower and an upper bound of the spectre for S^*S given by the expression:

$$S^*S = (\sum_{i=1}^{r} a_i \otimes Y_i^n)^* (\sum_{i=1}^{r} a_i \otimes Y_i^n),$$

where the a_i's are exact C^*-algebras and Y_i^n are $n \times n$ complex Gaussian Random matrices (definition follows shortly).

To obtain this result, the paper first finds an expression for the expectation of the trace of the product of $m \times n$ Gaussian Random matrices, B_i, $B_{\pi(i)}$, i.e., an expression for

$$E \circ Tr_n \left[B_1^* B_{\pi(1)} B_2^* B_{\pi(2)} ... B_p^* B_{\pi(p)}\right],$$

where E denotes expectation and Tr_n denotes the trace, i.e., the sum of the diagonal entries of the given matrix.

We shall consider certain aspects of this part of the paper. First we introduce the necessary definitions. The most important concept is that of a Gaussian Random matrix. Let there be given a probability space, (Ω, F, P). A complex $m \times n$ Gaussian Random matrix with variance σ^2, (GRM(m, n, σ^2)) is an $m \times n$ matrix with entries that are complex random variables defined on the probability space. The real and imaginary parts of the entries form a family of 2mn independent random variables with distribution $N(0, \frac{\sigma^2}{2})$. The matrix B^* denotes the transpose of the matrix B with entries replaced with complex conjugates.

In the following let $B_1, B_2, ... B_p$ be a set of independent Gaussian Random matrices with variance 1. Suppose that there is given a permutation π in S_p, pairing off matrices two and two. We are interested in an expression for $E \circ Tr_n \left[B_1^* B_{\pi(1)} B_2^* B_{\pi(2)} ... B_p^* B_{\pi(p)}\right]$.

Instead of writing the product in the form $\left[B_1^* B_{\pi(1)} B_2^* B_{\pi(2)} ... B_p^* B_{\pi(p)}\right]$ we could write it

as $C_{1*}C_2C_3^*...C_{2p-1}^*C_{2p}$. This entails that $C_{2i-1} = B_i$ and $C_{2i} = B_{2\pi(i)}$. Calculating to find which C_i's that correspond to each other in terms of the permutation π, one finds the following:
$C_{2i-1} = B_i = B_{\pi(\pi^{-1}(i))} = C_{2\pi^{-1}(i)}$ and $C_{2i} = B_{\pi(i)} = C_{2\pi(i)-1}$ We may thus define a permutation π^* on the indices of the C_i's, in the following way:

$$\pi^*(2i-1) = 2\pi^{-1}(i) \qquad \text{for } i = 1,2,...,p$$

$$\pi^*(2i) = 2\pi(i) - 1 \qquad \text{for } i = 1,2,...,p$$

Below are some diagrams illustrating how one obtains π^* from π. In the first example, π is the permutation (12)(34) in S_4, i.e., it is the case that $B_{\pi(1)} = B_2$ and $B_{\pi(3)} = B_4$.
This entails that $C_1 = C_4, C_2 = C_3$ and so on:

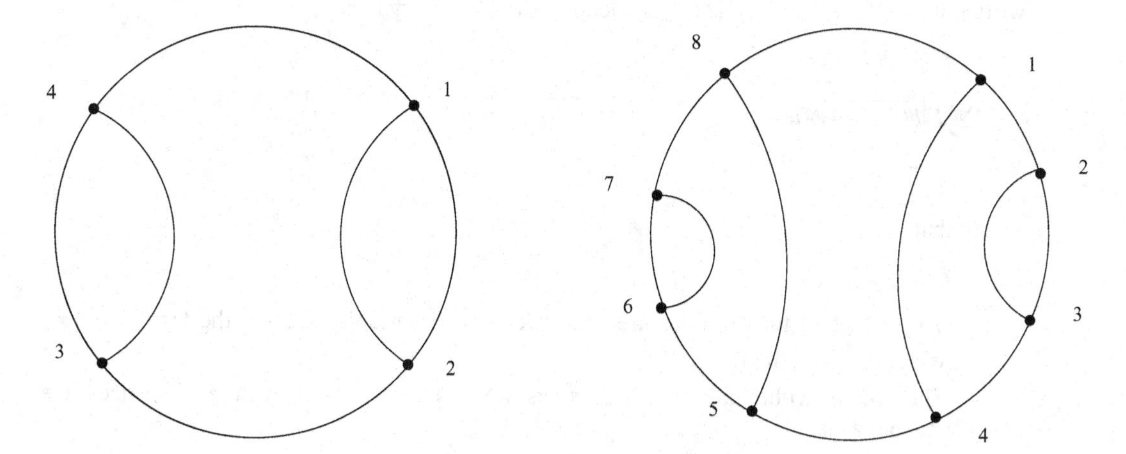

Picture 1. π to the left is the permutation (12)(34). π^* is pictured on the right.}

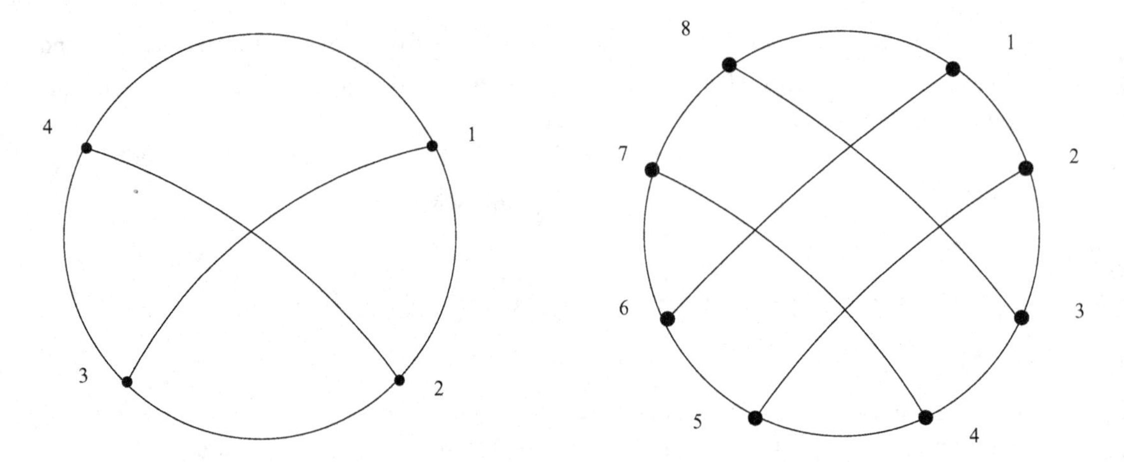

Picture 2. π to the left is the permutation (13)(24). π^* is pictured on the right.

By standard matrix calculations it is possible to show that the expression above can be written in the following form. $b(i,j,k)$ denotes the (i,j) entry of B_k :

$$\sum_{\substack{1\le u_1,u_3,\ldots,u_{2p-1}\le n \\ 1\le u_2,u_4,\ldots,u_{2p}\le m}} E\left[\overline{b(u_2,u_1,1)}b(u_2,u_3,\pi(1))\cdot \ldots \cdot \overline{b(u_{2p},u_{2p-1},p)}b(u_{2p},u_1,\pi(p))\right]$$

Recall that

- The entries of the matrices are complex random variables, i.e., of the form $f+i\cdot g$, where $f,g\sim N(0,1)$.
- The entries within a single matrix, as well as entries from different matrices are independent.

It is also the case that the expectation of products of independent random variables will be zero. However, if we have products of an entry with its complex conjugate, it will be different from zero. When the variance is 1, this product will also have value 1. From the above sum, it is thus possible to state the conditions for which terms of the sum are different from zero, namely when the terms in the product are pair wise conjugate, i.e.:

$$b(u_{2i},u_{2i+1},\pi(i)) = b(u_{2\pi(i)},u_{2\pi(i)-1},\pi(i)) \qquad \text{for } i\in\{1,2,\ldots p\}.$$

We thus have that $u_{2i}=u_{2\pi(i)}$ and $u_{2i+1}=u_{2\pi(i)-1}$. Working a bit on these equations, using the definition of π^* we obtain an equivalence relation on the set $\{1,2,\ldots,2p\}$. This equivalence relation is generated by the relation $\sim_{\pi^*}$ given by $j\sim_{\pi^*}\pi^*(j)+1$.

It is also possible to picture equivalence classes by diagrams:

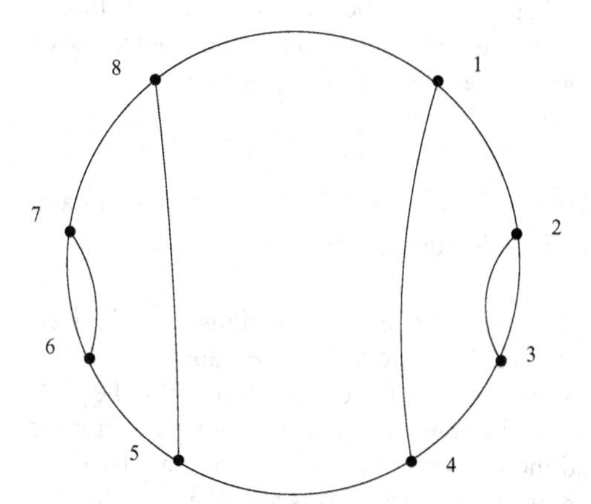

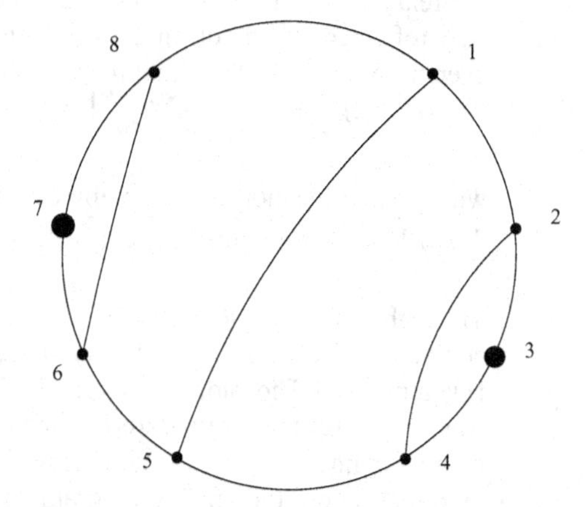

π^* from picture 1 is pictured on the left. Equivalence classes contain numbers that are joined by lines in the right diagram. In this case there are 5 equivalence classes.

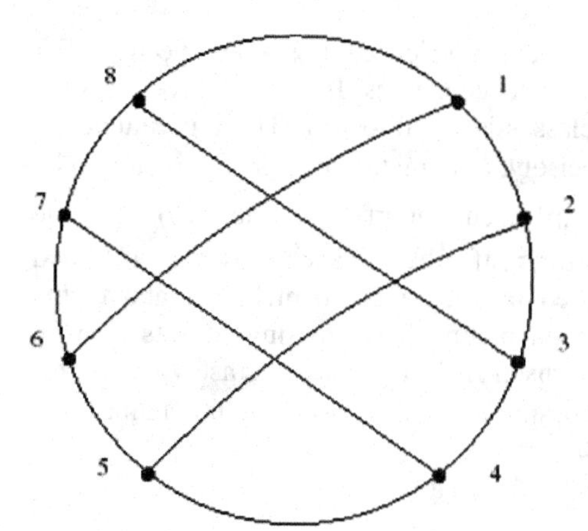

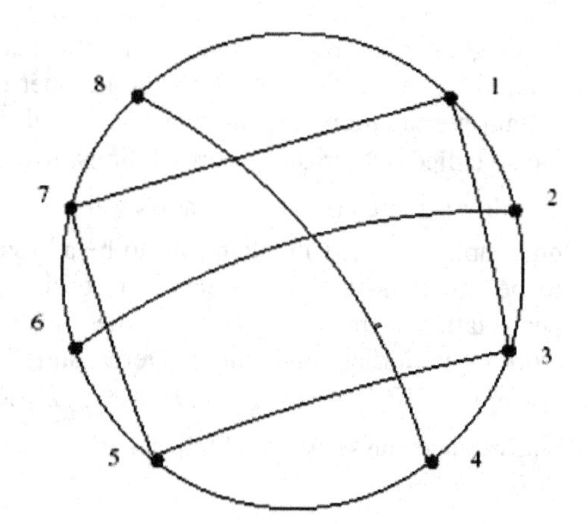

π^* from picture 2 is pictured on the left. Equivalence classes contain numbers that are joined by lines in the right diagram. In this case there are 3 equivalence classes.

Note that the equivalence classes of this equivalence relation will either entirely consist of even numbers or entirely of odd numbers. (π^* takes odd numbers into even numbers and even numbers into odd numbers. When taking the equivalence relation, one is added to the result).

We are now ready to express the value of the expression above. Note that the total number of terms will be $m^p \cdot n^p$ (there are m (n) choices for each even (odd) number up to 2p). But as we noted, some of these will be zero. We can express, in terms of the indices, the conditions for terms being equal to 1. These indices will be related by the equivalence relation that we presented above. That is, if we choose a value for, for example, u_1 then all u_i's where i is related to 1, are also determined. Thus we have to count the number of equivalence classes of the relation. In the expression, it is distinguished between odd and even indices. But this is also reflected when forming equivalence classes as classes either contain odd numbers or even numbers. It follows that the value of the expression above will be equal to

$$E \circ Tr_n\left[B_1^* B_{\pi(1)} B_2^* B_{\pi(2)} ... B_p^* B_{\pi(p)}\right] = m^{k(\pi^*)} \cdot n^{l(\pi^*)},$$

where $k(\pi^*)$ denotes the number of equivalence classes with even numbers and $l(\pi^*)$ denotes the number of equivalence classes with odd numbers.

Through out the presentation of this case study, I have illustrated permutations as well as an equivalence relation by diagrams. These diagrams cannot be found in the paper. But when Haagerup and Thorbjørnsen worked on these results, they first drew diagrams like these. It was only later that they were replaced by formal definitions. Returning to our definition of understanding, we could state that we understand the expression of, say the permutation π^*, by directing our thoughts to diagrams like the one presented earlier. It is also the case that the diagrams can be used when reasoning about the permutations (and as the permutations are used in expressing the expectation of the trace of the product of the B_i's, one may also claim that the diagrams can be used to obtain facts about this expression.)

In fact, later in the paper, certain properties of the equivalence classes are needed. For example, one property concerns the number of equivalence classes. It can be proven that for certain permutations the number of equivalence classes is equal to p+1. These permutations are so called non crossing permutations. More precisely, a permutation π in S_p is said to be crossing if there are numbers a<b<c<d in {1, 2, ...,p} such that $\pi(a) = c$ and $\pi(b) = d$. The quadruple (a, b, c, d) is then said to be a crossing for π. If π has no such crossings, it is said to be non-crossing. This concept is easily pictured by diagrams. In picture 1 above, the permutation is non-crossing, and the permutation in picture 2 is crossing. It was also by working with diagrams that Haagerup and Thorbjørnsen first convinced themselves that, for example: ``If a permutation $\pi^* \in S_{2p}$ is non-crossing it is the case that the number of equivalence classes is equal to p+1.

Understanding and the nature of mathematical objects

Finally, I wish to show how the description of understanding presented here connects to my previous work on characterizing mathematical objects.

In `Ontology and Mathematical Practice' (2004), I present a position on the nature of mathematical objects based on a case study in modern mathematics. According to this position mathematical objects are introduced or constructed by mathematicians. Claiming that mathematical objects are human constructions raises the question whether there is any difference between mathematical objects and other objects that are invented by human beings, such as fictional characters. I claim that there is a difference. One difference lies in the fact that for mathematical objects it is possible to distinguish between two acts of presenting a given object.[3] Take as an example, Mount Blanc. It is possible to describe this mountain by the expression `the highest mountain in Europe'. It is also possible to travel to this mountain and point to it. Thus there is a difference between describing the object and exhibiting the object. This is not possible for fictional characters.

[3] This distinction is taken from Panza (2003).

For some mathematical objects, it is possible to exhibit the object by a picture. Consider, for example, a circle. It can be defined as a curve fulfilling the equation $x^2 + y^2 = 1$, but it can also be drawn. Obviously, it is not always the case that a mathematical object can be visualized in such a way. However, I argue in the paper that it is still possible to distinguish between two different acts of presenting an object.

In the case study, we have discussed above, there are obvious exhibitions of the permutations and equivalence classes. Thus, one may state that in the case where there is a visual exhibition of a mathematical object, then one can understand the definition of this object by directing ones thoughts to the exhibition. In this way, it is possible to link the description of what it is to be a mathematical object to what it takes to understand a definition of the object.

References:

Carter, J. (2004): Ontology and Mathematical Practice. *Philosophia Mathematica 12*, 244-267.
Changeux, J.-P. and Connes, A. (1995): *Conversations on Mind, Matter, and Mathematics*, Princeton: Princeton University Press.
Haagerup, U. and Thorbjørnsen, S. (1999): Random Matrices and K-theory for exact C*-algebras. *Documenta Math. 4*, 341-450.
Hersh, R. (1993): Proving is convincing and explaining. *Educational Studies in Mathematics 24*, 389-399.
Mancosu, P. (2005): Mathematical reasoning and visualization (Visualization in logic and mathematics. In: Mancosu, Jørgensen and Pedersen (Eds.), *Visualization, explanation and reasoning styles in mathematics* (pp. 13-30). Dordrecht: Springer.
Panza, M (2003): Mathematical proofs. *Synthese 134*, 119-158.
Sierpinska, A. (1994): *Understanding in Mathematics*. London: Falmer Press.
Steiner, M. (1978): Mathematical explanation. *Philosophical Studies 34*, 135-151.
Tappenden, J. (2005): Proof Style and understanding in mathematics I: Visualization, unification and axiom of choice. In: Mancosu, Jørgensen and Pedersen (Eds.), *Visualization, explanation and reasoning styles in mathematics* (pp. 147-207). Dordrecht: Springer.

MATHEMATICAL CONSTRUCTS IN THE PHYSICAL REALITY

Gesche Pospiech
Chair of Didactics of Physics, TU Dresden

In mathematics concepts are being developed that serve as a tool or as a structuring element for applications be in mathematics itself be in other fields of knowledge. The main point is the mathematical modelling which in the core means a mapping between the phenomena of the real world and the mathematical world. The construction of this mapping is not easy - especially for beginners, since it requires several steps of abstraction or idealisation and the change between different representations of the same problem. Furthermore, equal-looking symbols often are connotated with different meanings. All these aspects constitute an additional difficulty for beginners besides the subject-specific knowledge. In this paper these will be discussed and proposals for dealing with them will be made.

INTRODUCTION

> *"As far as the laws of mathematics refer to reality, they are not certain; and as far as they are certain, they do not refer to reality."* (A. Einstein)

Mathematics and physics have a widely acknowledged relationship with each other. The application of mathematics to physics allowed for the step from philosophy of nature to modern science. Since the times of Newton and Galilei the importance of mathematics and its concepts for physics is undoubted.

> *"The miracle of the appropriateness of the language of mathematics for the formulation of the laws of physics is a wonderful gift which we neither understand nor deserve."* (E. Wigner)

Nowadays the whole of science and even beyond, e.g. to economical questions, cannot be thought of without mathematical modeling: it delivers predictions or provides theoretical models for explaining. The basic aspects of mathematical modeling in complex phenomena gain importance, especially as the computer power increases and more and more problems can be treated numerically in ever increasing precision. The role of mathematics, however, lies deeper than the numerics:

> "*Mathematical concepts turn up in entirely unexpected connections. Moreover, they often permit an unexpectedly close and accurate description of the phenomena in these connections. Secondly, just because of this circumstance, and because we do not understand the reasons of their usefulness, we cannot know whether a theory formulated in terms of mathematical concepts is uniquely appropriate.*" (E. Wigner)

Physics is an empirical science, heavily relying on experiments. But even the planning and the interpretation of experiments needs - qualitative or quantitative - predictions, which only can be given in the framework of a theory which in its turn needs mathematical structures. So the role of mathematics goes far beyond the evaluation of experiments and the use of mathematical tools. Therefore, today the cooperation between experimental and theoretical

B.Sriraman, C.Michelsen, A. Beckmann & V. Freiman (Eds). (2008). *Proceedings of the Second International Symposium on Mathematics and its Connections to the Arts and Sciences* (MACAS2). Information Age Publishing, Charlotte: NC , pp.233-240

physicists is more needed than ever before. Mathematical models in physics have to be constructed, evaluated and tested.

MATHEMATICS IN PHYSICS EDUCATION

The goal of physics education at school consists not only in presenting facts, laws and the scientific methodology but also learning about nature, about doing physics, the nature of science and the insight into the range of physics in everyday life. Scientific literacy is the key word for these goals. Because of the most important role of mathematics in physics the students should gain the following two insights:

1. *Mathematics is a valuable tool for physics.* The physicist in general does not care much about precise proofs or the mathematical structure as such. But computations are needed to make predictions, to determine values of physical quantities and eventually to verify whether the assumptions of an explanation might be correct.
2. *Mathematics provides the underlying structure of a theory.* As seen for instance with the Noether Theorems or the principle of least action mathematical results can be of immense relevance for the fundamentals of physics. Mathematical concepts reveal an inner structure and give hints for possible phenomena. Analogies between different fields can be derived or lead to new discoveries.

It may not be forgotten that the mathematical tools need interpretation in terms of physics: A famous example is the Lorentz-Transformation of special relativity: its mathematical shape was known as the transformation between to relatively moving electromagnetic systems. The interpretation however, which led to the astonishing consequences in time and space was only found by Einstein a few years later.

The same mathematical structure may apply to several distinct physical phenomena: e.g. the equation for the electric potential, the equation of heat transport or diffusion have the same mathematical shape. Hence, the semantics of equal looking structures may be very different. (s. Table 1)

Mathematics	***Physics***
Numbers	Numbers with units
Fraction	Relation
Function in an abstract sense	Functional dependencies
Geometrical objects	Symbolic representations of physical units
Differentiation	Rate of change
Integration	Integration

Table 1: Differences between the semantics of similar constructs in mathematics and physics

This leads to differences in language between mathematics and physics lessons which make the transfer of knowledge or tools quite difficult. E.g. in mathematics functions are treated from a different view point and have special writings: f: x ---> f(x), there is only f(x) and g(y) whereas in physics we have s(t) or (t) or F(s) with their special meaning stemming from conventions.

This central difficulty might be the reason that most students have their problems with the mathematical side of physics: they dislike it and they lack experience in applying it, even if they attribute mathematics an important role not only in physics lessons but also in physics. So they often learn the formulas by rote without really understanding their meaning. The problems students have consist mainly in translating between the physical objects with their relations and the mathematical formalism. They have to switch between different representations of the same problem, (Friege 2001). Nevertheless, one of the most important methods in physics – apart from experimenting – is mathematical modelling requiring just this ability. Most curricula expect at least that the students at the end of their school career are able to model physics phenomena. However, this seems to be a crucial difficulty and hence needs special attention in design of physics courses from a long term perspective. To design appropriate problems requires an deeper analysis.

The development of the relationship between mathematics and physics may comprise the following steps and is strongly connected to the scientific method:

1. Physical phenomena are observed (surely on the basis of hypotheses or with a theory in mind) and they are analysed in a precise experiment or observations.
2. They are described: What happens?
3. An explanation is thought: Why does this happen?
4. What could be the relevant quantities and which relations do they have: the more .... the less, or vice versa ?
5. The influences of different parameters have to be separated and analysed, either experimental or theoretical. Idealisation takes place and gives numerical predictions.
6. These predictions have to be tested and may lead to further research.
7. The model has to be analysed and interpreted in all its derivations: Are all derivations consistent with other experiments or observations? Are further experiments necessary?

We concentrate now on the connection between mathematics and physics at school.

Establishing the connection between mathematics and physics

First of all the students have to clarify for themselves the structure of the discussed topic, be it for instance electricity, mechanics or thermodynamics. Among the different possibilities are: writing explanatory texts, verbal explanation and drawing a concept map (e.g. Darmofal 2002). All these techniques imply additional representations of the physical structure, which means that the students build a mental model before going on to the mathematisation. On the other hand, the teacher has to be aware of the mathematical abilities of his students on a technical and conceptual level. It might be helpful to show the functioning of a mathematical tool in a non physics context, possibly known to the students. Then in a third step the connections between mathematical formulas and physical structure have to be constructed by the students. (For a simple example from biology see Köppl, 2006) This proves to be the critical step for bridging the gap between the "world of objects" and the "world of formulas" (Gerstberger 1999) and needs sufficient time which is often not given (Monk 1994).

The process of modelling is mirrored in steps 2 through 4 from the preceding section. The steps form the ladder from daily life to the abstraction in mathematical formulation. The goal is the understanding of the process leading to a formula. Furthermore, also step 6 is essential for anchoring the meaning of a formula, but too often is neglected in lessons. Students, therefore, tend to simple learn the characters of a formula, hardly remember their meaning and – not interpreting it in their own words – they may just interchange the different chracters: the formula is useless and meaningless to them. Cultivating verbal discussion or writing and then viewing mathematics as kind of codification of the insight which has been gained in prior analysis.

Mathematical tools

In order that the students recognize the meaning of formulas, a basic understanding of certain mathematical concepts is substantial:

- *Functions*: They are most important because they are used for describing dependencies between physical quantities. Examples for time dependent functions are the accelerated motion or the radioactive decay.
- *Integration*: This tool is – at least in its most simple form – used for modelling the accumulation of physical quantities along a path.
- *Differentiation*: One of the most important features in modelling consists in describing the rate of change of a physical quantity in time or space. The earliest example pupils encounter is the velocity.
- *Geometry/vectors*: Geometrical properties are often needed for analysing the motions of objects in space, especially in adding forces or velocities, superposition of motions planetary motions and optics, too.
- *Algebra.* Students need algebraic abilities for manipulating formulas on different levels and in different contexts.

Often the problem is mentioned that the pupils do not yet know the appropriate mathematical tools for treating the physics contents. Partly this may be a problem of synchronizing the lessons in mathematics and physics, partly it is a principal problem because of the different goals and views in those two subjects.
But with some compromise both subjects may help each other (Beckmann, 2003). I give some examples from the saxonian curriculum in Germany:

Grade	*Mathematics*	*Physics*
6	proportionalities	Velocity, density
7	rational numbers	Ohm's Law
8	different types of functions	Meaning of parameters
9	quadratic functions	Laws of accelerated motion
10	trigonometric functions	Periodic motions, oscillations

Table 2: Parallelism between mathematics and physics in saxonian curriculum

SPECIFIC ROLE OF MATHEMATICS AND MATHEMATICAL MODELLING

A short overview over the thinking of a small number of 11th graders indicates that they mostly view mathematics as a tool for physics (data by a questionnaire), but do not appreciate its structuring role. However, since students should see the structure of physics and recognise analogies, the structuring role of mathematics has to be stressed during the whole education. I want to show some possibilities for this purpose.

Structuring elements in physics education

Examples for showing similar structures throughout the curriculum:

- *Similar equations:* the resistance of a long wire, the energy needed for heating a body, air resistance of a moving body
- *Similar laws:* Ohm's law, Hooke's law, law of
- *vector fields or field lines:* gravitational field, electric field, magnetic field
- *Integration* in the sense of summing up along a distance or summing up in time. The most used examples are the integration of a constant function: uniformly rectilinear motion, uniformly accelerated motion, work of a constant force along a path; or the integration of a linear function: uniformly accelerated motion, he kinetic energy, the potential energy of a spring and the energy of magnetic field of a solenoid.

At least three examples will be necessary in order that the students recognise the application of a mathematical method not only as a trick but as a method and hence as a structuring element for physics. They gain insight into scientific methodology and get aware of analogies.

Art of Modelling

The presupposition for modelling is knowledge of the physics background, making the appropriate idealisations and then making the transition to the mathematical representation (s. also Table 3). To achieve this ability with students a careful path from the everyday experience to the formula has to be observed.

Modelling requires several steps:

- The analysis of the present physical situation: Here studies show that beginners and experts differ in assigning meaning to the objects and their relationship. This implies the following:
  - The decision which parameters or measurable values are important and which could be safely neglected
  - The identification of the relevant principles or equations and the way they have to be applied
- The transition to the idealized situation: the physical situation has to be represented in a graphical or iconic way. This step should be supported by verbal descriptions and explanations, (see also Table 3).
  - Drawing of vectors e.g. force diagrams
  - Plotting measuring results e.g. temperature dependence on time
  - use of a graphical calculator
- The translation into a mathematical formulation
  - Recognizing functional dependencies

Mathematical Concept	***Idealization***	***Physical Object***
point	mass point	(rigid) body
vector	force arrow	force
function		trajectory
equation	law	

Mathematical Concept	*Idealization*	*Physical Object*
vector field	field lines	magnetic field

Table 3: The relation between physical objects and corresponding mathematical concepts

Mathematical modelling

To introduce the modelling process and train some of the related activities it is necessary to identify key techniques and often used methods.

- the inductive-deductive derivation of physics laws both from theoretical considerations and from experiments. This implies recognizing functional dependencies and condensing the measuring values into mathematically formulated laws. A problem consists in that it is by no means unique to derive a functional dependence or even a law from three or four pairs values. However, this is often suggested by the procedere at school. Hence this inductive Ansatz has to be complemented by deductive reasoning.
- analysis of a given situation: everyday situations have to be stripped off their hiding "clothing" and their physical core to be laid open. Herewith an important feature is to encourage creativity: Mostly, students are used to be given problems with exact numbers and one and only one correct answer. Suggesting reasons, recognizing applicable laws, estimating numbers and deriving consequences should be important part of the "Aufgabenkultur".

For illustration I give a few examples which might serve as an introduction into modelling.

1. contextual problem (with relation to thermodynamics, 7th or 8th grade):

How many gas cartouches will be necessary for a five-day hike through the wild if you want to have a hot coffee and a cooked meal every day ?

This problem requires mainly the decision which quantities are necessary or important and have to be known and the courage to estimate numbers (amount of water, efficiency the the gas burner, decision on cooked meal). It has no significant difficulties beyond the standard problems in physics lessons. It stimulates the creativity of students.

2. everyday situation (with relation to thermodynamics, 8th grade):

How much cooling can reached on a hot day by wrapping a bottle with a soft drink in wet cloth?

If the precise change of temperature with time is not in question, there is only the energy balance to be taken into account: The evaporating water takes its energy mainly from the inside (the drink), because it has stronger heat conducting contact then the surrounding air. There have to be estimates about the amount of evaporating water, the heat capacity of a soft drink and of glass, knowledge of evaporation energy of water (about 2350 J/g) and the law of energy conservation. The students can gain insight in the power of balancing: they can treat the experimental design mainly as a black box which simplifies the estimation a great deal.

Furthermore, the result can be controlled by an experiment: Besides the connection between theory and experiment, the students learn to control the parameters and side conditions: temperature of soft drink, evaporating water and surroundings, amount of water and so on.

3. physics problem (applying standard techniques, mechanics 11th grade)

How much energy is needed to get a satellite into a geostationary orbit?

The first Ansatz will be to determine the potential energy of a satellite. In this example the approximation for small heights can not be applied so that the energy has to be calculated with a variable force along the path. This could even be analytically integrated. Students will learn a certain precision of thought by comparing with the standard formula for the potential energy, (Lotze 2006).

An additional difficulty could be taken into account by gifted students: the velocity dependent air resistance and the specific weight of the air during the flight through the atmosphere.

As is seen with the last two examples the emphasis of modelling may lie either on knowing the result of a process or on analysing the whole process in detail. Both approaches have their own difficulties, advantages and disadvantages. Since processes are difficult to model - they require at least difference equations - it may be useful to apply bilancing, especially in lower grades. Often it is proposed to use a spreadsheet for calculation of difference equations. It has the advantage of being easily accessible to teachers and students, but the great disadvantage of quite uncomfortable handling. A computer program - written e.g. in PASCAL or DELPHI - could be much more convenient for a more variable use. An alternative would be a modeling system like PAKMA or STELLA which allows for visual modelling - similar to a concept map - and the calculation happens in the background.

CONCLUSION

Developing competencies in scientific thinking is the main task of physics education. The knowledge of mathematical tools is an important part. To achieve that students are able to model phenomena from daily life they need training with suitable problems. Therewith the balance has to be found between the necessary guidance and the students' self directed learning and inquiring. The guidance consists in giving the students a direction how to proceed, which steps to take; the free inquiry needs that the students learn to structure their knowledge and to find themselves suitable idealizations. So a task of physics education research is the development of suitable problems and learning environments in order to engage the students in appropriate activities.

The deliberate use of mathematics can help to achieve better insight in physics laws and their common structures and to further physical thinking.

References

Beckmann, A (2003). Fächerübergreifender Mathematikunterricht, Teil 2: Mathematikunterricht in Kooperation mit dem Fach Physik, Hildesheim:Franzbecker.

Darmofal et al (2002). Using concept maps and concept questions to enhance conceptual understanding, 32nd ASEE/IEEE Frontiers in Education Conference. Boston.

Friege, G (2001). Wissen und Problemlösen - Eine empirische Untersuchung des wissenszentrierten Problemlösens im Gebiet der Elektrizitätslehre auf der Grundlage des Experten-Novizen-Vergleichs. Berlin: Logos Verlag.

Gerstberger, H (1999). Sprache und Mathematisierung beim Lernen von Physik. In: Zur Didaktik der Physik und Chemie, Alsbach:Leuchtturmverlag.

Köppl. R (2006). Hyänenhunde - kühle Taktiker beim Energieeinsatz in heißen Savannen. *MNU, 59(7), 416-422.*

Lotze, K-H (2006). Wie gut bereitet der Mathematikunterricht auf ein Studium der Physik vor?. *Praxis der Naturwissenschaften Physik/ Physik in der Schule 55(5), 8-17.*

Monk, M (1994). Mathematics in physics education: a case of more haste less speed. *Phys. Educ.28 (1994), 209-211.*

Pospiech, G (2006). Modellierung und mathematische Kompetenz im Physikunterricht. Beitrag zur Frühjahrstagung des Fachverbandes Didaktik der Physik in der DPG, Kassel

Wigner, E (1960). The Unreasonable Effectiveness of Mathematics in the Natural Sciences. *Comm. Pure Appl. Math., 13(1).*

USING MATHEMATICS AND STATISTICS TO ANALYZE WHO ARE THE GREAT SLUGGERS IN BASEBALL

Randy Taylor
Las Positas College, Livermore, CA, USA,
&
Steve Krevisky
Middlesex Community College, Middletown, CT, USA

*In this presentation, we share the results of statistical work that we have done over several years, in order to determine who are the best sluggers in the game of baseball in the US? Using z scores (z = [**score – mean**] / **standard deviation**), we examined yearly home run and slugging average figures, so as to analyze which batters were the most SD's above the mean. We used cutoffs of 200 at bats or 250 plate appearances before expansion, and increased this by about 5 %, to account for the increased # of games played after the expansion in the early 1960's. Since real data are involved, we feel that this would be a very good application for students in a basic statistics class, and we will present various charts in the following discussion, and summarize our findings. This is a shortened version of what we have presented at numerous conferences in the US.*

INDIVIDUAL YEAR HOME RUN RESULTS

Gathering the statistics was much more time consuming than imagined. Data was obtained for all years 1920-2003 from ***The Sports Encyclopedia: Baseball 2004*** **by David S. Neft, Richard M. Cohen, and Michael L. Neft**. This book contained all the information required, but the print was very small. It also used numbers to indicate players who were traded, which was important in counting players totals in both leagues if they played in both leagues. We now look at Z-Scores for various Home Run Champions. The Z-Score is the number of standard deviation above the mean a player's total is and is given by:

Z-Score = (individual HR total – year's mean HR)/year's standard deviation

Next we calculate some American League results:

Home Run Champions 1920 - 2003					
American League					
Year	**HR Mean**	**HR Standard Deviation**	**Name**	**HR**	**Z-Score**
1920	4.85	7.27	Babe Ruth	54	6.76
1921	6.05	8.87	Babe Ruth	59	5.97
1927	5.42	10.05	Babe Ruth	60	5.43
1932	8.59	10.32	Jimmie Foxx	58	4.79
1938	11.60	11.88	Hank Greenberg	58	3.91
1956	13.34	9.39	Mickey Mantle	52	4.12

B.Sriraman, C.Michelsen, A. Beckmann & V. Freiman (Eds). (2008). *Proceedings of the Second International Symposium on Mathematics and its Connections to the Arts and Sciences* (MACAS2). Information Age Publishing, Charlotte: NC , pp.241-244

1961	15.01	12.34	Roger Maris	61	3.73
1967	11.87	8.99	Carl Yastrzemski	44	3.57
1990	11.12	8.74	Cecil Fielder	51	4.56
2002	15.83	10.51	Alex Rodriguez	57	3.92

The top five Z-Scores are: (1) Babe Ruth, 1920, z = 6.76, (2) Babe Ruth, 1921, z = 5.97, (3) Babe Ruth, 1927, z = 5.43, (4) Jimmy Foxx, 1932, z = 4.79, and (5) Cecil Fielder, 1990, z = 4.56.

Next we calculate some National League results:

Home Run Champions 1920 – 2003 National League					
Year	**HR Mean**	**HR Standard Deviation**	**Name**	**HR**	**Z-Score**
1922	6.31	7.18	Rogers Hornsby	42	4.97
1930	10.86	11.20	Hack Wilson	56	4.03
1949	10.87	9.78	Ralph Kiner	54	4.41
1954	14.01	11.70	Ted Kluszewski	49	2.99
1998	15.44	12.48	Mark McGwire	70	4.37
2001	18.03	13.37	Barry Bonds	73	4.11

b

The top five Z-Scores are: (1) Rogers Hornsby, 1922, z = 4.97, (2) Ralph Kiner, 1949, z = 4.41, (3) Mark McGwire, 1998, z = 4.37, (4) Barry Bonds, 2001, z = 4.11, and (5) Hack Wilson, 1930, z = 4.03.

INDIVIDUAL SLUGGING AVERAGES

Most batters who hit many Home Runs normally have high Slugging Averages, but some don't, like Harmon Killebrew. Also, many Slugging Average leaders were good all-around hitters and never lead in Home Runs, like Stan Musial. Using this category will give us another way to analyze who the top sluggers/power hitters are. We will be using Z scores so that each hitter is compared to the time that he played in.

- **Z SCORE = (PLAYER'S SLA – MEAN SLA OF THE SEASON) / YEAR'S SD**

First of all, we define Slugging Average, and show how it is computed.

- **Slugging Average = Total Bases divided by Total at Bats.**

SLA = TB / AB.

Example: You get 10 hits in 30 at bats and you have 3 doubles, 2 triples and 2 home runs. This is the data listed would be listed in ***The Sports Encyclopedia: Baseball 2004.*** First, you need the # of singles. How do we do this? There is a formula for this: We take the following approach:

- **# Singles = # Hits – (# 2B + # 3B + # HR).**

In this case, # S = 10 – (3 + 2 + 2) = 10 – 7 = 3.

Next, you get the # of totals bases as follows:

- **# Total Bases = 1(#S) + 2(#2B) + 3(#3B) + 4(#HR)**

This can be viewed as the dot product of 2 vectors, as follows:

- **#TB = [1, 2, 3, 4] *[3, 3, 2, 2]**

#TB = (1*3) + (2*3) + (3*2) + (4*2)

#TB = 3 + 6 + 6 + 8 = 23

Then, SLA = 23/30 = .767, which is quite good? It should be noted that this decimal number is often just call 767, just as a hitter batting .310 is frequently said to be hitting 310.

Now we calculate some American League results:

Slugging Average Champions 1920 - 2003					
American League					
Year	**SLA Mean**	**SLA Standard Deviation**	**Name**	**SLA**	**Z-Score**
1920	400.76	95.07	Babe Ruth	847	4.69
1921	431.29	84.98	Babe Ruth	846	4.88
1932	422.46	82.80	Jimmie Foxx	749	3.94
1937	437.91	77.45	Joe DiMaggio	673	3.04
1941	414.57	84.30	Ted Williams	735	3.80
1956	422.51	81.45	Mickey Mantle	705	3.47

The top five Z-Scores are: (2) Babe Ruth, 1921, z = 4.88, (2) Babe Ruth, 1920, z= 4.69, (3) Jimmy Foxx, 1932, z = 3.94, (4) Ted Williams, 1941, z = 3.47, and (5) Mickey Mantle, 1956, z = 3.47

Next we calculate some National League results:

Slugging Average Champions 1920 - 2003					
National League					
Year	**SLA Mean**	**SLA Standard Deviation**	**Name**	**SLA**	**Z-Score**
1922	437.68	80.84	Rogers Hornsby	756	3.94
1930	467.46	116.51	Hack Wilson	723	2.19
1948	405.03	76.71	Stan Musial	702	3.87
1949	416.11	75.31	Ralph Kiner	658	3.21
1954	431.84	86.72	Willie Mays	667	2.71

1998	432.25	98.95	Mark McGwire	752	3.23
2001	451.64	98.42	Barry Bonds	863	4.18

The top five Z-Scores are: (1) Barry Bonds, 2001, z = 4.18, (2) Rogers Hornsby, 1922, z= 3.94, (3) Stan Musial, 1948, z = 3.87, (4) Mark McGwire, 1998, z = 3.23, and (5) Ralph Kiner, 1949, z = 3.21.

CONCLUSIONS AND IMPLICATIONS FOR FURTHER STUDY

We hope to eventually have a complete data base for all the years from 1920-2006. This will take time, but we think that this work will shed more light on who the great sluggers in baseball are. We think that students and professors alike would enjoy this work, and many a good project could stem from this activity. We could also do this for individual players' careers ! Finally, we note the following:

Even using lower cuts offs for At Bats of 250 before 1961/1962 and 265 afterward, some significant Home Runs totals were still not used in the calculations. For example:

1960 Johnny Blanchard, New York Yankees, 243 AB, 21 HR
1963 Johnny Blanchard, New York Yankees, 218 AB, 16 HR
1966 Joe Adcock, California Angels, 231 AB, 18 HR
1996 Gary Sheffield, Florida Marlins, 218 AB, 16 HR
2000 Mark McGwire, St. Louis Cardinals, 236 AB, 32 HR

REFERENCES

Taylor, R. and Krevisky, S. (2005). Using Mathematics to Catch the Wave of Great Baseball Sluggers. Paper presented at annual meeting of the American Mathematical Association of Two Year Colleges (AMATYC), San Diego.

Cohen, R., Neft, D, and Neft, M. (2004) The Sports Encyclopedia Baseball .

GIVING SENSE TO THE MATH FORMULA

Marina Rugelj, Tine Golež

St. Stanislav Institution for Education, Diocesan classical gymnasium, Ljubljana

A math and a physics teacher can find some common topics in both subjects. We present an example of collaboration between the two teachers. Experiments in physics may give the sense to the math formula. This is also true for trigonometric functions in mathematics and oscillations in physics. The papers shows how both teachers can together executed such a combined lesson.

THE MATHEMATICS TEACHER'S INTRODUCTION...

With the use of coupled pendulum in physics we give sense to the math formula

$$\sin\alpha + \sin\beta = 2\sin\frac{\alpha+\beta}{2}\cdot\cos\frac{\alpha-\beta}{2}.$$

In the third year of High school (Gymnasium) we teach students the "angle addition formulas":

$$\sin(\alpha \pm \beta) = \sin\alpha\cos\beta \pm \sin\beta\cos\alpha$$

$$\cos(\alpha \pm \beta) = \cos\alpha\cos\beta \mp \sin\alpha\sin\beta$$

From these formulas we also derive the transformation of a sum of two trigonometric functions to a product:

$$\sin\alpha + \sin\beta = 2\sin\frac{\alpha+\beta}{2}\cdot\cos\frac{\alpha-\beta}{2}$$

$$\sin\alpha - \sin\beta = 2\cos\frac{\alpha+\beta}{2}\cdot\sin\frac{\alpha-\beta}{2}$$

$$\cos\alpha + \cos\beta = 2\cos\frac{\alpha+\beta}{2}\cdot\cos\frac{\alpha-\beta}{2}$$

$$\cos\alpha - \cos\beta = -2\sin\frac{\alpha+\beta}{2}\cdot\sin\frac{\alpha-\beta}{2}$$

These are very technical lessons for students. It is hard to present these formulas as meaningful and applicable. Also exercises in textbooks are given without any motivation. For example:

Calculate (without the use of calculator): $\sin 75^0 + \sin 15^0 =$

Or simplify the fraction: $\dfrac{1+2\cos x}{2\sin x+\sqrt{3}} =$

B.Sriraman, C.Michelsen, A. Beckmann & V. Freiman (Eds). (2008). *Proceedings of the Second International Symposium on Mathematics and its Connections to the Arts and Sciences* (MACAS2). Information Age Publishing, Charlotte: NC , pp.245-254

These formulas are also found on the list of important formulas which students are expected to be familiar with on the high school finals.

Mathematics teachers very often hear such questions: "Why should we know these formulas? Where can we use them?" A math teacher option to answer might be: "You have to know them to pass the final exam." A smarter way would be answer such questions within collaboration with physics teachers. Namely, these formulas can be given clear meaning by the use of physics.

Before we turn to physics let us look at the following three graphs of functions as mathematicians.

The first: $f(x) = \sin(10x)$

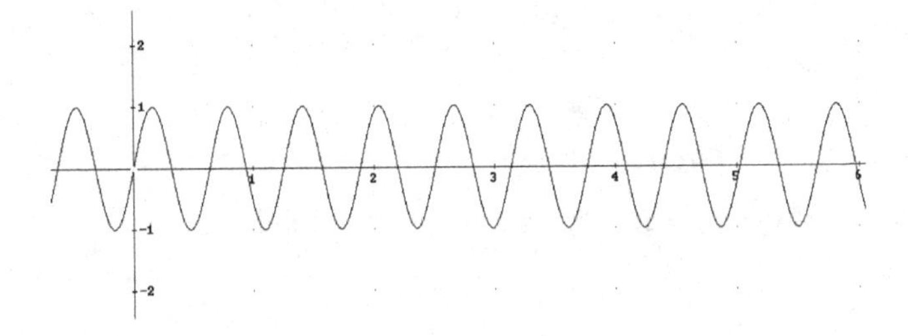

Let the second be $f(x) = \sin(12x)$

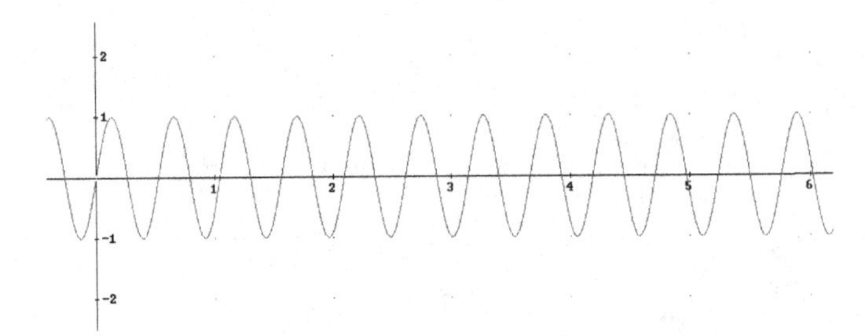

If we sum up these two functions by the use of formula $\sin\alpha + \sin\beta = 2\sin\frac{\alpha+\beta}{2}\cdot\cos\frac{\alpha-\beta}{2}$

we get

$f(x) = \sin(10x) + \sin(12x) = 2\cos x \sin(11x)$

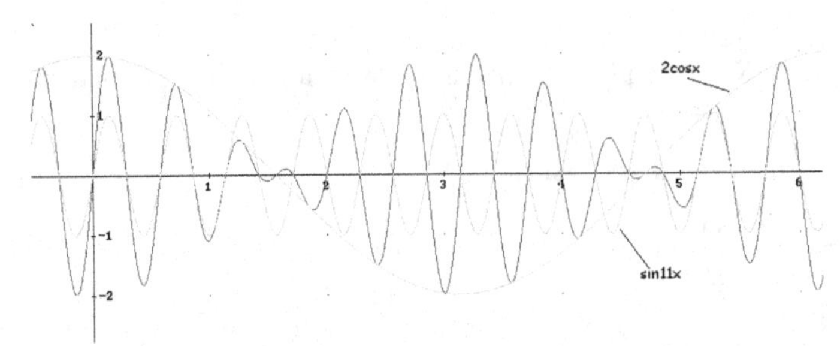

On the graph we sketched our function and the two factor functions. Regarding this graph and the meaning of a product on a graph can be discussed further as a purely mathematical question, but now we want to turn our attention to the meaning of the above graph from the physicist point of view.

THE PHYSICS TEACHER'S EXAMPLE OF USE...

In the third year of High school (Gymnasium) in Slovenia we also teach about oscillations. Coupled pendulum is rarely mentioned, but we regard the coupled pendulum as the essential topic because the basics of this phenomenon lead us to understand waves, sound, and light.

Students master already fundamental knowledge about simple pendulum: graphs $s(t)$, $v(t)$, $a(t)$, period of different types of simple pendulums, energy changes... We proceed by experiment: two simple pendulums are coupled by a spring of weak stiffness (Fig. 1). It is easy to observe two facts: this is not a sinusoidal oscillation, but it is still a periodical one. One assumes the oscillation depends on the initial conditions. Can we choose such initial condition that the oscillation is a sinusoidal one?

Fig.1. Home made coupled pendulum.

Students usually correctly suggest the two initial conditions which lead the system to oscillate not only periodically but also sinusoidal.

If the initial position is the same, the two pendulums will always have same phases. Therefore, it is a *sin* oscillation. The second option is to begin with the same position but different sign (Fig 2., Fig 3.)

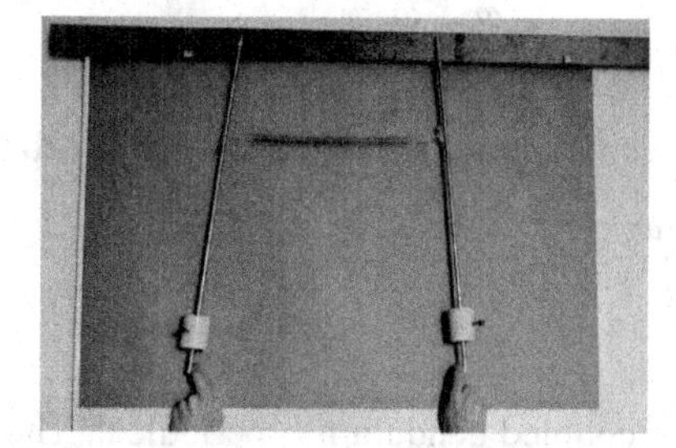

Fig. 2., 3. The two initial conditions which lead to the sinusoidal oscillations.

There are only two sinusoidal oscillation modes which we can describe by two functions:

$s_1(t) = A\sin\omega_1 t$ and $s_2(t) = A\sin\omega_2 t$.

The amplitude is the same in both cases. Angular frequencies ω_1 and ω_2 tell us how many times the pendulum oscillate in the first and in the second case in 2π seconds. We know that $\omega = \frac{2\pi}{t_0}$, therefore period is $t_0 = \frac{2\pi}{\omega}$ and frequency is $\nu = \frac{1}{t_0} = \frac{\omega}{2\pi}$.

The periods of the two oscillations described above are very similar, one can easily predict that the first initial condition will cause slightly longer period. We measure both periods by observing five oscillations.

The results are: $t_{01} = 1{,}46$ s and $t_{02} = 1{,}37$ s.

Physics is able to foretell many outcomes of different experiments. Can these two measurements tell us something about the first experiment (periodical, non sine oscillation)?

The oscillation from the first experiment is a sum of the two sinusoidal oscillation modes:

$$s(t) = A\sin\omega_1 t + A\sin\omega_2 t = A(\sin\omega_1 t + \sin\omega_2 t)$$

Before we conclude what this means, we define a *beat* – that is what we observe by looking at oscillating coupled pendulum. The period of a *beat* is the time between the two rest positions (the velocity is 0) of the left (or right) pendulum.

The frequency of the beat is the absolute value of the difference of the two frequencies (of the two normal modes, the two sinusoidal modes of oscillation). Why?

AND MATHEMATICS AGAIN…

At this moment the applicability of the formula becomes evident:

$$\sin\alpha + \sin\beta = 2\sin\frac{\alpha+\beta}{2}\cdot\cos\frac{\alpha-\beta}{2}$$

$$s(t) = A\sin\omega_1 t + A\sin\omega_2 t = A(\sin\omega_1 t + \sin\omega_2 t) = 2A\sin\frac{\omega_1+\omega_2}{2}t\cdot\cos\frac{\omega_1-\omega_2}{2}t$$

Rearranging this formula, we get

$$s(t) = 2A\sin\frac{\omega_1+\omega_2}{2}t\cdot\cos\frac{\omega_1-\omega_2}{2}t = \left(2A\cos\frac{\omega_1-\omega_2}{2}t\right)\cdot\sin\frac{\omega_1+\omega_2}{2}t$$

Let us denote $A_1(t) = 2A\cos\frac{\omega_1-\omega_2}{2}t$.

Coupled pendulum oscillation can now be described by the formula:

$$s(t) = A_1(t)\sin\frac{\omega_1+\omega_2}{2}t$$

What does it mean?

The oscillation of the coupled pendulum is periodical. The frequency of coupled pendulum is the average of the two frequencies: $\omega = \frac{\omega_1+\omega_2}{2}$.

But the beat frequency is hidden in the amplitude $A_1(t) = 2A\cos\frac{\omega_1-\omega_2}{2}t$.

We have already defined: the period of a beat is the time beginning at the moment when e.g. left pendulum is at rest and ending at the moment when the same pendulum is at rest position again.

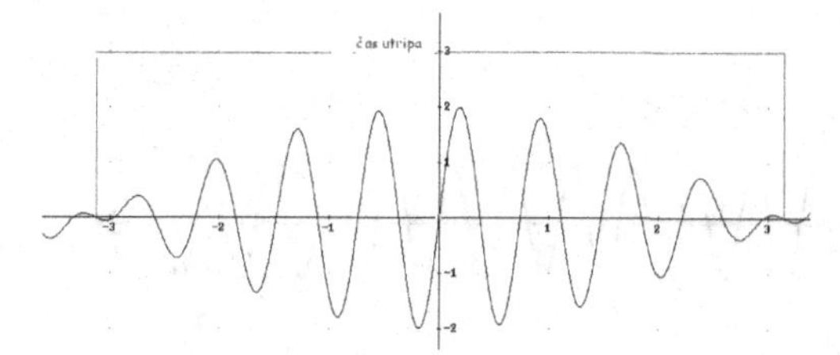

Between one and another zero of the function $f(x) = \cos x$ is π therefore we can calculate the period of beat t_B :

$$\frac{\omega_1 - \omega_2}{2} t_B = \pi \Rightarrow t_B = \frac{2\pi}{\omega_1 - \omega_2}$$

The frequency of a beat ν_B $\nu_B = \frac{1}{t_B} = \frac{\omega_1 - \omega_2}{2\pi} = \frac{\omega_1}{2\pi} - \frac{\omega_2}{2\pi} = \nu_1 - \nu_2$ is therefore implied by calculation.

AND PHYSICS…

Let us go back to our experiment. We know the two periods: t_{01} = 1,46 s and t_{02} = 1,37 s.

Frequencies are $\nu_1 = \frac{1}{t_{01}} = 0{,}685 s^{-1}$ and $\nu_2 = \frac{1}{t_{02}} = 0{,}730 s^{-1}$ hence the calculated beat frequency is:

$$\nu_B = \nu_1 - \nu_2 = 0{,}730 s^{-1} - 0{,}685 s^{-1} = 0{,}045 s^{-1}$$

From this data we can calculate the beat time:

$$t_B = \frac{1}{\nu_B} = \frac{1}{0{,}045 s^{-1}} = 22{,}2 s$$

By experiment we measure the beat period and get 22,4 s. The results are quite similar.

Let us look at graphs. Bellow are four graphs (Fig. 4) . The first two are $s(t)$ of the two normal modes. And the last two describe $s(t)$ for left and right simple pendulum which is connected to an equal pendulum in order to form a coupled pendulum together.

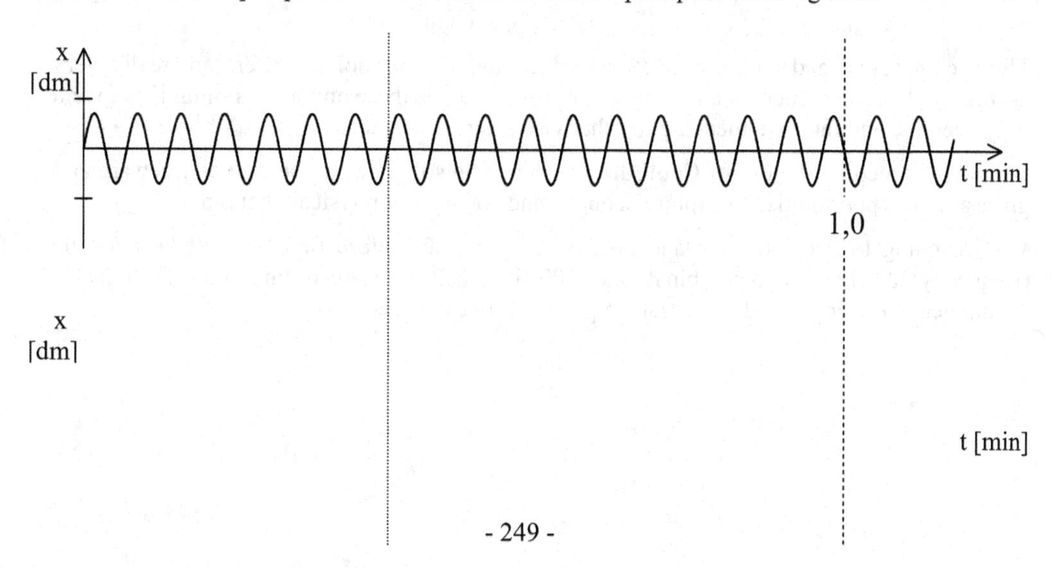

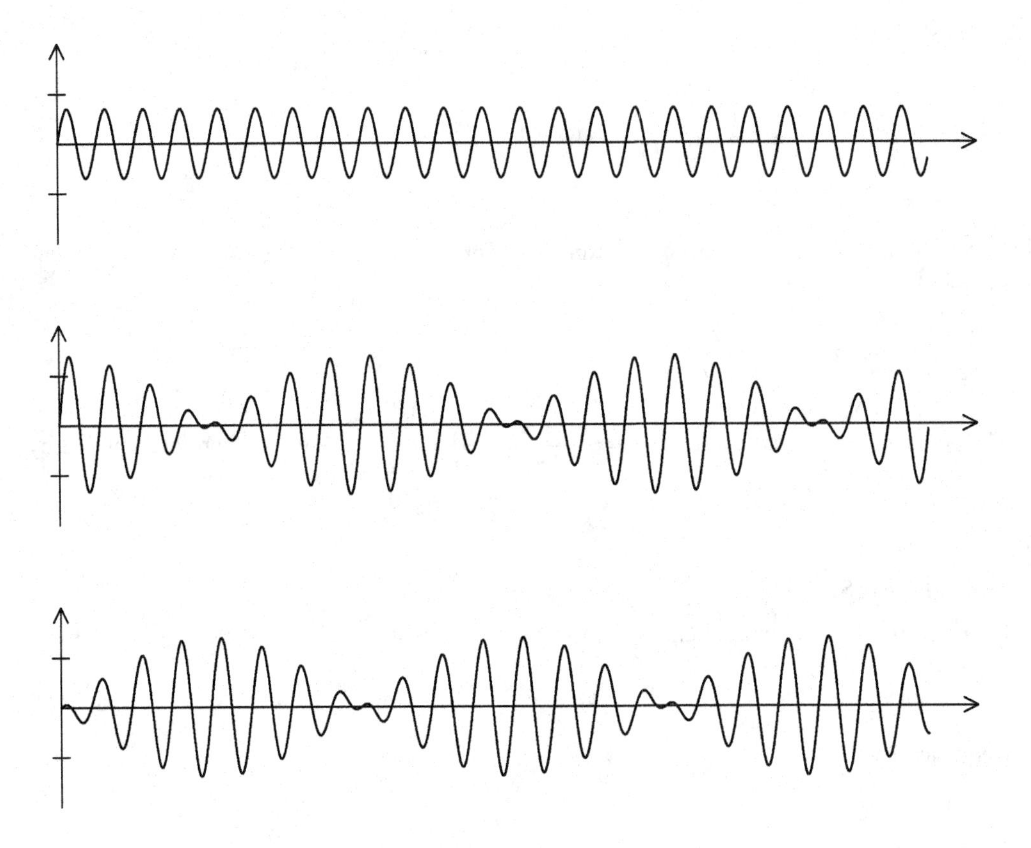

Fig. 4. The graphs of a coupled pendulum are displayed.

We should know that the frequency of the first sin oscillating mode is 17,5 min^{-1}And the frequency of the second is 20 min^{-1}. So we can predict the frequency of the beat: the result is the absolute value of the difference of the two frequencies. It is clear that the two lower graphs give beat frequency 2,5 min^{-1}.

One can also explore this experiment as an applet simulation. Many wonderful applets can be find on the net. (http://www.walter-fendt.de/ph11e/cpendula.htm)

The use of visual and audio simulations which today's technology offers are really very useful. Students and teachers alike can both profit a lot, as these animated simulations give a truly accurate and intuitive meaning to otherwise abstract formulas.

We now proceed by the use of CoolEdit program. The software can be find as freeware and gives a lot of opportunities to explore acoustic and some other physical phenomena.

We are going to generate tone and also a beat tone. Therefore first we will hear a tone (frequency 200 Hz) latter a combination of 200 Hz and 202 Hz. According to our findings we should hear a beating sound with beat frequency 2 times per second.

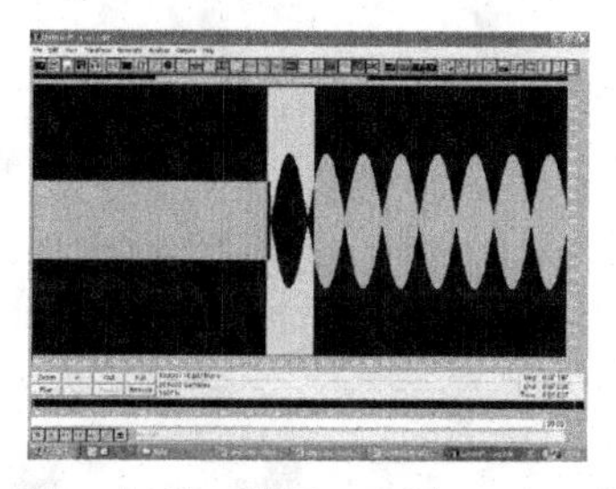

Fig. 5. CoolEdit enables us to generate tones and combined sound as explained in the text.

One can simply magnify a part of the graph by choosing and zooming it.

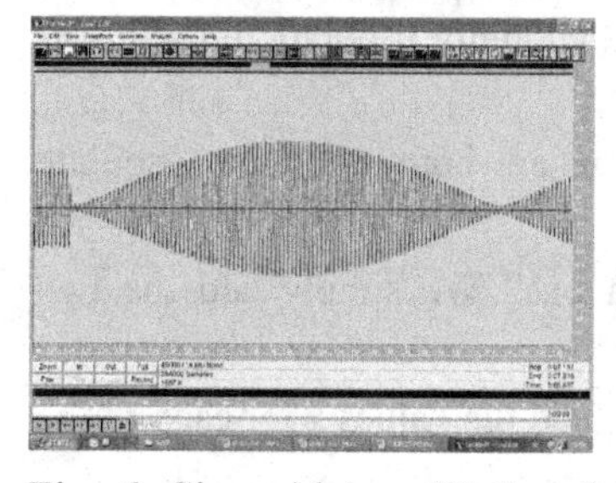

Fig. 6. Sinusoidal oscillation followed by coupled oscillation. It is easy to read the beat period from the time (horizontal) axe.

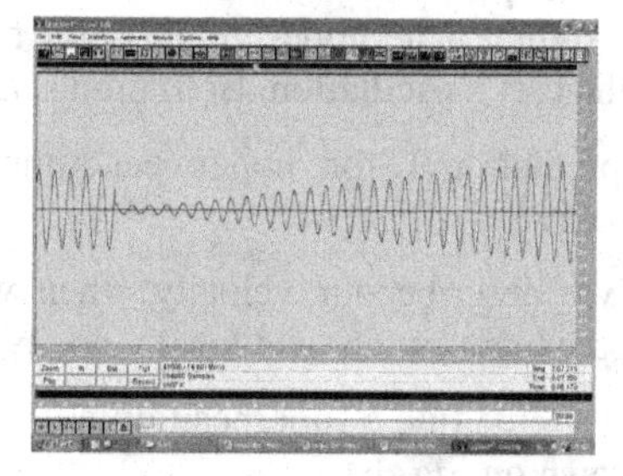

Fig. 7. Detail from Fig. 6. The mathematics teacher will want to zoom this part even deeper in order to show, that the frequency of the right part of the signal is 201 Hz. This is the average value of the two used frequencies.

Let us examine a double pendulum. It consists of the two simple pendulums but these pendulums are not equal as it was in the case of the coupled pendulum. (Fig. 8). The cart is attached by two springs to the clamps. This (the cart) is one simple harmonic oscillator. The second one is a simple pendulum made of a rope and a small weight. It is easy to plot both graphs x(t).

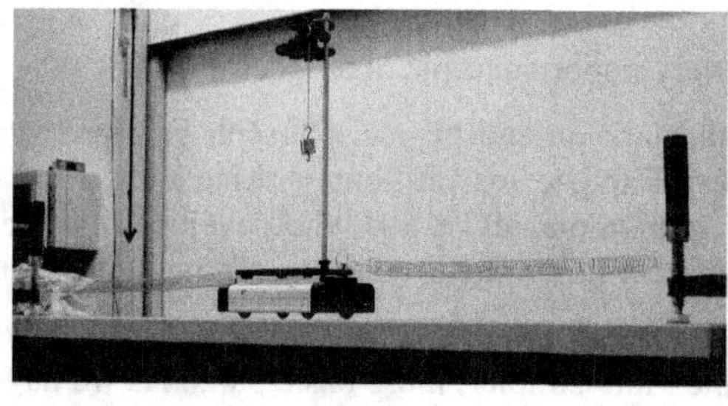

Fig. 8. A cart and a small simple pendulum. They form a double pendulum.

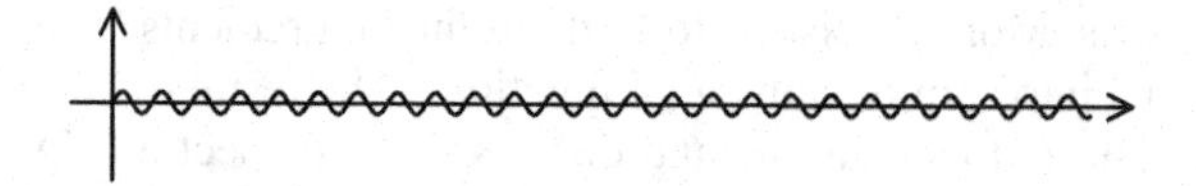

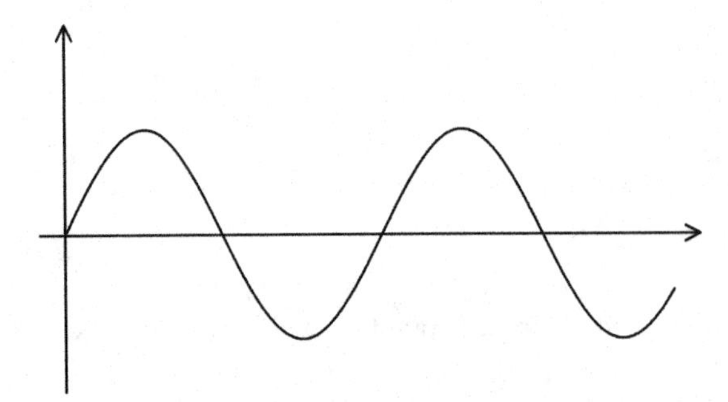

Fig. 9. The cart-pendulum system has two sinusoidal modes. The frequencies and amplitudes are quite different. The upper graph is for the small simple pendulum oscillation, the lower describes the cart's oscillation. Both graphs are $x(t)$.

Can we predict *x(t)* for upper pendulum in the case when both of them oscillate simultaneously?

How can we describe our velocity when walking inside a bus? We simply add the two velocities: the bus velocity and local velocity inside the bus.

Similarly we must find a superposition of the two graphs. We add the values x for each t. The result is shown on Fig. 10.

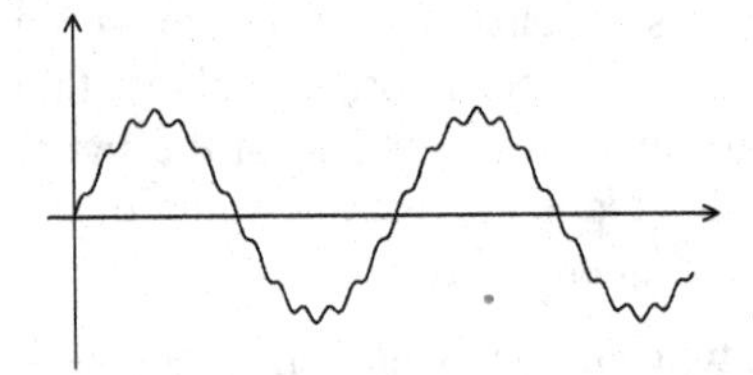

Fig. 10. The $x(t)$ graph for the simple pendulum when it oscillates simultaneously with the cart's (defining simple pendulum's coordinate system) oscillation.

...TOWARDS UDERSTANDING SPECTRA

These experiments and discussion lead us to a very important topic of a Spectrum analysis.

This is certainly not a new topic. French mathematician and physicist Joseph Fourier lived almost two centuries ago. By his full name Jean Baptiste Joseph Fourier (March 21, 1768 - May 16, 1830) is best known as a pioneer in harmonic analysis and who developed Fourier analysis and their application to the problems of heat flow. The Fourier transformation is also named in his honour.

Analysis means decomposing something complex into simpler, more basic parts. As we have seen, our x(t) graphs are quite complex (Fig. 4 the two lower graphs and Fig. 10). They are not sinusoidal curves but they are still periodical. Let us assume, we do not know the two curves which correspond to these oscillations. Fourier invented a mathematical procedure which enables us to find out the basic parts. More precisely, to find out the "ingredients" of a periodical curve. It decomposes the signal into sum of sinusoidal functions. The result of such an analyses is often displayed as spectrum. Let us examine and discuss the two spectra (Fig. 11)

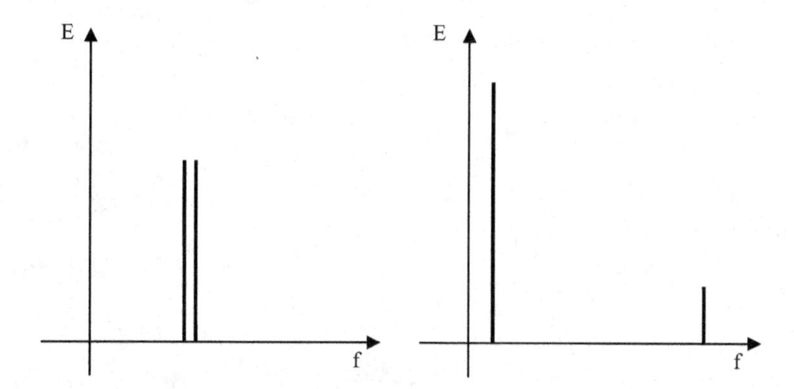

Fig. 11. The left spectrum is the coupled oscillation spectrum, the right one describes double pendulum. The frequencies correspond to the frequencies of the sinusoidal modes, and energies too.

Now we can find out what a spectrum shows: It tells us not only what are the frequencies of the normal (basic) modes but it also shows the (amount of) energy which corresponds to each frequency. The frequencies of the two sinusoidal oscillation modes of the coupled pendulum are very similar. The energy is also almost equal. But the second example, the double pendulum shows two quite different frequencies and also amplitudes.

A vibrating string can be presented as a huge number of coupled pendulums. As it is "made up" of "*N*" simple pendulums there are "*N*" sin oscillating modes. The frequencies of these modes can be either calculated either measured when a string is plucked. The results of a measurement are usually plotted as spectrum. But after investigation of the coupled pendulum such a spectrum should be more understandable for students.

One can examine many sound spectra. Sound spectrum is one of the determinants of the timbre or quality of a sound or note. It is the relative strength of pitches called harmonics and partials (collectively overtones) at various frequencies usually above the frequency of the main *A* (440 Hz which is the standard for musical pitch). But this would already be the start of another long and interesting topic…

LITERATURE:

RUGELJ, Marina, ŠPAROVEC, Janez, KAVKA, Dušan, PAVLIČ, Gregor. *Spatium = Prostor : matematika za 3. letnik gimnazij*. 3. izd. Ljubljana: Modrijan, 2005.

http://www.walter-fendt.de/ph11e/cpendula.htm

A CULTURAL VISIT IN MATHEMATICS EDUCATION

Annica Andersson
School of Teacher Education, Malmö University

In the Swedish curriculum for the social science education programs mathematical education there is a paragraph which states that

> *...the students shall deepen their insight into how mathematics has been influenced by people from many different cultures, and how mathematics has developed and still continues to develop (Skolverket,2000).*

How do we achieve this goal in mathematics education? Inspired by ethnomathematics this paper describes a teaching sequence containing two parts. The students in an upper secondary class in Sweden received an introduction to ethnomathematics subsequent discussion about mathematics and the global questions that it raised. This introduction was followed by a study visit where the students had the opportunity to study ethnomathematics in Australian Aboriginal art. The students' encounted geometry, functions and arithmetical progressions. Some students chose to discuss what mathematics is and what post constructions in art are.

INTRODUCTION

In the Swedish curriculum for the social science education programs mathematical education there is a paragraph which says that "the students shall deepen their insight into how mathematics has been influenced by people from many different cultures, and how mathematics has developed and still continues to develop"(Skolverket, 2000). How do we achieve this goal in mathematics education? My experience is that we, as teachers, usually don't pay any special attention to this goal. We are very clever in teaching all the goals in mathematical skills, and these skills are also the ones that are tested in national mathematical tests. I asked myself if ethnomathematics, defined by D'Ambrosio (1985) as the mathematics you find in different identified culture groups e.g. Indigineous peoples mathematics (e.g.Ascher, 1998; Lean, 1992 and Owens, 2001) and working place mathematics (e.g.Wedege, 2000) but also ethno mathematics as described by Bishop (1991), Archer (1998) and Barton (1996) could be a way to achieve the above described goal and at the same time bring discussions about global fairness and justice issues into the mathematical classroom. These discussions can be seen as an attempt to work with an critical mathematical discourse (following the ideas of e.g. D'Ambrosio, 2001; Ernest, 1998 and Skovsmose, 1994).

A scholarship gave me the opportunity to travel to Auckland, New Zealand and participate in the Third International Conference on Ethnomatematics, ICEM-3 in February 2006. This conference inspired me to put an ethnomathematics and cultural perspective into my mathematical teaching in an upper secondary school in the south of Sweden. One part was to give an introduction to ethnomathematics in the classroom. The second part was to make a study trip with the students in the second year on the international/social science program. It

B.Sriraman, C.Michelsen, A. Beckmann & V. Freiman (Eds). (2008). *Proceedings of the Second International Symposium on Mathematics and its Connections to the Arts and Sciences* (MACAS2). Information Age Publishing, Charlotte: NC , pp.255-264

is the experiences from this study trip I analyse in this paper. The study trip went to Arken Museum of Modern Art, in Copenhagen, Denmark, where the students studied the exhibition Dreamtime, Aboriginal art from the Ebes Collection. The exhibition contained more than 100 Australian Aboriginal art works and showed examples of Aboriginal maps (for more information about the exhibition please look at www.arken.dk/exhibition).

"Dreamtime" is a conception within the Australian Aboriginal culture describing a spiritual dimension of the Australian Aboriginals' existence and connects present time with the ancestors' time when the world was created. Dreamtime is represented by songs, dances and paintings, but also with maps showing traces in nature of the ancestors in forms of lakes, rivers and mountains (the Oxford Companion to Australian History, 2006).

Wedege (2006) argues quality in mathematics education when teachers declares what he/she sees as mathematical knowledge and at the same time express why students should learn mathematics and why mathematics should be taught. This is one of the reasons for my interest that from a mathematics education point of view see students expressed opinions of mathematics in our own culture and as a result of this see if there are possibilities to learn and understand mathematics from a cultural perspective (both our own and other cultures' perspective). As a researcher in teaching, the goal is to find the answers to the questions why; reasons for mathematical education, what; the goals for education and learning mathematics, who; the agents in teaching and learning and how: the context for teaching and learning (Wedege, 2000).

ETHNOMATHEMATICS AND THE ETHNOMATHEMATICAL RESEARCH AREA

Ethnomathematics is a research area describing how mathematics develops within different cultures and subcultures. The International Study Group of Ethnomathematics/ISGM was established about 15 years ago with a wide international participation and interest. The first international congress in ethno mathematics (ICEM) was held in Granada 1998 and continues since then every fourth year (D'Ambrosio, 2001). Researchers in the ethnomathematical field discuss Indigenous' people's mathematics; which is the focus in this paper, but also other cultural groups' mathematics. An assumption for this is that mathematics is developed in all cultures but not necessarily in the same way (Bishop, 1991).

D'Ambrosio (1985) defined the concept of ethnomathematics as the mathematics we find in different identified culture groups. Examples of different culture groups are groups of Indigenous people as the Australian Aboriginals and the Maori, but they can also be defined in different workplace groups and in groups of children and young people. The mathematical ideas and how they are expressed varies between different cultures and are found in different contexts as art, games and religion. One example of ethnomathematics is calendars. Even if since 1582 most people use a common calendar, we today have about 40 different calendar periods around the world. Calculations and time registration are some examples of ethnomathematics (D'Ambrosio, 2001). Other examples are the concepts of space, space orientation and map drawing (Harris, 1991).

D'Ambrosio discusses the concept of creativity and he claims that creativity and the students' learning will increase if they sense a feeling of cultural belonging. He discusses an alternative pedagogical discourse that clears the student's cultural background and makes it the base in teaching. Culture shall here be seen in a wide sense:

> The main fact of interest ... is the recognition that culture is a broad concept and goes beyond the traditional view of regarding it as associated with ethnic or geographic parameters (D'Ambrosio, 1985:42).

D'Ambrosio also shows examples of social groups, for example different age groups, who have their own culture referring to symbols, jargons and codes for behaviour and expectations. The different groups develop their own mathematics, which D'Ambrosio describes with the term "mathematizing". This is in contrast to the academic mathematics, the one taught in schools. Adults' mathematics in the workplace might also be regarded as ethnomathematics (see fore example Wedege, 2000). This alternative pedagogical discourse is described by D'Ambrosio with a comparison to a plant: understanding and creativity needs good soil and needs to be taken care of: "Here, fertility is determined by motivation, curiosity and initiative" (D'Ambrosio, 1985:80). D'Ambrosio questions if the academic and formal mathematics really develop these qualities within the students.

Bishop (1991) has studied the connections and the resemblance between different cultural groups referring to mathematical ideas and activities. According to him, the resemblances teach us about the cultural phenomena mathematics and facilitate an understanding of the roots of mathematical thinking. The mathematical ideas and activities are described by Bishop within the areas of counting, space, measuring, design and games and to find models of explanation for mathematical phenomena. Ethnomathematics can also be understood as reconstructions of Indigenous people's mathematical thoughts, ideas and principles. Mathematics is sought by ethnomathematicians in products from different cultures as handicraft, utility goods, tools, architecture and art. Bishop (1991) points out different research approaches within the ethnomathematical research area. The two of interest in this paper are the mathematical knowledge in traditional cultures with an anthropological perspective and mathematical knowledge in non western communities with an historical perspective.

THE AUSTRALIAN ABORIGINAL MAPS

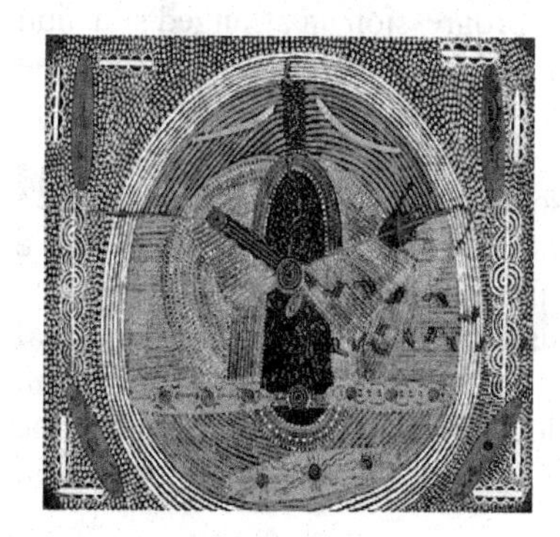

Rover Thomas Joolama

Clifford Possum Tjapaltjarri

Fig 1. Examples of Australian Aboriginal maps/paintings (www.arken.dk).

Harris (1991), an Australian researcher, studied the mathematical area for many years within a cultural context and especially within Australian Aboriginal cultures. She describes the Aboriginals' important sense for room and space in the following way:

> If Aboriginal people were to set up their own mathematics programs uninfluenced by white Australian traditions of what is important in a school mathematics program… the highest priority would certainly go to the space strand. (Harris, 1991:19)

Harris describes the Aboriginals' spatial capacity for space and directions. Where ever they are able to describe the directions in space even in, for them, new places and environments, both out in the deserts or inside buildings. When we (Westerners) should use the words right and left Aboriginals use the compass directions as direction markers. Harris uses the Australian Aboriginal maps as examples when she explains the Aboriginals' expressions/symbols for abstractions. The western maps are full of symbols and abstractions. The Aboriginal maps are also filled with abstractions - depending on whether the map is "only a map" for travels and trips or if it also has a mythical or sacred purpose. Aboriginal maps are often filled with mythical symbols for different places.

The art pieces shown at the Dreamtime exhibition were figurative pictures with specific landscapes, sacred places, vegetation and traces of animals with both symbolic and practical meaning for the Australian Aboriginals' way of life in the wilderness. Aboriginal maps were shown as art with symbols for places, but also for the walker, man and time (Ebes, 2006). E.d. the number of circles in a map can also be seen as time. A picture of five circles in front of one big circle can be seen as five days later you come to a big tree. A picture of five men can tell a story of one man's five day travelling in the landscape (Ebes, 2006).

A PIECE OF ACTION RESEARCH

The study reported in this paper can be placed in the field of action research. Skovsmose and Borba (2000) describe in a model for action research in mathematics education the connection between "current situation", "arranged situation" and "imagined situation". In my work, the current situation can be described as formal academic mathematics education. I had a different idea of mathematical education where students and mathematics were placed in a cultural context (imagined situation) and in order to achieve progression an arranged situation was arranged with an introduction to ethnomathematics and the visit to the Dreamtime exhibition.

The sequence of education reported in this paper can also be seen as a small attempt as a teacher to work with a Critical Mathematics Education perspective. According to Skovsmose and Nielsen (1996) is this kind of education depending on the goal in the curricula saying the students shall be prepared to take part in a political life outside school and that mathematical education shall provide them tools to analyse the world around them both locally and global. The culture is seen as important and the mathematical dialogue between the students and the teacher involves, according to this approach, a reflective way in connection to the world around us.

THE INVESTIGATION

THE STUDENTS AND THEIR BACKGROUND

The 16 students (14 girls and 2 boys) in this study were in the second year of an upper secondary school (a Swedish gymnasium) and they were 17-18 years old. The education program chosen by these students is an international social science program which prepares for future university studies. The students have communication and languages as their main subjects.

In Sweden all students in upper secondary school have to start their mathematical education with the course Mathematics A which consists of arithmetical problems and geometry. This course is followed by Mathematics B including mainly probability and functions. Mathematics C includes differentiation and exponential functions and equations and the Mathematics D course contains integrals and trigonometry and gives the student formal competence to study natural science subjects and programs at university. The students in this group were all in the middle of the Mathematics C course.

INTRODUCING THE STUDENTS TO ETHNOMATHEMATICS

In a 2-hour lesson two weeks prior to the exhibition visit, I introduced the students to the ethnomathematical field. They saw pictures of Maori art and handicraft and pictures of interesting architecture from New Zealand and Australia. The students also had the time and opportunity to discuss global issues, questions of justice and injustice and what Indigenous people are. Issues discussed were mathematics as power, who can/should learn mathematics and whose mathematics is accepted or right. The debate took place in the whole group but parts of it were arranged in smaller groups who reported their conclusions to the whole group.

THE STUDENTS' ASSIGNMENT AT THE EXHIBITION

The student's assignment was to answer the following three questions during their time at the exhibition:

1. What kind of mathematics do you think lies behind the different works of art? Choose some of the artworks and reflect more on them.
2. Can you see any possibility to use Indigenous or other cultures' art and crafts in our (Swedish) mathematical education? Which parts of mathematics could be understood better in this perspective?
3. Can you see any possibilities to use our own Swedish art and crafts traditions in mathematics education?

The students wrote their answers during the time spent at the exhibition and handed them in to me. They had to answer the questions individually, but they were free to discuss the maps/paintings in small groups. At the museum, notices gave information about the different works of art and what patterns and symbols in the respective works of art meant.

THE STUDENTS' ANSWERS

> A world full of ochre red plains, purling rivers, villages full of life and dreams about life and death pressed forward, in front of our eyes. We saw topographical maps showing roads and ways to villages and important waterholes as well as more abstract maps showing different paths through life. (A Student's description of the Dreamtime exhibition)

The art prompted the students to think of different aspects of school mathematics. The students' answers to the first question about mathematical knowledge behind the respective works of art diverged in three different tracks.

A. The majority of the students reflected on education in geometry and sought geometrical figures and patterns in the paintings. They found patterns of circles, rectangles, symmetrical patterns and graphs. One student verified the painting "After the fires" that "in some way he must have calculated three uniform sized circles containing different dots inside them". Other students counted for pictures of functions. For example Amanda who wrote about the painting "Kangaroo men dreams": "It looks similar to x^2 – graphs in a system of coordinates".

One student found details in the painting similar to tangents to curves with different gradients. I overheard a conversation between two students who discussed the conceptions of speed and differentiation. It was clear to me that these students understood the concept of differentiation clearly and here applied it in a new context.

B. One group of students found arithmetical progressions when they decided to count the circles and the dots:"...and then we discovered a pattern in the pictures. The numbers 4, 5, 8 and 10 seemed to pop up everywhere. Was it just a coincidence or not?" wondered Lisa and went on counting and then found that the progression went on: "To our delight we found another connection. After seven steps the first number is 10 times bigger" and they also found the connection" If one multiplies two numbers after each other in the progression as example 4x5=20 5x8=40 you get a number further on in the progression" and concludes lively: - Within this area more research is needed!!

This group of students gained a new experience of mathematics during their time at the exhibition. It was interesting for me as a teacher to observe them counting and testing. In general, during ordinary lessons at school one of these students is an ambitious girl, solving all the problems in the textbook and she really wants to understand everything properly – a relational learning student. One of the others is a student who usually is not especially interested in mathematics and she prefers instrumental learning (Skemp, 1976). The latter student showed a high mathematical activity during their testing of different progression theories while observing the painting. This was very interesting for me as a teacher to observe.

C. The last theme is characterized by students' reflecting over if it really is any mathematics in the paintings. Linda wrote: "Is this maths or is it copied patterns from nature?" She answered the question herself like this: "I don't know really, but man is borne as a logical being and maths is logic so in one sense one knows a kind of maths before one learns "our" mathematics". Anneli reasoned in the following way: "The question is if there is mathematics behind the maps. Mathematics is there but when the paintings were made one probably didn't think of maths..." While studying the painting "Women figures", Ann wrote:" I see mostly accuracy and precision, more than special mathematical knowledge". Tora gave this comment: "I stared at them (the maps) and tried to find out the mathematical in the paintings and that was quite hard to do" and then continues "the Aboriginal people were inspired of the forms and patterns they saw in nature, but what is mathematical about that?" She concludes with this remark:" Is it me being so chained to traditional thinking that mathematics demands a question, a solution and an answer?"

The second question was about the possibility to use Indigenous or other cultures' art and handicrafts in our (Swedish) mathematical education and which parts of mathematics could be understood better in this perspective. Here geometry is reflected by the students who sought geometric forms and patterns earlier. They think students would understand better when searching geometrical forms, patterns and symmetries in other ways than "just looking in a mathematical textbook or at the whiteboard".

One student suggests that experiencing mathematics in daily activities and in other places, not only in the mathematics classroom would make mathematical understanding easier. For this girl it was important that we left the ordinary school environment and sought mathematics in more unusual places. Another student, Mia, writes that as the Aboriginals are able to seek mathematics in nature, we ought to be able to do so too. We have our lovely and beautiful nature where we can seek forms and inspiration she argues. Linda gives the following comment: For the students with difficulties in understanding mathematics it must be good for their progress to see geometry through art or "through baskets and stuff from other places because then you get more practical examples".

The third question about possibilities for us to use our own Swedish art- and handicrafts traditions in mathematical education was answered shortly and concrete by the students. The following suggestions were given: knitting, crochet and embroidery, making patchwork and folk dresses, produce handicrafts, things and nesting boxes, paint pictures with mathematical thinking behind, look at and make pictures with dimensions and discuss them. The student, Stefan, with a deep historical interest wants to see "how mathematics have used and changed over time and study art with a mathematical perspective from cave paintings to the baroque art, the renaissance art and today's modern art". As a summary the students tend to see a connection between mathematics and the handicraft subjects at school. A more frequent collaboration between these subjects in school could accord these students an increase in their mathematical interest and mathematical understanding.

DISCUSSION AND CONCLUSIONS

The goal with this paper has been to describe an educational sequence with two parts. The first part was a classroom discussion where students were introduced to the conception of ethnomathematics followed by a discussion about such issues as global justice and questions about Indigenous people, power and mathematics. The second part was a study trip to the Dreamtime exhibition.

Art and handicrafts contain some mathematics components, which Stieg Mellin-Olsson (1987) defined as folk mathematics. Material calculations and economical calculations can be seen as examples of this. But there is also a part of the manufacturing process that is not mathematical, but can be seen as a post construction of mathematics and/or mathematical applications, forms and abstractions. Post-constructions are done in the mathematics classroom when we seek mathematics in existing art and handicraft. It may be interesting to reflect on what section of mathematics the artist and artisan apply (concretely use) and which parts we post construct together with the students.

The answers to the third question indicate that students in secondary school see a value for mathematical education to connect with handicraft subjects. It can also be understood as if there are possibilities to work across subjects and cooperate for example with the art and handicraft subjects and teachers. From a students' perspective the mathematic gets concrete and direct if they get the opportunity to discuss mathematical questions during the manufacturing process.

A question that I have asked myself when I have studied my student's answers to the three questions is this: What answers *didn't* they give?
For example, I did not find any answers similar to the problems we find in their mathematical textbooks. Arithmetical problems such as these: How much does it cost to make the picture? Calculate how much wood was needed to make the frame? How many percent of the picture is red? These types of questions/answers were not described by any student at all which might have been expected as that is the kind of mathematics the students have been used to work with in their textbooks in their mathematics education at school. Maybe I would have got these answers if the students had not listened to the presentation first. But I can also see alternative explanations. Boaler (1993) discusses the context in the mathematical classroom and shows examples of the dichotomy between so called everyday/general mathematical problems that often are constructed in a context ignoring the complexity and range of the student's experiences and convictions and because of this not always regarded as everyday problems in the student's world. When we now found ourselves outside the ordinary school environment with its culture and context, the students sensed a freedom to use other thoughts and methods as an alternative. It was most interesting for me as a teacher to overhear and note usually instrumentally learning students deepen their mathematical conversation. "... if students are encouraged to use their own methods and explore their usefulness, general mathematical understanding will be deepened." (Boaler, 1993:16).

Benn (2001) describes the mathematical classroom and the mathematical education in a wide social, cultural, political and economical context. From a mathematical education point of view education should be enriched with discussions of interest for the students with the goal to put mathematics in a context outside school, and by that widen their view on mathematics and its applications. I sence ethnomathematics can be one way of doing this.
I find that ethnomathematics is one way to achieve the goal in the Swedish curriculum for the social science education programs mathematical education which says that "the students shall deepen their insight into how mathematics has been influenced by people from many different cultures, and how mathematics has developed and still continues to develop"(Skolverket, 2000).

The education sequence presented in this paper shows an attempt to discuss global issues in a context like Critical Mathematical Education described by Skovsmose and Nielsen (1996). The result of this sequence showed a high student activity during the introduction lesson. It also showed students with usually not so high ambition levels in mathematics presenting a higher interest and activity both during the introduction and during the exhibition visit. Benn (2001) argues for a mathematical education that consequently is built on a social, economical, political and cultural consciousness. An education of this kind would according to Benn transform mathematical education and make mathematics accessible for more people. This was my experience from this exhibition visit. I think that ethnomathematics might be one way to achieve these goals. As a teacher I was able to register a new, or different, positive interest in mathematics in some of the students which will inspire me in the future both as a teacher and a researcher.

REFERENCES

Benn, R. (2002). Secret knowledge: Indigenous Australians and learning mathematics. In L. Oestergaard and T. Wedege (red), *Numeracy for Empowerment and democracy? ALM 8.* (pp.50-59).Roskilde: Roskilde University printing.

Bishop, A. J. (1991). *Mathematical Enculturation. A cultural perspective on Mathematics education.* Dordrecht: Kluwer Academic Publishers.

Boaler, Jo (1993). The role of contexts in mathematical classrooms. *For the learning of Mathematics, 13(2),*12-17.

D'Ambrosio, U. (1985). *Socio-cultural bases for Mathematics Education.* Unicamp, Campinas, Brazil.

D'Ambrosio, U. (2001). *Ethnomathematics. Link between Traditions and Modernity.* Rotterdam: Sence Publichers.

Ebes, H (2006). *Arken Museum of Modern Art. Dreamtime-Aboriginal art from the Ebes collection 11 february-11 june 2006.* Ishöj: Arken.

Harris, P. (1991). *Mathematics in a cultural context. Aboriginal perspectives on space, time and money.* Geelong: Deakin University Press.

Mellin-Olssen, Stieg. (1987). *The Politics of Mathematical Education.* Dordrecht: Kluwer Academic Publishers.

Skemp, Richard R. (1976). Relational and instrumental understanding. *Mathematics teaching, Bulletin of the association of Teachers of Mathematics,77,* 20-26.

Skolverket (2000). Upper secondary school. Goals to reach. Retrieved 2006-12-11 from www.skolverket.se

Skovsmose, O; Borba, M. (2000). *Research Methodology and Critical Mathematics Education.* Roskilde: Centre for Research in Learning Mathematics.

Skovsmose, O; Nielsen, L. (1996). Critical Mathematics Education. In Bishop, A J. *International Handbook of Mathematics Education, part 2.* Dordrecht: Kluwer Academic Publishers.

The Oxford Companion to Australian History. Retrieved 2006-10-25 from www.oxfordreference.com

Wedege,T. (2000). *Matematikviden og teknologiske kompetencer hos kortuddannede voksne.* Roskilde: IMFUFA, tekst nr. 381.

THE INFINITE IN SCIENCES AND ARTS

W. Mueckenheim

University of Applied Sciences Augsburg

Actual infinity in its various forms is discussed, searched and not found.

INTRODUCTION

The infinite or, speaking precisely, the *actual infinite*[1], has been suspected in many domains. G. Cantor was convinced that the infinite exists in God, in many domains of nature and in mathematics. "Dementsprechend unterscheide ich ein 'Infinitum aeternum increatum sive Absolutum', das sich auf Gott und seine Attribute bezieht, und ein 'Infinitum creatum sive Transfinitum', das überall dort ausgesagt wird, wo in der Natura creata ein Aktual-Unendliches konstatiert werden muß, wie beispielsweise in Beziehung auf die, meiner festen Überzeugung nach, aktual-unendliche Zahl der geschaffenen Einzelwesen sowohl im Weltall wie auch schon auf unserer Erde und, aller Wahrscheinlichkeit nach, selbst in jedem noch so kleinen, ausgedehnten Teil des Raumes"[2] [1, p. 399]. In a letter to D. Hilbert he wrote about the applicaton of transfinite set theory "auf die *Naturwissenschaften*: Physik, Chemie, Mineralogie, Botanik, Zoologie, Anthropologie, Biologie, Physiologie, Medizin etc. ... Dazu kommen aber auch Anwendungen auf die sogenannten 'Geisteswissenschaften', die meines Erachtens als Naturwissenschaften aufzufassen sind, *denn auch der 'Geist' gehört mit zur Natur*"[3] [2, p. 459].

---

[1] A potentially infinite quantity is always finite though not bounded, like a function. An actually infinite quantity is static, complete, i.e., *finished* and larger than any finite quantity of the same kind. Cantor calls it "finished infinite" (vollendet-unendlich) [1, p. 175], [2, p. 148, p. 181].

[2] Accordingly I distinguish an eternal uncreated infinity or absolutum which is due to God and his attributes, and a created infinity or transfinitum, which is has to be used wherever in the created nature an actual infinity has to be noticed, for example, with respect to, according to my firm conviction, the actually infinite number of created individuals, in the universe as well as on our earth and, most probably, even in every arbitrarily small extended piece of space.

[3] to the *natural sciences*: physics, chemistry, mineralogy, botany, zoology, anthropology, biology, physiology, medicine etc. ... In addition there are the applications to the so-called 'arts' which, in my view, also have to be considered natural sciences, *because also the 'mind' belongs to nature.*

B.Sriraman, C.Michelsen, A. Beckmann & V. Freiman (Eds). (2008). *Proceedings of the Second International Symposium on Mathematics and its Connections to the Arts and Sciences* (MACAS2). Information Age Publishing, Charlotte: NC, pp.265-272

Cantor's first proofs of actual infinity had partially religious character - based on his firm belief in God and encouraged by the Holy Bible which says: "'Dominus regnabit in infinitum (aeternum) *et ultra*'. Ich meine dieses '*et ultra*' ist eine Andeutung dafür, dass es mit dem ω nicht sein Bewenden hat, sondern dass es auch *darüber hinaus* noch etwas giebt"[4] [2, p. 148]. "Ein Beweis geht vom Gottesbegriff aus und schließt zunächst aus der höchsten Vollkommenheit Gottes Wesens auf die Möglichkeit der Schöpfung eines Transfinitum ordinatum, sodann aus seiner Allgüte und Herrlichkeit auf die Notwendigkeit der tatsächlich erfolgten Schöpfung eines Transfinitum"[5] [1, p. 400].

In the following we will search high and low for evidence of actual infinity, be it as a reality or as an idea. With respect to *reality*, only physics including cosmology is concerned, while the question of actual infinity in the realm of theology (is it reality or idea?) is a difficult matter. Nevertheless, it belongs to the general topic and will be discussed. Actual infinity *as an idea* is to be suspected mainly in mathematics. In addition there is frequent mentioning of infinity in aesthetics and art but those effusive statements[6] rarely lay claim to cover the meaning of actual infinity and will not further be considered. Also subjective statements of historical scholars[7] as well as merely formal applications like the use of infinite vector spaces in theoretical physics do not belong to our topic.

DOES THE INFINITELY SMALL EXIST IN REALITY?

Quarks are the smallest elementary particles presently known. Down to 10^{-19} m there is no structure detectable. Many physicists including the late W. Heisenberg are convinced that there is no deeper structure of matter. On the other hand, the experience with molecules, atoms, and elementary particles suggests that these physicists may be in error and that matter may be further divisible. However, it is not divisible in infinity. There is a clear-cut limit.

---

[4] 'The Lord rules in infinity (eternity) and beyond.' I think this 'and beyond' is a hint that ω is not the end of the tale.

[5] One proof is based on the notion of God. First, from the highest perfection of God, we infer the possibility of the creation of transfinity, then, from his all-grace and splendor, we infer the necessity that the creation of transfinity in fact has happened.

[6] "Silently one by one, in the infinite meadows of the heaven ..." (Henry W. Longfellow). "The Development of the aesthetics of the infinite" (Marjorie Hope Nicholson). "Above, infinite space, illimitable emptiness, with only the sun shining brazenly, eternally ..." (Sir Francis Chichester). " ... empty land stretching silently into infinity" (Alan Moorehead). "Infinity is where things happen that don't." (Anonymous). "I am painting the infinite" (Vincent van Gogh). "Towards the Infinite" (Joan Miró). [3]

[7] "Actual infinity does not exist in mathematics" (Aristoteles). "The number of points in a segment one ell long is its true measure." (R. Grosseteste). "Actual infinity exists in number, time and quantity" (J. Baconthorpe). "Je suis tellement pour l'infini actuel" (G. Leibniz). [4]

Lengths which are too small to be handled by material meter sticks can be measured in terms of wavelengths λ of electromagnetic waves, for instance.

$$\lambda = c/\nu \qquad (c = 3*10^8 \text{ m/s})$$

The frequency ν is given by the energy E of the photon

$$\nu = E/h \qquad (h = 6{,}6*10^{-34} \text{ Js})$$

and a photon cannot contain more than all the energy of the universe

$$E = mc^2$$

which has a mass of about $m = 5*10^{55}$ g (including the dark matter). This yields the complete energy $E = 5*10^{69}$ J. So the unsurpassable minimal length is $4*10^{-95}$ m.

DOES THE INFINITELY LARGE EXIST IN REALITY?

Modern cosmology teaches us that the universe has a beginning and is finite. But even if we do not trust in this wisdom, we know that theory of relativity is as correct as human knowledge can be. According to relativity theory, the accessible part of the universe is a sphere of $50*10^9$ light years radius containing a volume of 10^{80} m^3. (This sphere is growing with time but will remain finite forever.) "Warp" propulsion, "worm hole" traffic, and other science fiction (and scientific fiction) does not work without time reversal. Therefore it will remain impossible to leave (and to know more than) this finite sphere. Modern quantum mechanics has taught us that entities which are non-measurable in principle, do not exist. Therefore, also an upper bound (which is certainly not the supremum) of 10^{365} for the number of elementary spatial cells in the universe can be calculated from the minimal length estimated above.

HOW LONG LASTS ETERNITY?

We know that the universe has been expanding for about $14*10^9$ years. This process will probably continue in eternity. According to newer astronomic results the over-all energy density is very close to zero, suggesting an Euclidean space. Eternity, however, will never be completed. So time like space has a potentially infinite character. Both are of unbounded size but always finite.

Will intelligent creatures survive in eternity? One constraint is the limited supply of free energy which is necessary for any life. This problem could be solved, however, by living for shorter and shorter intervals with long hibernation phases. By means of series like 1/2 + 1/4 + 1/8 + ... = 1, a limited amount of energy could then last for ever - and with it intelligent life [5].

Alas, there is the risk of sudden death of a creature by an accident. If we assume that in our civilisation one out of 200 lives ends by an accident, then we can calculate that the risk to die by an accident during this very minute is roughly 10^{-10}. The risk that 6,000,000,000 people will die during this minute is then $10^{-60,000,000,000}$ (not taking into account cosmic catastrophes, epidemic diseases etc.). This is a very small but positive probability. And even if in future the risk of accidental death can be significantly reduced while the population may be enormously increased[8], the risk of a sudden end of all life within a short time interval is not zero and, therefore, will occur before eternity is finished [6, p. 135].

So, after a while, nobody will be present to measure time - and it may well be asked if an entity does exist which in principle cannot be measured.

IS THE HEAVEN INFINITE?

The existence of a creator has become more and more improbable in history of mankind. Copernicus, Bruno, and Darwin contributed to remove mankind from the centre of the universe, the position chosen by God for his creatures. The development of the character of God himself closely reflects the social development of human societies. G. C. Lichtenberg observed: "God created man according to his image? That means probably, man created God according to his."

The discovery of foreign cultures, in America and Australia, showed that people there had not been informed in advance about the God of Jews, Christians or Moslems - a highly unfair state of affairs in case belief in this God was advantageous before or after death.

The results of neurology and cerebral surgery show that characteristic traits and behaviour usually retraced to the human soul can be arbitrarily manipulated by electric currents, drugs, or surgery while an immortal soul cannot be localized.

Of course it is impossible to prove or to disprove the existence of one or more Gods, but it is easy to disprove the absolutum, as Cantor called it, i.e., the infinity of every property of a God. Medieval scholastics already asked, whether God could make a stone that heavy that he himself was incapable of lifting it. God cannot know the complete future unless the universe is deterministic. But in this case, there could be no free will and no living creature could prove itself suitable or unsuitable to enter paradise or hell - and the whole creation was meaningless.

Therefore, actual infinity, as being inherent to theological items, cannot be excluded but is at best problematic.

ARE THERE INFINITE SETS IN MATHEMATICS?

Set theory contains an axiom postulating the existence of an actually infinite set, and its most important theorem states that there are even different infinities. In particular the cardinal number of the infinite set of real numbers is larger than the cardinal number of the "smallest" infinite set, namely the set of natural numbers. This result has raised differing opinions. While most mathematicians were pleased about Cantor's theory "diese erscheint mir als die bewundernswerteste Blüte mathematischen Geistes und überhaupt eine der höchsten

---

[8] but at most to 10^{78} individuals because this is the number of hydrogen atoms in the universe, and life without at least one atom seems impossible.

Leistungen rein verstandesmäßiger menschlicher Tätigkeit"[9] [7, p. 167], "aus dem Paradies, das Cantor für uns geschaffen, soll uns niemand vertreiben können"[10] [7, p. 170], others remained sceptic. "In der Renaissance, besonders bei Bruno, überträgt sich die aktuale Unendlichkeit von Gott auf die Welt. Die endlichen Weltmodelle der gegenwärtigen Naturwissenschaft zeigen deutlich, wie diese Herrschaft des Gedankens einer aktualen Unendlichkeit mit der klassischen (neuzeitlichen) Physik zu Ende gegangen ist. Befremdlich wirkt dem gegenüber die Einbeziehung des Aktual-Unendlichen in die Mathematik, die explizit erst gegen Ende des vorigen Jahrhunderts mit G. Cantor begann. Im geistigen Gesamtbilde unseres Jahrhunderts ... wirkt das aktual Unendliche geradezu anachronistisch"[11] [8]. "Infinite totalities do not exist in any sense of the word (i.e., either really or ideally). More precisely, any mention, or purported mention, of infinite totalities is, literally, meaningless" [9].

In fact, there is some evidence that actual infinity, at least if understood as a number larger than every natural number, leads to problems. One simple observation is that every initial segment of even positive integers has a cardinal number

$$|\{2, 4, 6, ..., 2n\}| = n$$

which is less than some of the elements. There is no reason to expect that this discrepancy will disappear for the union of all initial segments, because it increases with increasing size of the segment. But this union is the set of all even positive integers (because there is no positive integer outside of the union). Its cardinal number should be larger than any element of the set.

The most important theorem of set theory states that the set of real numbers of the interval [0, 1] has a greater cardinal number than the set of natural numbers, i.e., the continuum [0, 1] is uncountable.

The infinite binary tree *T*, consisting of nodes, namely the numerals 0 and 1, contains the binary representations of all real numbers of the interval [0, 1] in form of infinite paths, i.e., sequences of nodes, starting at the root node and running through the levels of the tree, guided by the edges, as indicated below. Some real numbers, like 0.1000... = 0.0111...., are present even in duplicate.

---

[9] This is, in my opinion, the most admirable blossom of the mathematical mind and altogether one of the outstanding achievements of purely intellectual human activity.

[10] No one shall expel us from the paradise Cantor created for us.

[11] During the renaissance, particularly with Bruno, actual infinity transfers from God to the world. The finite world models of contemporary science clearly show how this power of the idea of actual infinity has ceased with classical (modern) physics. Under this aspect, the inclusion of actual infinity into mathematics, which explicitly started with G. Cantor only towards the end of the last century, seems displeasing. Within the intellectual overall picture of our century ... actual infinity brings about an impression of anachronism.

```
0        0.
         / \
1       0   1
       / \ / \
2     0  1 0  1
...     .....
```

Every *finite* binary tree T_n contains less paths than nodes. Down to level n there are $2^{n+1} - 1$ nodes but only 2^n path. In the infinite binary tree T the set of nodes remains countable. But the set of paths must be uncountable, because for each real number of the interval [0, 1] there is at least one path in T representing it.

The union $T_{\mathbb{N}}$ of *all finite* binary trees *covers all levels* enumerated by natural numbers. With respect to nodes and edges it is identical with the infinite binary tree

$$T_{\mathbb{N}} = T.$$

According to set theory (including the axiom of choice) a countable union of countable sets is a countable set. The set of paths in the union tree $T_{\mathbb{N}}$ is merely a countable union of *finite* sets, and, therefore, $T_{\mathbb{N}}$ contains only a countable set of paths. But does $T_{\mathbb{N}}$ contain only finite paths?

An *index* denotes the level to which a node belongs. The union of all indexes of nodes of finite paths is the union of all initial segments of natural numbers

$$\{1\} \quad \{1, 2\} \quad \{1, 2, 3\} \quad \dots \quad \{1, 2, 3, \dots, n\} \dots \quad \dots = \{1, 2, 3, \dots\}.$$

This is also the set of all last elements of the finite segments, i.e., it is the set of all natural numbers. This is the set of *all* indexes - there is no one left out. With respect to this observation we examine, for instance, all finite paths of the tree $T_{\mathbb{N}}$ which always turn right: 0.1, 0.11, 0.111, ... If considered as sets of nodes, their union is the *infinite* path representing the real number 0.111... = 1. From this we can conclude that also *every* other infinite path belongs to the union $T_{\mathbb{N}}$ of all finite trees.

The trees $T_{\mathbb{N}}$ and T are identical with respect to all nodes, all edges, and all paths (which would already have been implied by the identity of nodes and edges). But the set of all paths is countable in the tree $T_{\mathbb{N}}$ and uncountable in the same tree T.

Such problems of set theory can be avoided if we replace the fundamental concept of a one-to-one correspondence or bijection (leading to different transfinite cardinal numbers) by the concept of *intercession* [6, p. 116] briefly outline below.

Two infinite sets, A and B, *intercede* (each other) if they *can be* put in an intercession, i.e., if they can be ordered such that between two elements of A there is at least one element of B and, vice versa, between two elements of B there is at least one element of A.

The intercession includes Cantor's definition of equivalent (or equipotent) sets. Two equivalent sets always intercede each other, i.e., they can always be put in an intercession. (The intercession of sets with nonempty intersection, e.g., the intercession of a set with itself, requires the distinction of identical elements.[12]) The intercession is an equivalence relation, alas it is not as exciting as the bijection. *All* infinite sets (like the integers, the rationals, and the reals) belong, under this relation, to *one and the same equivalence class*. The sets of rational numbers and irrational numbers, for instance, intercede already in their natural order. There is no playing ground for building hierarchies upon hierarchies of infinities, for accessing inaccessible numbers, and for finishing the infinite. Every set which is not finite, is simply infinite, namely potentially infinite.

This point of view is in much better accordance with reality than the belief in actual infinity which has no application outside of mathematics at all [10, 11]. There are no numbers to be taken from an infinite set or from a Platonist shelf. "Die Zahlen sind freie Schöpfungen des menschlichen Geistes"[13] [12, p. III]. "A construction does not exist until it is made; when something new is made, it is something new and not a selection from a pre-existing collection" [13, p. 2]. When the objects of discussion are linguistic entities ... then that collection of entities may vary as a result of discussion about them. A consequence of this is that the 'natural numbers' of today are not the same as the 'natural numbers' of yesterday" [14, p. 478]. And we may add, what differs is not constant, i.e, it is not a constant set.

CONCLUSION

We have investigated actual infinity in all domains which reasonably could be suspected to sustain it. But the result is such that we can give the last word to D. Hilbert, who, at the end of his famous paper praising actual infinity of set theory, comes to a plain and astonishing conclusion:

"Zuletzt wollen wir wieder unseres eigentlichen Themas gedenken und über das Unendliche das Fazit aus allen unseren Überlegungen ziehen: Das Gesamtergebnis ist dann: das Unendliche findet sich nirgends realisiert; es ist weder in der Natur vorhanden, noch als Grundlage in unserem verstandesmäßigen Denken zulässig - eine bemerkenswerte Harmonie zwischen Sein und Denken"[14] [7, 190].

---

[12] As an example an intercession of the set of positive integers and the set of even positive integers is given by {1, 2, 3, ...} {2', 4', 6', ...} = {1, 2', 2, 4', 3, 6', ...}.

[13] Numbers are free creations of the human mind.

[14] Finally, let us return to our original topic, and let us draw the conclusion from all our reflections on the infinite. The overall result is then: The infinite is nowhere realized. Neither is it present in nature nor is it admissible as a foundation of our rational thinking - a remarkable harmony between being and thinking.

References

[1] E. Zermelo (Ed.): Georg Cantor, Gesammelte Abhandlungen mathematischen und philosophischen Inhalts, Springer, Berlin (1932); reprinted by Olms, Hildesheim (1966).

[2] H. Meschkowski, W. Nilson (Eds.): Georg Cantor Briefe, Springer, Berlin (1991).

[3] E. Maor: To infinity and beyond, Birkhäuser, Boston (1986).

[4] W. Mueckenheim: Die Geschichte des Unendlichen, Textbook, Univ. of Appl. Sciences, Augsburg (2004). http://www.fh-augsburg.de/~mueckenh/MR/Publ.mht

[5] F. J. Dyson: Time without end: Physics and biology in an open universe, Rev. Mod. Phys. **51** (1979) 447-460.

[6] W. Mückenheim: Die Mathematik des Unendlichen, Shaker-Verlag, Aachen (2006). http://www.shaker.de/Online-Gesamtkatalog/details.asp?ID=1471993&CC=21646&ISBN=3-8322-5587-7

[7] D. Hilbert: Über das Unendliche, Math. Ann. 95 (1925) 161-190. http://gdzdoc.sub.uni-goettingen.de/sub/digbib/loader?did=D26816

[8] P. Lorenzen: Das Aktual-Unendliche in der Mathematik, Philosophia naturalis 4 (1957) 3-11. http://www.sgipt.org/wisms/geswis/mathe/ulorenze.htm#Das Aktual-Unendliche in der Mathematik

[9] W.A.J. Luxemburg, S. Koerner (Hrsg.): A. Robinson: Selected Papers, Vol. 2, North Holland, Amsterdam (1979) p. 507.

[10]W. Mückenheim: "Physical Constraints of Numbers", Proceedings of the First International Symposium of Mathematics and its Connections to the Arts and Sciences, A. Beckmann, C. Michelsen, B. Sriraman (eds.), Franzbecker, Berlin (2005) 134-141. http://arxiv.org/pdf/math.GM/0505649

[11] The site of MatheRealism. http://www.fh-augsburg.de/~mueckenh/MR.mht

[12] R. Dedekind: Was sind und was sollen die Zahlen?, Vieweg, Braunschweig (1888). http://dz-srv1.sub.uni-goettingen.de/sub/digbib/loader?did=D46393

[13] Edward Nelson: Predicative Arithmetic, Princeton University Press (1986) 2. http://www.math.princeton.edu/~nelson/books/pa.pdf

[14] David Isles: What evidence is there that 2^65536 is a natural number?, Notre Dame Journal of Formal Logic, Volume 33, Number 4, (1992) 465-480.

http://projecteuclid.org/Dienst/UI/1.0/Summarize/euclid.ndjfl/1093634481?abstract

PROMOTING STUDENTS´ INTERESTS IN MATHEMATICS AND SCIENCE THROUGH INTERDISCIPLINARY INSTRUCTION

Claus Michelsen
Center for Science and Mathematics Education
University of Southern Denmark

Abstract: *This paper presents the German-Danish research project IFUN*[1] *- Interest and Interdisciplinary Instruction in Science*[2] *and Mathematics. The aim of the project is to investigate on how upper secondary students' interest in the subjects of mathematics, physics, chemistry and biology might be improved by increased instructional interplay and integration between the subjects. The individual student's interests in interdisciplinary domains of mathematics and science are studied within a two-dimensional framework: (i) Content context: The student's interest in a particular interdisciplinary domain of science and mathematics (ii) Identity: The student's affiliation with and valuation of mathematics and science. We present the main results from a questionnaire study which was carried out in the first phase of the project. Focus is on three issues: (i) The students' interests in mathematics and science at lower and upper secondary school (ii) The students' attitudes and expectations to mathematics and science at upper secondary school (iii) Making instruction in mathematics and science more interesting. The results of the study show that students have a very stable and high interest in mathematics and as a rule are positive towards interdisciplinary instruction. We conclude that the results indicate it is possible to expand interest in one subject to another subject through interdisciplinary instruction.*

1. INTRODUCTION

Research studies indicate that students' interest in science and mathematics declines during lower secondary school level (Hoffmann et al 1998), but international surveys as well as other indicators do not indicate any general decrease of interest in science among the population in most countries (Sjøberg 2003). A simple assumption is that many young people find various aspects of science and mathematics interesting, but it is beyond the capacity of the school to catch and hold this interest. Until recently research in interest has not to any great extent addressed the relation between interest and education at the upper secondary level. The reason for this could be the assumption that students have developed relatively stable individual interest profiles during lower secondary school. According to Prenzel (1998) there are several factors that lead to questioning this assumption. The students are confronted with an abundance of topics, and although many of these do not correspond to the students' individual interest they may lead to new areas of interest. At Danish upper secondary education the students' can choose between different subject packages and a number of

---

[1] The acronym IFUN refers to *Interesse og Fagoverskrindende Undervisning i Naturvidenskab* and *Interesse und Fächerübergreifender Unterricht in den Naturwisseschaften* which is Danish respectively German for Interest and Interdisciplinary Instruction in Science and Mathematics.
[2]We use the term science as a common denominator for the subjects of physics, chemistry and biology.

B.Sriraman, C.Michelsen, A. Beckmann & V. Freiman (Eds). (2008). *Proceedings of the Second International Symposium on Mathematics and its Connections to the Arts and Sciences* (MACAS2). Information Age Publishing, Charlotte: NC , pp.273-284

optional subjects. The choices of students might among other things be based on interest. Also the students' at upper secondary education have to make decisions about future education and profession. Personal interest is a key factor behind modern youth's educational choice and career aspirations are used as indicators of interest in scientific fields (Gardner 1998).

It is characteristic for research in students' interests that the domains, which have been studied, are ones that belong to academically defined curricula with focus on a single subject, for instance physics or mathematics. In the light of current tendency towards increased interplay between different subjects, interest research in the field of mathematics and science should also include broader and interdisciplinary domains, which are associated with applied, environmental, technological, ethical and socio-scientific content. It is the intention of the IFUN-project to fulfill these gaps in interest research.

2. INTEREST AND LEARNING

Dewey (1913) advocated that interests are the most important motivational factors in learning and development, and stressed the necessity of interest for the maintenance of learning as a self-initiated, content related activity, the mastery of which produces pleasure and satisfaction. The positive contributions of the idea of interest are twofold. It protects us from a merely internal conception of mind and from a merely external conception of subject matter. Thinking of the mind as an inner world of itself is not compatible with the conception of interest as an activity that moves toward an end and search for means. To avoid externality of subject matter, the presented matter must mean something complete to the students in its readiness and fixed separateness. This means that interest it not obtained by thinking about it and aiming at it, but by considering and aiming at the conditions that lie back of it, and compel it. Pointing on the normative aspect of education Krapp (2003) argues that teachers must take into account, which contents or objects are chosen for teaching and learning and what the long-term goals of development are to be. Therefore a theory of interest must state how contextual preferences develop and what the conditions are for a learner to have a more or less lasting interest in certain contents and to be ambivalent or even decline others. According to Cobb et al (2003) processes of learning should be interpreted broadly to encompass what is typically thought of as knowledge, the evolution of learning-relevant social practices and constructs as identity and interest.

The current state of interest research is the result of a fruitful exchange between researchers from psychology and educational science. Interest is conceptualized as a relational construct between a person and an object. An interest represents a more or less enduring specific relationship between a person and an object of interest in her or his life-space. The object of interest can refer to concrete things, a topic, an abstract idea, or any other content of the person's life-space. Three major views are reflected in interest research (1) interest as a characteristic of the person (individual interest) (2) interest as a characteristic of the learning environment (interestingness) (3) interest as a psychological state (actualized individual interest/situational interest) (Renninger et al 1992, Krapp 2002). The conceptual differentiation between situational and individual interests entails a focus on the process by which externally stimulated situational interest is stabilized and maintained and finally becomes integrated into a person's self as an individual interest. The overall implication of the literature in interest research is that situational interest can facilitate cognitive processing and improve students' learning. Mitchell (1993) proposes that situational interest is multifaceted and distinguishes between catching interest and holding interest. The shift from catching to holding a person's situational requires a learning environment that makes the content of learning meaningful for the students according to their actual goals. According to Krapp (2002, 2003) the transition from catch to hold will only occur if the student's activity is related to personal goals and the affective experience is positive and emotionally satisfactory. In the sense of situational interest an interest relationship depends on situational conditions, such as the specific context the object is presented or the activities the student is allowed to engage in. This dependency on favorable environmental conditions suggests an analysis of the content, the characteristics of the learning environment and the student's dispositional interest.

The design of interesting educational tasks encompasses studying alternatives existing to an actual practice in mathematics and science teaching, which demands an understanding of the characteristics of attitudes to the potential alternatives of the students who will engage in them. Identity and self-esteem have a key role in formulating personal and professional classroom identity (Middleton, Lesh & Heger 2003). We therefore broaden the analysis to include the ways that students' think about themselves in relation to mathematics and science, and the extend to which they have developed a commitment to and have come to see relevance and value in mathematics and science as it is realized both in and outside the classroom. To provide the study with this aspect the notion of students' personal identities is included in our analysis. The study of the individual student's interests in interdisciplinary domains of mathematics and science is thus conducted within a framework consisting of two dimensions:

- Content context: The student's interest in a particular interdisciplinary domain of science and mathematics
- Identity: The student's affiliation with and valuation of mathematics and science

3. INTEREST AND INTERDISCIPLINARY TEACHING

The choice of focus on broad and interdisciplinary domains of mathematics and science is motivated by research from both mathematics education and science education pointing at active and student-centered activities centered on applications of mathematics and science in real world settings as a way to facilitate increased interests in the subjects (e.g. Schiefele & Csikszentmihalyi 1995) An interest based approach to teaching raises the issue of the mathematics and science curriculum's content as problematic. Most students experience their curriculum as a series of courses in different disciplines rather than as an integrated interdisciplinary experience. Looking at the traditional mathematics and science curricula, one could say that in general the concepts taught are the basic concepts of mathematics and science. Only a tiny fraction has been concerned with cross-curricular, technological, and socio-scientific content of the mathematics and science curricula, and one might ask the question if contemporary education prepare the students to think mathematically and scientifically beyond school. To prepare future citizen for future life competencies are needed for the individual to cope with the complex world and some specific competencies can be acquired better in domains interdisciplinary contexts of mathematics and science than in others (Michelsen 2005). Also a proper image of the current scientific activities must include the interdisciplinary aspect – mathematicians, physicist, biologists, economists etc. are involved in the study and modeling of complex interdisciplinary systems.

4. RESEARCH DESIGN AND METHODOLOGY

The IFUN-project emphasizes the ways in which students' interests and learning styles can be capitalized upon a resource to ensure that all students have access to significant interdisciplinary ideas in mathematics and science. In this approach the focus is on practices of learning and understanding, including changes in resources and activities of practices that would strengthen the learning environment. The project involves students and teachers from 6 upper secondary schools - 3 from the southern part of Denmark and 3 from Schleswig-Holstein in the northern part of Germany. 2 of the Danish schools and 2 of the German schools are so-called general high-schools (gymnasium) while the remaining 2 schools are technical high-schools. The participating students are all grade 11 students and they have all opted for a science branch. The project is divided into two phases: (i) a descriptive/explanative phase and (ii) an intervention phase.

The first phase of the study is designed to provide a stimulus for interesting instruction in mathematics and science. The aim of this is to give empirical evidence on which kinds of interdisciplinary topics the students are interested in learning about. Focus is on situational conditions, such as the specific context the topic is presented in or the activities the student is allowed to engage in, that causes a situational interest in the topic and promote a shift from catching interest to holding interest. A major point here is that the development of situational interest is an individual process that cannot be understood separately from the environment in which it emerges and develops and from the individual student's identity. Self-reports – questionnaires and interviews - from the students are used for assessing the students' interests. In the second phase an intervention project will be initiated based on the insight gained in the first phase. The intention is to develop, implement and evaluate a didactic concept for interdisciplinary instructional units in mathematics and science. The intervention is planned as an implementation of 6-8 different instructional sequences developed in close cooperation between educational researchers and teachers from upper secondary school with the aim of producing a change in the classroom practice towards an interest based and interdisciplinary instruction.
In the following section we present the main results from a questionnaire study which was carried out in the first phase.

5. MAKING INSTRUCTION IN MATHEMATICS AND SCIENCE MORE INTERESTING

In the period from December 2005 to February 2006 grade 11 students from Danish upper secondary school in the region of Southern Denmark were given a 143-item questionnaire that took up one class period. A total of 255 students (147 girls and 118 boys) answered the questionnaire. To shed light on the to a great extend general results from the questionnaire interviews with 21 grade 11 students from Danish upper secondary school were conducted in the period from March 2006 to May 2006. The 143-items of IFUN questionnaire focus on seven domains, which shed light on different aspects of the three dimension of the IFUN-study. In this paper focus is on the results from three of the seven domains of the questionnaire: (i) *The students' interests in mathematics and science at lower and upper secondary school* (ii) *The students' attitudes and expectations to the subjects of mathematics and science at upper secondary school* (iii) *Making instruction in mathematics and science more interesting.* The goal of this part of the IFUN-study is to shed light on the students' views on changing mathematics and science education towards a more interdisciplinary approach with the aim of improving it more interesting and relevant.

5.1 Interest in mathematics and science

The domain *Interest in mathematics and science* focuses on the students' attitudes towards and interests in science and mathematics in basic school and in general and technical upper secondary school. The basic school (Folkeskolen) of the Danish education system is a 9-year comprehensive school. Prior to this, there is a voluntary pre-school class, and after there is a voluntary 10th school year. About 50% percent of a year group continues after completion of basic school to the academically oriented 3-years general or technical upper secondary school. The shift from basic school to general upper school is tremendous for many students. While the teachers of basic school are educated at teacher colleges the teachers of general upper secondary school are educated at universities. This first part of the questionnaire provides information about the students' interest in relation to the subjects of mathematics, physics, chemistry and science in lower and upper secondary school, and it consist of 18 questions, where the student rates their interest in and the relevance of the subjects of basic and general upper secondary and one question, that provides information about what the students finds interesting an uninteresting in mathematics, Danish language, sports, physics, biology, English, chemistry and history. From this part of the questionnaire we get the following information about the students' attitudes towards mathematics and science education:

- According to the students mathematics is among the three most interesting and important subjects at lower and at upper secondary school.
- The students' interest in mathematics does not change from lower to upper secondary school.
- There are no significant gender differences concerning students' interest in mathematics.
- The students believe that their parents consider mathematics as one of the most important subjects in school.
- The instruction in mathematics and science at upper secondary school is more interesting than the instruction at lower secondary school.
- Very few students have suffered defeat in mathematics lessons. Considerable more students have suffered defeat in physics lessons.
- Around 15% of the students think physics is interesting and around 15% think it is not interesting.

5.2 Attitudes and expectations to mathematics and science

This domain consist of 16 questions formulated articulated as 4 identical questions to each of the 4 subjects of mathematics, physics, chemistry and biology. We ask the students if it is relevant for them to be occupied with the subject, if they expect the subject to be important for their future education and profession, if their knowledge of the subject have impact on their attitude towards nowadays global challenges, and if their interest for the subject has changed within the latest months. Response to the each question is marked on 3-points scale:

1 = yes
2 = no
3 = don't know

Figure 1 below shows the correlation between questions S3.5 *Is it relevant for you to be occupied with mathematics?* and S3.6 *Can you imagine that you in a future education or future occupation will make use of something related to mathematics?*

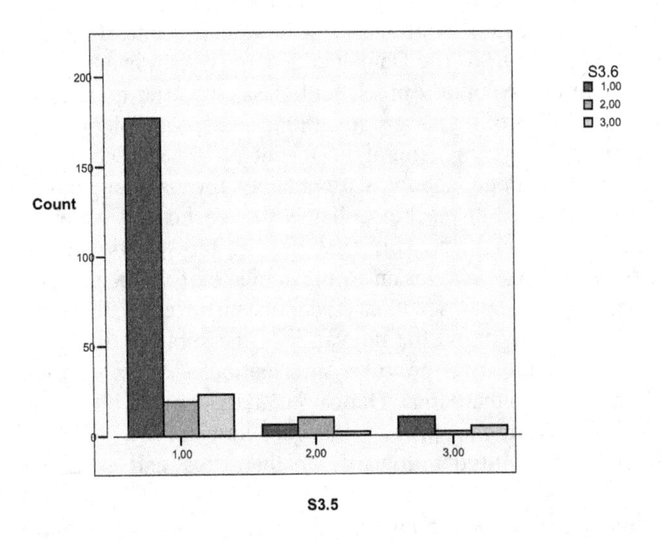

Figure 1 (N = 255)

Compared with the other subjects mathematics have the highest correlation between relevance for the student and the expectation of future applicability in education and profession. The subjects of biology and chemistry have the highest correlation between the question about relevance and the question *Does your knowledge about the subject have importance for your attitudes towards the global challenges, e.g. environmental problems, poverty, epidemics, terror.* Table 2 below shows this correlation for mathematics.

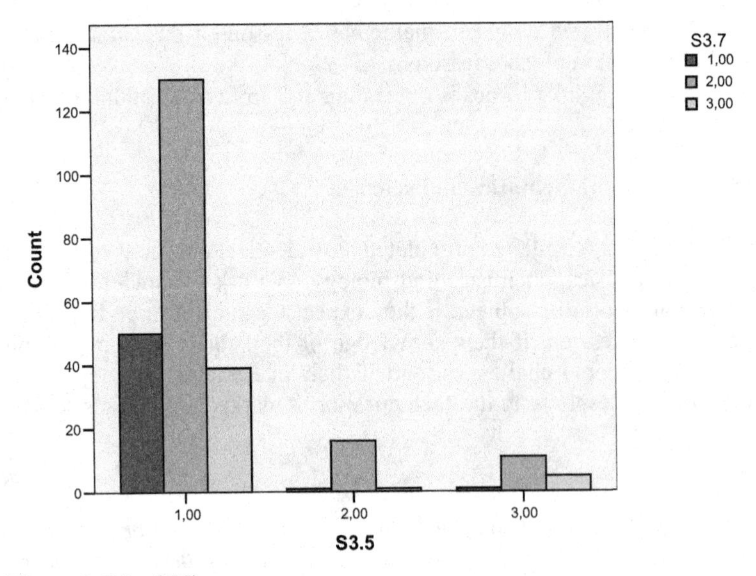

Figure 2 (N = 255)

It is clear from Figure 2 that the students do not relate their conception of mathematics as a relevant subject to their attitudes towards global problems.
Only a tiny fraction of the students report a change in interest for the subject within the latest month. For those who have changed interest, in positive or negative direction, there are two main reasons:

- Change of teacher.
- Change in the students' perception of level of understanding of the subject.

5.3 Making instruction more interesting

The domain of the questionnaire consists of an introduction *Instruction in mathematics and science would be more interesting if the following change was made* followed by 15 statements about possible changes in mathematics and science education, each with a 5-point Likert scale and a field for comments. The statements cover a broad spectrum ranging from an increased interplay between the subjects of mathematics and science over focusing on career opportunities related to mathematics and science to including ethical and moral questions and the human aspect in the teaching of science and mathematics. The responses to the 15 statements were marked on a five-point scale as follows:

5 = strongly agree
4 = agree
3 = neutral
2 = disagree
1 = strongly disagree

This enabled mean responses to the statements to be calculated as shown on Table 1.

Statement	Mean, girls	Std deviation, girls	Mean, boys	Std deviation, boys
Inclusion of examples of professions, where mathematics and science are applied	3,8015	0,83284	4,0339	0,92387
Instruction is more often organized as around an interdisciplinary theme	3,2117	1,01772	3,3162	1,11146
Cooperation between mathematics, science and humanistic subjects	3,2044	1,06508	3,2797	1,08514
Cooperation between mathematics, science and subjects of arts	3,2482	1,14263	2,6154	1,22366
Cooperation between mathematics, science and subjects of social sciences	3,6131	0,99446	3,6017	0,99692
Cooperation between the subjects of mathematics, physics, chemistry and biology	3,4745	0,95548	3,5254	1,01873
Cooperation with local companies	3,9781	0,79951	3,9576	0,91888
Cooperation with universities and other research institutions	4,1825	0,69894	4,1271	0,84273
Using alternative representations of subject matter, e.g. narratives or drama	3,5182	1,09210	3,3305	1,10210
Inclusion of the human dimension of the scientific activity	3,3650	0,96159	3,6102	1,01303
Inclusion of moral and ethical issues	4,0803	0,88333	3,8644	0,96001
Inclusion of societal perspectives	3,5547	0,89047	3,2966	1,04029
More focus on the learning of the abstract concepts of mathematics and science.	3,4380	0,88169	3,5254	0,97588
Modern media are used to visualize the concepts	3,6277	0,82254	3,6949	0,85243
Visits to companies, universities is included in instruction	4,1606	0,77870	3,9746	0,95597

The responses from the students are divided up according to gender. The reason for this is that a tremendous amount of empirical work focuses on gender differences in interest (Renninger et al 1992, Hoffmann et al 1988). We notice the significant positive attitude towards increased interdisciplinary in the teaching of mathematics and science, and that gender differences are only observed for only few of the statements.

One of the main issues of the IFUN-study is the students' interest in increased interdisciplinary instruction. To shed more light on the students' attitudes towards interdisciplinary activities we show with the two figures below the correlation between the students' conception of the relevance of being occupied with mathematics and their attitudes towards increased cooperation between mathematics, science and subjects of social sciences (Figure 3) and towards increased cooperation between the subjects of mathematics, biology, chemistry and physics (Figure 4).

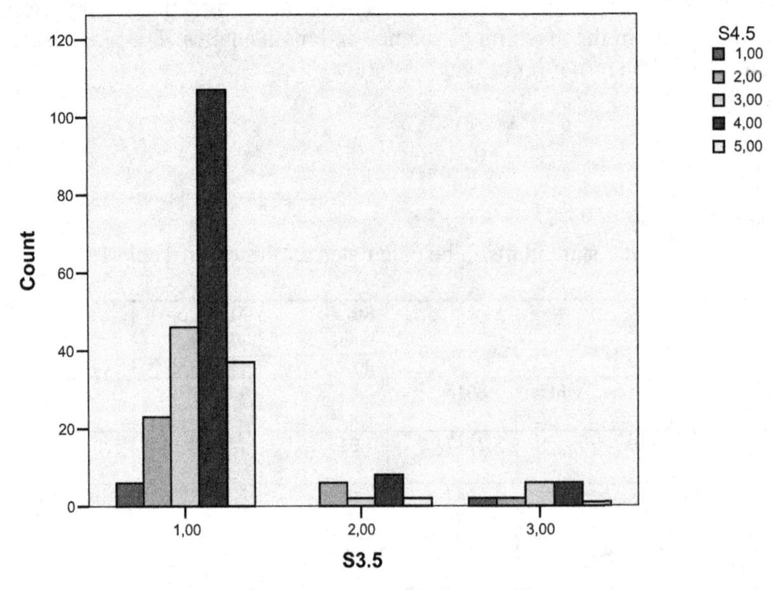

Figure 3 (N = 255)

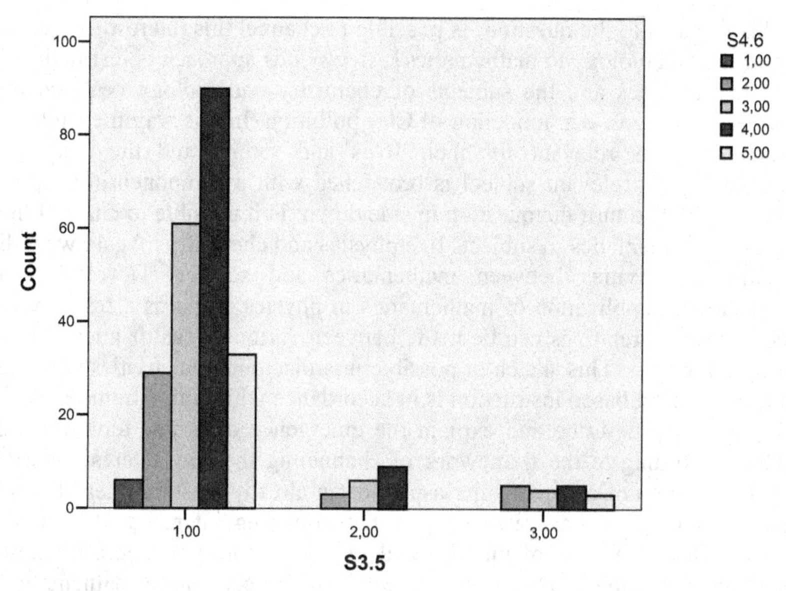

Figure 4 (N = 255)

The figures show a positive correlation between the students' conception of mathematics as a relevant subject and their attitudes to increased cooperation between mathematics, science and social sciences.

6. CONCLUDING POINTS

Students who have opted for a science branch at upper secondary school have a very stable and positive interest in mathematics. Although the type of mathematics instruction varies substantially between lower and upper secondary school there is no significant change in the students' interest in mathematics when the students shift from lower to upper secondary school. Our results show that as a rule the students are positive towards interdisciplinary instruction. The only significant gender difference is related to cooperation between mathematics, science and subjects of arts and to including moral and ethical issues in the instruction of mathematics and science. In both cases the girls have more positive attitudes than the boys. From the findings we can also conclude that inclusion of societal issues will increase the students' interest.

Dewey (1916) emphasizes in his famous work *Education and Democracy* that an occupation is the only thing which balances the distinctive capacity of an individual with his social service. Clearly issues related to study and profession opportunities have a relatively high mean in the study. Our interviews with students revealed that it is very difficult for them to explain why mathematics is important for their future profession. The study thus indicates that the students' interest in mathematics and science will increase if career oriented aspects are included in instruction.

The results of our study open up for some interesting perspectives for the second phase of the IFUN-study. We notice that although mathematics is considered as a relevant and interesting subject by the students the subjects of chemistry and biology have a higher correlation between the students' conception of the relevance of the subject and their attitudes towards global challenges. This leads us to the question: Is possible to channel this interest for global challenges from chemistry and biology to mathematics? An obvious approach is instructional cooperation between mathematics and the subjects of chemistry and biology centered on themes related to global challenges, e.g. modeling of CO_2 pollution. In this way the students experience that mathematics is relevant for their lives and future, and the students' conception of mathematics as a relevant subject is broadened with a component related to global challenges. We might also turn the question upside down: Is it possible to channel the students' stable interest in mathematics to subjects like physics and chemistry? Again we will point on interdisciplinary activities between mathematics and science. Developing a mathematical concept through application of mathematics in physics might as a result have that the student recognizes that relations can be made between various areas of interest that until then did not belong together. This sketch of possible interdisciplinary activities with the aim of promoting a more interest based instruction is in accordance with Fink's framework of channeling and overlapping to describe and explain the emergence of a new topic related interest (Krapp 2002). According to the framework of channeling the new interest emerge through increased differentiation of one particular aspect of an already existing interest. And according to the overlapping framework a detection of overlapping interest in two areas which until then were different objects of interest lead to a new object of interest. These considerations concerning a possible expansion of the interest, for example of mathematics, into other subjects will play a major role in phase two of the project, where focus is on the development of interest based instructional sequences.

7. FINAL COMMENTS

Empirical results clearly show that interest-based learning has many benefits (Renninger et al 1992). The first phase of the IFUN-study suggests that upper secondary students' interest for mathematics and science can be improved by putting more weight on interdisciplinary activities. A reform of the content of mathematics and science at upper secondary education towards an interdisciplinary approach that includes applied, environmental, technological, career opportunities and socio-scientific aspects have the potential to appeal to the minds of upper secondary school students. Effective instruction will depend on a curriculum that introduces important topics in both science and mathematics in a suitable order with appropriate linkages between them. In order for students to gain an interest and appreciation in science and math they need to be aware of how these subject areas affect their lives.

The IFUN-project is granted by the EU Interreg IIIa Programme in the regions of Fyn/KERN and Sønderjylland/Scleswig

References:

Cobb, P., Confrey, J., diSessa, A., Lehrer, R., & Schauble, L. (2003) Design Experiments in Educational Research. *Educational Researcher, Vol 32, No. 1,* 9-13.
Dewey, J. (1913). *Interest and Effort in Education.* Boston, New York and Chicago: Houghton Mifflin Company.
Dewey, J. (1916). *Education and Democracy.* The Macmillan Company.
Gardner, P.L. (1998). The Development of Males' and Females' Interests in Science and Technology. In Hoffman, L., Krapp, A., Renninger, K.A. & Baumert, J. (eds.) *Interest and Learning.* Kiel: IPN, 41-57.
Hoffman, L., Krapp, A., Renninger, K.A. & Baumert, J. (eds.) *Interest and Learning.* Kiel: IPN.

Krapp, A. (2003) Interst and human development: An educational-psychological perspective. *Development and Motivation, BJEP Monograph Series II (2),* 57-84.
Krapp, A. (2002) Structural and dynamic aspects of interest development: theoretical considerations from an ontogenetic perspective. *Learning and Instruction, 12,* 383-409.
Middleton, J.A., Lesh, R. & Heger, M (2003) Interest, Identity, and Social Functioning: Central Features of Modeling Activity. In Lesh & Doerr (eds.) *Beyond Constructivism. Models and Modeling Perspectives on Mathematical Problem Solving, Learning, and Teaching.* Mahwah: Lawrence Erlbaum Associates, 501-517.
Michelsen, C. (2005). Expanding the domain: Variables and functions in an interdisciplinary context between mathematics and physics. In Beckmann, Michelsen & Sriraman (eds.) *Proceedings of The First International Symposium of Mathematics and its Connection to the Arts and Sciences.* Hildesheim & Berlin: Verlag Franzbecker, 201-214.
Mitchell, M. (1993) Situational Interest: Its Multifaceted Structure in the Secondary School Mathematics Classroom. *Journal of Educational Psychology, 85*, 424-436.
Prenzel, M. (1998) Interest Research Concerning the Upper Secondary Level: An overview. In Hoffman, L., Krapp, A., Renninger, K.A. & Baumert, J. (eds.) *Interest and Learning.* Kiel: IPN.
Renninger, K.A., Hidi,S. & Krapp, A. (eds.) (1992) *The Role of Interest in Learning and Development.* Hillsdal: Lawrence Erlbaum Associates.
Schiefele, U. & Csikszentmihalyi, M. (1995) Motivation and Ability as Factors in Mathematics Experience and Achievement. *Journal for Research in Mathematics Education, 26(2),* 163-181.
Sjøberg, S. (2003). Krise! Hvilken krise? Myter og realiteter om naturfagene i Norge. In Jorde & Bungum (eds.) *Naturfagdidaktikk. Perspektiver. Forskning. Udvikling.* Oslo: Gyldendal Akademisk.

MULTI-DISCIPLINARY PROJECTS IN UPPER SECONDARY SCHOOL – NEW ROLES FOR MATHEMATICS?

Mette Andresen, Lena Lindenskov

Danish University School of Education - Aarhus University, Denmark

It is a central part of the Danish 2005-reform of the general upper secondary school that for the first time ever multi-disciplinary courses have become compulsory for all teachers and all students, as well as part of final examination for all. Before the reform only some teachers were actively involved in inter- and multi-disciplinary projects. This paper reports from a case study of such a new triple/four-disciplinary project in mathematics, physics, chemistry and 'general study preparation' [Danish: almen studieforberedelse] performed by experienced teachers. The aims of the case study is to inquire the relations the students establish between mathematics and other subjects and to look for possible influences on the students' conceptions and use of methods. The paper ends by considering some methodological issues in follow-up studies of how multi-disciplinarity is practiced in school and perceived by teachers, students and in official documents.

BACKGROUND OF THE CASE-STUDY – INTERESTS AND QUESTIONS

This paper is based on experiments in a 'Reform-2006' classroom in upper secondary school in Denmark, as we find it highly relevant to produce knowledge about what actually happens when new pedagogy, which is developed, used and admired by few fiery souls, are adjusted and becomes compulsory for all by educational laws and regulations. Mathematics from now on will in some teaching sequences be an element of a new type of projects in upper secondary school.

It is an official aim of the new pedagogy elements of 'inter- and multidisciplinary projects' to provide students with possibilities to recognize and practice new and unusual connections/relations between mathematics and the sciences.

We are interested in issues in the reform process concerning teachers' and students' actions and concerning teachers' and students' perception of this reform element. It is our interest to explore how fiery souls act and think; how 'average' teachers – whatever 'average' in this sense may mean – act and think; what happens in the teacher planning and coaching practice, which is being developed; how students' learning may be affected; and how reform elements – like 'multi-diciplinary projects' – may be understood and perceived by teachers and students?

In a case-study of a triple/four-disciplinary project in mathematics, physics, chemistry, and 'general study preparation' we explore the following questions concerning students: which relations do the students seem to establish between mathematics and other subjects; and how may students' mathematical conceptions and use of methods be influenced?

We collect and analyse qualitatively data from several sources, describe and analyse episodes of classroom teaching and learning. We try to identify traces of school mathematics in the

B.Sriraman, C.Michelsen, A. Beckmann & V. Freiman (Eds). (2008). *Proceedings of the Second International Symposium on Mathematics and its Connections to the Arts and Sciences* (MACAS2). Information Age Publishing, Charlotte: NC , pp.285-292

students' activities by analysing what content and experience the teachers and the students refer to, and in which situations.

MANY KINDS OF COOPERATION IN SCHOOL - WHAT MAY MULTI-DISCIPLINARITY MEAN?

Six forms of cooperation between teachers in school have been distilled from what has actually happened in Danish primary and lower-secondary schools:

1. A teacher - let's say a teacher in English - asks a colleague - let's say a teacher in Geography - to help by giving a specific Geography course element relevant to the course in English

2. Two teachers cooperate because of a common theme – and learners will probably learn better each subject – a symbiosis may be created – although the teachers think in subject before they think inter/cross-curricular

3. A teacher chooses a theme or problem area independently of school subjects but because of relevance for students' actual and future lives. In order to qualify analyses and handling of the theme or problem, some relevant concepts, approaches, methods from one or more subjects are used. This is called functional inter-curriculum.

4. Students choose a theme or problem – students find information, analyse and communicate results. Students come to be more knowledgeable, learn about the theme or problem, and learn some methods

5. A group of teachers from different subjects choose a theme with a potential of being analysable in common. All subjects who wish to participate do so. This form is called formal inter-curricular.

6. A teacher and/or students choose a problem and use knowledge from everyday experiences. (Kristensen, 1998)

Use and discussions of these kinds of curricular elements are not new, but the main ideas have shifted through history (Kristensen, 2001). For a further discussion of some of the conflicts in the area, see Jensen (1995). For a pragmatic view of the area, see Undervisningsministeriet (2000).

LITERATURE FROM MACAS1

In MACAS1 Claus Michelsen emphasises that "the lack of coordination between the curricula of physics and mathematics is in one of the primary cause of students' difficulty of application of mathematics in physics. It is difficult for the students to transfer concepts, ideas and procedures learned in mathematics to a new and unanticipated situation in physics lessons." Michelsen points to an alternative approach that stresses the importance of modelling activities in an interdisciplinary context between the two subjects.

Lllk Victor Freiman and Nicole Lirette-Pitre focus in MACAS1 on how a new approach to the school curriculum with emphasis on linking different subjects and making interdisciplinary connections, requires that teachers have clear visions of the interconnections of the multiple disciplines and can develop different teaching approaches to help attain the transdisciplinary learning results in communication, information and communication technology (ICT), critical thinking, personal and social development, study and work habits, and culture and heritage.

R. Filo and M. Yarkoni, M. analyse in MACAS1 interdisciplinary project integrating geometry and art as interdisciplinary learning of parallel concepts that are expressed differently in each field yet complement each other. They "assume that combining art with math should cause the students to feel more positive towards the study of math. As well as, including art as an integral part of math studies will strengthen and 'elevate the status' of the art classes."

INSTITUTIONAL MILIEU AND TEACHERS' ATTITUDES TO REFORM

The school is situated in a suburb of Copenhagen with a diverse population. The school has been – and still is - one of the leading upper secondary school in the country in regard to teaching and curriculum experiments. Reform processes are relevant to study both as grounded in experienced teachers and as well as grounded in newly appointed teachers. In this case study we study experienced teachers only. The three teachers all have experiences from many years of teaching and experimenting in this school.

What meets us when we visit the school and the experienced teachers is a friendly and warm environment for learning in the classroom. Asked about the reform situation the teachers talk about more demands for documentation besides the new demands for compulsory cooperative teacher work. From media and from professional discussions it seems that many teachers feel they neither get appropriate time nor appropriate further training and counselling to cope with the new demands. Also some teachers are concerned that reforms ideas, which in the proper sense they are in congenial company with, will turn out to be impossible to implement. Some teachers feel that some of the reform intentions are mutually contradictory. Especially the concerns turn on single disciplines towards multi-disciplinarity.

THE CASE – A TRIPLE-FOUR DISCIPLINARY PROJECT

The triple/four-disciplinary project involved a second-year class in mathematics, chemistry and physics, and focused on rockets: the theory and practice of rocket construction as well as military and civil use of satellites and rockets. The three teachers plus a prospective teacher in mathematics planned, managed and evaluated the teaching sequence in common. Each subject spent about three weeks' lessons on the project. Besides, the new school subject 'general preparation for study' spent about 25 lessons. In chemistry, the preceding sequence of lessons concerned with pyrotechnics led up to the rocket project. The rocket project was succeeded by a five-day study visit in Munich by the students, which included a visit to the Technical Museum. According to the teachers' plan, the content parts of the project were related to the subjects like this:

Mathematics: Deduction and calculations of the rocket equation: $v_{final} = u \cdot \ln\left(\frac{M + m_0}{M}\right)$

(Here, the rocket's velocity is expressed by the exhaust velocity and the masses of the empty rocket and of the launch amount of fuel).

Chemistry: Generation of energy, temperatures of combustions, liquid and solid fuels and different types of rockets.

Physics: Newton's laws, circular orbits and gravitation, conservation of momentum, and ballistic motion.

General preparation for study: History of rockets, Satellites and GPS (=Global Positioning System).

The students are requested to write individual chapters to a common report and to each give a five-minutes oral presentation of the theme of their chapter in front of teachers, class-mates and students from another class.

The project seems to provide conditions for students to improve their understandings and competences on two levels. Opportunity to learn about rockets, i.e. understand some principles, investigate by experiments and calculations different quantities of rockets and their movements, and opportunity to get informed about historical developments and some civic uses of rockets. Furthermore students get opportunity to understand and practice mathematical, physics and chemistry concepts and to further their understanding on some connections between some of the concepts.

Data for our case study analysis consist of film recordings and field notes from about eight lessons, teaching materials, and notes from informal talk with the teachers and the students' written reports, power point presentations and written homework tasks. Further, it is our plan to make interviews with a group of students some months after the teaching sequence.

In this paper, the analyses are based on two selected episodes from the data: one episode takes place during a lesson in chemistry and the other occurs while the students do group work in the classroom. The episodes are selected to enlighten how the students perceive school mathematics and relate it to physics, to chemistry and to their everyday experiences.

Episode 1

This episode takes place in the project's first chemistry lesson. Introducing the chemistry part of the project to the students, the teacher states that most likely, the students will not feel any distinction between the chemistry parts of the project and the parts, relating to general preparation for study. He stresses that there would be no intended distinction between these parts.

The teacher goes through a note on rocket engines, presuming that the students have prepared for the lesson by reading it in advance.

The concept of specific momentum is introduced and explained in the note (*Raketmotorer* page 2, mea translation):

> The specific momentum, which is in fact no momentum, is defined by "kg pressure power per kg consumed fuel per second". That is, if a given fuel may deliver a 200 kg pressure power by consumption of 1 kg fuel each second, then $I = \frac{200kg}{1kg/s} = 200s$.
>
> Notice, that the unit of specific momentum is second. Notice also that the specific momentum gets the same value and the same unit if it is defined by "pounds pressure power per pound consumed fuel per second".
>
> It is easy to show that $I = \frac{u}{g}$ where u is the exhaust velocity and g is the gravitation.

After a short discussion about types of chemical reactions that give a huge amount of energy very fast, the teacher continues, referring to the note:

Teacher: Then we have a short chapter on specific momentum – have you worked with this together with G (the physics teacher)?

Student: We have talked about the concept of momentum

Teacher: Have you also talked about the concept of specific momentum?

(Confirmative mumbling)

Student: And in mathematics, too

Teacher: Do we have to do it again, then?

Student1: I would like you to explain this again

Student2: Will you please do it again?

Teacher: Let's do that

A few minutes later, the teacher explains that the higher velocity of the exhaustion, the higher velocity of the rocket. He refers to the note's table of graphs of pressure, density, temperature and speed as functions of the position in the rocket nozzle. One of the students interrupts the teacher:

Student1: A (name of the teacher)? What is this: *I* equals *u* divided by *g*, what is it?

Teacher: *I* is the specific momentum, it is a property of the fuel, isn't it?

Student1: But is it a formula, or what? That is, I have never seen it before.

Teacher: Then you see it now!

Student2: We have seen it before in a different way

Student1: Yes, we have seen it before in a different way

Student3: It is the same

Apparently, student1 uses the term 'a formula' in the meaning of 'shared knowledge in the class', known from previously taught lessons. That is, in our interpretation, that student1 wants to know whether the expression $I = \frac{u}{g}$

is supposed to be well known. The students 2 and 3 agree that it is part of the shared knowledge but give student1 an excuse for not recognizing it ('we have seen it in a different way').

In parallel, the question 'what is it?' in our interpretation reveals that student1 also feel unsure of the meaning of the concept symbolized by *I*. The teacher follows up with a question, apparently with the intention to find out whether the students understand the expression and the concept of specific momentum or not:

Teacher: In what way have you seen it?

Student: We know *I* equals … *p* times *v*, wasn't it so?

Student: I equals p times vm

Teacher: The u is the same as…

Student (interrupts): or u times v

Teacher: …is the same as momentum divided by mass, isn't it?

Student1: But is it a formula for momentum, or what?

Teacher: It is a formula for what is called the specific momentum of the fuel. It is related with the exhaust velocity so you can say that these things – it is easy to see, based on the conservation of momentum, that the higher speed the better, isn't it? That is, the higher specific momentum the better.

The problem might be that the concept of specific momentum is distinct from the concept of momentum. The first one is rather new to the students, whereas the latter, supposedly, is part of their shared knowledge. The students might not be aware of the distinction.

The episode demonstrates two levels of questions: one level concerns the pull of shared knowledge, and negotiations about which formulas and expressions the teacher and the students are allowed to use, when they want to refer to this. The other level concerns the meaning of the concepts, referred to by the symbols.

Episode 2

The second episode reports on one group's work with a task, which forms the basis for one of the individual presentations. The episode serves to demonstrate how the students choose appropriate mathematical tools when faced with a problem. The teacher presents the problem: which height does the rocket reach? One of the students refers to a ballistic experiment, previously done with the class. The experiment concerned with a ball but the teacher argues, that they then calculated the height of its path based on the mass of the ball. Then the teacher and one of the students in common refer to another experiment on height, which involved a tall house in the neighbourhood. The teacher makes a sketch on the blackboard (fig 1).

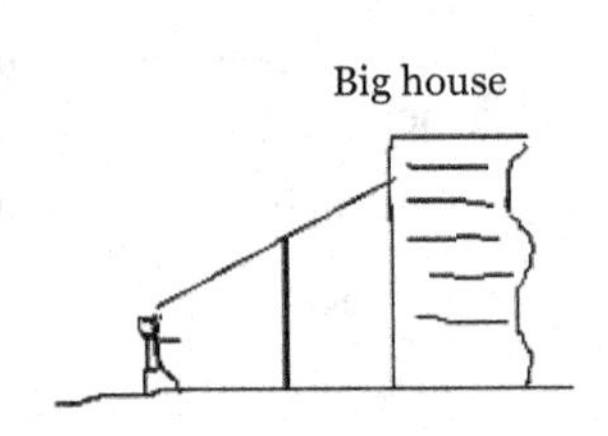

Fig1. The teacher's sketch One student asks whether it would be appropriate in the actual case to use trigonometry, and the teacher answers yes. The teacher then introduces the use of laser to measure the height.

In the following lesson, the group work starts. A group of four students are requested to design an experiment, which could serve to determine the height of the rocket's path. The students make a drawing on the computer, resembling the teacher's sketch (fig 2.)

Fig 2. The students' sketch

Based on their drawings, the students discuss the design of the experiment. They neither refer to the previous experiments with the ball nor to the one with the big house. This may be because the teacher has already discussed these two in plenum. Neither do they refer to the textbook nor to the mathematics lessons. They ask the teacher if it would be all right to use similar triangles, and the teacher answers yes:

Teacher: You are allowed to use Pythagoras and similar triangles – even if the math teacher loves cosines and sinus, you are not obliged to do the same!

Student: But how can we determine this distance…we just use similar triangles.

The short excerpt indicates how the students manage to compare the actual problem with previous experiences and to consider different strategies before they choose an appropriate one to solve the problem. The students, seemingly, do not feel obliged to use their most advanced school mathematics (in the form of trigonometry). On the contrary, the teacher (humorously) encourages them to be critical towards their math teacher's motives for working with trigonometry.

FURTHER DISCUSSIONS

It is by no means easy to inquire the relations the students establish between mathematics and other subjects and to look for possible influences on the students' conceptions and use of methods. It raises methodological concerns: as a first rough indicator of traces of multi-disciplinary thoughts, acts, and competences we choose to use references made by teachers and learners in different situations to classes in mathematics, in physics, and general study preparation and to their everyday situations. We choose to conceptually understand the traces by means of authenticity: Which kinds of authenticity are played out in classroom social interaction? Which kinds of authenticity are played out in students' written reports and oral presentations of these? And what does authenticity mean to the robustness of research results? See Lindenskov (2003).

This first case study cannot stand alone, so we plan for follow-up studies during the next years with teachers and students, whereupon we will compare their perceptions and the traces of multi-disciplinarity with discourses on multi-disciplinarity, authenticity, and references in the relevant legal documents from the Ministry of Education.

Issues of generalizations surely are demanding and fragile. How can it be asserted our analyses may bring something of value for other schools, teachers, or students? We wish to point at the on-going discussion of how relations between practice and research can be described and enhanced. See i.e. K. Ruthven (1999, 2000a,b).

References:

Beckmann, A., Michelsen, C., Sriraman, B. (Eds.) (2005). *Proceedings of The First International Symposium of Mathematics and its Connections to the Arts and Sciences.* 19th – 21st May 2005, Schwäbisch Gmünd, Germany. Berlin: Euro Verlag Franzbecker.

Jensen, B.E. (1995). Tværfaglighed. *Undervisningsministeriets tidsskrift Uddannelse,* 9. København: Undervisningsministeriet.

Kristensen, H. J. (1998). Tværfaglighed - flop eller noget af betydning. Skolebiblioteket, 1. Localised February 2007 at world wide web at www.emu.dk

Kristensen, H. J. (2001). Hvad er tværfaglighed? – reportage fra et foredrag på Århus Dag- og Aftenseminarium. Localised February 2007 at world wide web at www.emu.dk

Lindenskov, L. (2003a). Tutors' approaches to Mathematics to live and work: A model of four types. In: J. Maasz & W. Schloeglman (Eds.), Mathematics to Live and Work in our World (pp. 146-153). Linz, Austria: Universitätsverlag Rudolf Trauner.

Michelsen, C. (2005). Expanding the domain - Variables and functions in an interdisciplinary context between mathematics and physics. In: Beckmann, A., Michelsen, C., Sriraman, B. (Eds.) (2005). Proceedings of The First International Symposium of Mathematics and its Connections to the Arts and Sciences. 19th – 21st May 2005, Schwäbisch Gmünd, Germany. Berlin: Euro Verlag Franzbecker.

Michelsen, C., Glargaard, N. and Dejgaard, J. (2005) Interdisciplinary competences – Integrating mathematics and subjects of natural sciences. In M. Anaya and C. Michelsen (Eds.), *Relations between mathematics and other subjects of science or art* (pp 32-37)

Rachel Filo, R.; Yarkoni, M. (2005). A.Geomart"* – Geometry Reflected Through Art. In: Beckmann, A., Michelsen, C., Sriraman, B. (Eds.) (2005). *Proceedings of The First International Symposium of Mathematics and its Connections to the Arts and Sciences.* 19th – 21st May 2005, Schwäbisch Gmünd, Germany. Berlin: Euro Verlag Franzbecker.

Ruthven, K. (1999) Reconstructing Professional Judgment in Mathematics Eudcation: from Good Practice to Warranted Practice. In C. Hoyles, C. Morgan and G. Woodhouse (Eds.), Rethinking and Mathematics Curriculum, Studies in Mathematics Educations Series (Vol. 10, pp 203-216).

Ruthven, K. (2002a) Linking researching with teaching: towards synergy of scholarly and craft knowledge. In L. English (Ed.) Handbook of International Research in Mathematics Education (Lawrence Erlbaum, Mahwah NJ) pp. 581-598. Available online at http://www.icme-organisers.dk/dg02/Ruthven.doc

Ruthven, K. (2002b) Towards synergy of scholarly and craft knowledge. In H-G. Weigand et al. (Eds.) Developments in Mathematics Education in German-speaking Countries: Selected Papers from the Annual Conference on Didactics of Mathematics, Potsdam, 2000 (Franzbecker Verlag, Hildesheim) pp. 121-129.Available on-line at http://webdoc.gwdg.de/ebook/e/gdm/2000/ruthven_2000.pdf

Smith, D. S. (1999) Mathematics and Scientific Literacy. In C. Hoyles, C. Morgan and G. Woodhouse (Eds.), *Rethinking the Mathematics Curriculum*, Studies in Mathematics Educations Series, 10, pp 130-139).

Thorup, O. (2003) Erfaringer med ledelse og organisation i en elektronisk skole. I O. Thorup, A. Witzke (red). *"Stik mig en bærbar" IT-undervisning på Avedøre Gymnasium* (pp 3-8).

Undervisningsministeriet (2000). *De fire tværgående dimensioner på htx* - Uddannelsesstyrelsen temahæfteserie som nr. 8 under temaet den pædagogiske proces. s, København: Uddannelsesstyrelsen. Området for gymnasiale uddannelser.

Victor Freiman, V. ; Lirette-Pitre, N. (2005). Innovative approach of building connections between science and math didactics in pre-service teacher education using wiki-technology. In: Beckmann, A., Michelsen, C., Sriraman, B. (Eds.) (2005). *Proceedings of The First International Symposium of Mathematics and its Connections to the Arts and Sciences.* 19th – 21st May 2005, Schwäbisch Gmünd, Germany. Berlin: Euro Verlag Franzbecker.

Vilhelmsen O. (2003) Naturfag og matematik i klasser med bærbare computere. I O. Thorup, A. Witzke (red). "Stik mig en bærbar" IT-undervisning på Avedøre Gymnasium (pp 17-18).

Wittmann, E. (1998). Mathematics education as a 'deign Science'. In A. Sierpenska & G. Kilpatrick (Eds). *Mathematics Education as a research domain: a search for identity.* Kluwer Academic Publishers, Dordrecht, p. 87-103.

COOPERATION BETWEEN MATHEMATICS AND PHYSICS TEACHING – THE CASE OF HORIZONTAL LAUNCH

Tine Golež, St. Stanislav Institution for Education,
Diocesan Classical Gymnasium, Ljubljana

The study of horizontal launched object is a topic which can serve as an example how to connect the teaching of mathematics and physics. It does not require unusual and expensive equipment. But such a connection is possible only if the two teachers are willing to cooperate. A detailed instruction about the interdisciplinary teaching of this topic is the theme of this presentation

INTRODUCTION

Following the idea I presented at MACAS1 Symposium my work is focused on how to connect the teaching of physics and mathematics. The fact that a physics teacher master mathematics quite in details while a mathematics teacher is less acquainted with physics emerges from the essential role which mathematics play in physics: it is the language of physics. Therefore the cooperation between the two teachers and suggestions for concrete activities could be much easier initiated by a physics teacher while fully developed by both of them. The case of a horizontally thrown object (horizontal launch) which is presented here is a good theme for the cooperation. Having very cooperative math teachers colleagues it was easy for me to invite them to use/test/comment my suggestions in order to enhance the teaching. In our case the common use, testing and commenting on the practically performed experiment greatly enhanced our cooperation and improved the quality of teaching.

FROM PHYSICS ...

Introductory experiment

Kinematics is the branch of classical mechanics concerned with describing the motions of objects without considering the factors that cause or affect the motion. After mastering the linear motion we focus on the horizontal launch. This plane movement is studied quite in detail. The first experiment is well-known two coins experiment (Fig. 1).

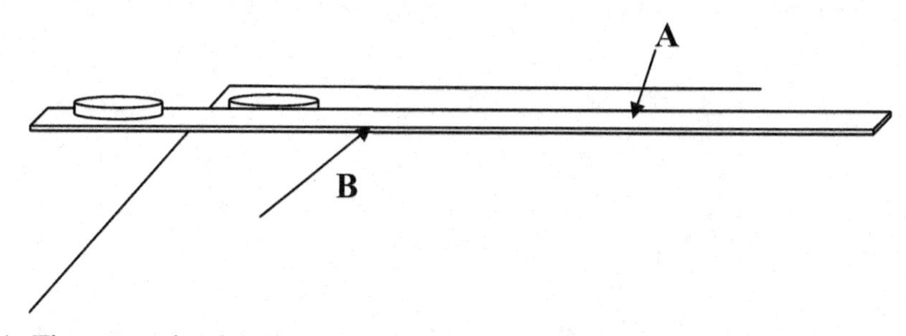

Fig. 1. The set-up is placed on a high object (for example a 2 meter high wardrobe). One finger is pressing at point A while the other will rapidly push (snap) the ruler at point B.

B.Sriraman, C.Michelsen, A. Beckmann & V. Freiman (Eds). (2008). *Proceedings of the Second International Symposium on Mathematics and its Connections to the Arts and Sciences* (MACAS2). Information Age Publishing, Charlotte: NC , pp.293-298

Sudden movement of the ruler will cause the left coin to fall down as a freely falling object while the right one will be launched in horizontal direction. Before carrying out this experiment, students are asked to discuss and guess the experiment's possible outcome. Very few of them predict that the coins will hit the floor at the same time. After a few repetitions of the experiment, they agree, that one can hear only one "clap", therefore that the two coins are falling downwards with equal speed.

We conclude: movement of a horizontally thrown object can be split into two independent movements: a vertical one which is equal to a freely falling body movement (the two coins began their path at the same level and time and they simultaneously hit the floor). But this simple experiment can not tell us what is happening regarding the horizontal part of the coin's movement. The vertical movement can be calculated by the same equations we have previously used for a freely falling object. That is:

$$y = \frac{gt^2}{2} \quad (1)$$

and

$$v_y = gt \quad (2)$$

Note, that the y-axis is contrary to the mathematics common use oriented downwards while x-axis is the usual horizontal direction.

Improved experiment

A physics experiment and the accompanied measurements should be as simple as possible. But when they cannot give us enough data to explain the phenomenon, one must enhance or redesign the experiment. That is the reason to continue this experiment using a spring-gun (Projectile launcher produced by PASCO) fixed on the with-board pillar (Fig. 2). It enables us to repeatedly shoot a plastic ball. Such an enhanced experiment enables us to measure quantities which lead us to the conclusions about the horizontal part of this movement. First, students would not realize, how using this gun could help us to analyse the case of a horizontally thrown object. They say: "The velocity of the ball is very high. How can we measure this movement?"

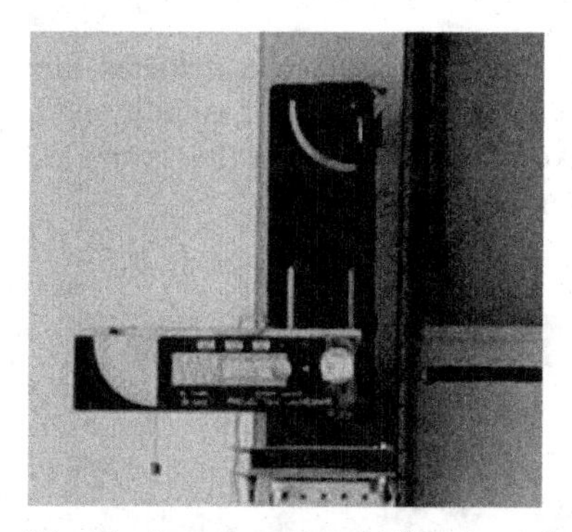

Fig. 2. A spring gun is fixed on the white-board pillar. Its position in the classroom is more visible on Fig. 3.

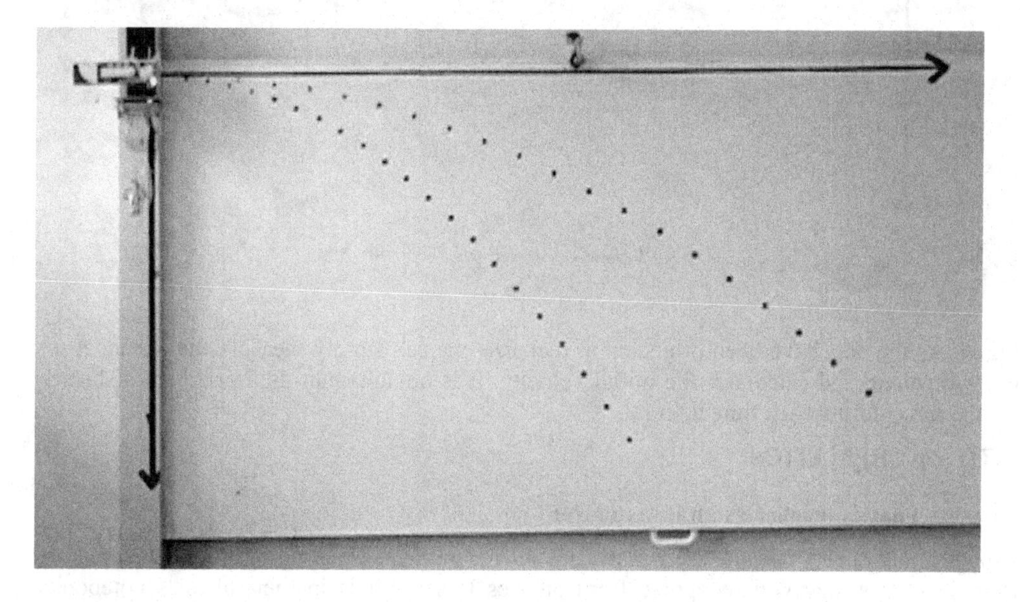

Fig. 3. This is photography of the whiteboard in the classroom during physics lesson. The dots have been drawn while we projected the horizontal throw in slow motion mode.

The simplest way is to record the experiment using a digital camera. Later we can analyse the slow motion of the ball's path. It is even not difficult to place the movie projector at the right distance from the whiteboard in order to get a 1:1 scale projection. The movie is projected in slow motion; so slow, that the teacher can mark the positions of the ball at the time equidistant points. The gun enables us to shoot with three different initial velocities. The two of them are recorded and projected. Teacher's trajectory's marks as shown on Fig. 3.

Some students notice that the distances between marks in horizontal direction are the same when we consider one shoot (with particular velocity). But the majority of students will realize that, when we point it out on the board (Fig. 4).

It is obvious (Fig.4) that the horizontal component of velocity is constant. We can write the two equations for the horizontal movement:

$v_x = v_0$ (3)

$x = v_0 t$ (4)

The physics teacher simply measures the distances on the whiteboard and calculates the initial velocity. Furthermore he explains the instantaneous velocity as a vector quantity and its components. He also carries out "A hunter and a monkey experiment" but this is not so essential for mathematics teacher, so this is not described here.

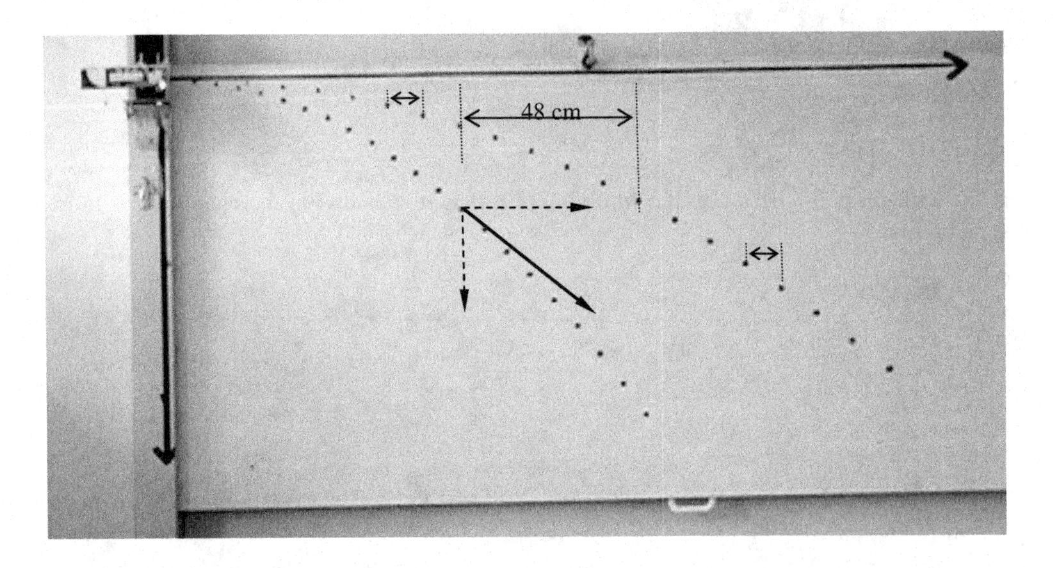

Fig. 4. As the dots have been projected in real size one can simply measure the distances on the whiteboard and calculates the initial velocity. It is obvious that the horizontal distances are the same during each time interval.

...TO MATHEMATICS

"A mathematics teacher as an investigator"

I suggest that mathematics teacher motivates his/her students by bringing the photocopies of (Fig. 3) scenes they will recognize from physics lessons. It is in their physics notebooks already! He invites the students to follow the mathematical investigation of the picture. This investigation will lead them to calculate some important facts about shooting without looking at the data which are collected in the physics notebook. First the teacher writes down the equations which are certainly known to mathematics teachers as well. Therefore:

$$y = \frac{gt^2}{2} \quad (5)$$

and

$$x = v_0 t \quad (6)$$

The mathematics teacher can proceed by acknowledging that he doesn't know the initial velocity of the ball and also, he does not know the time intervals between two marked positions of the ball. But he claims he could find out the missing data by analyzing the picture. He asks the students how to plot a y (x) graph (expressed by the parametric expressions (5) and (6)) which fits the dots on the whiteboard. Combining the two equations

(t is expressed from the equation (6) and used in the equation (5)) one derives the parabola equation:

$$y = \frac{g\left(\frac{x}{v_0}\right)^2}{2} \quad (7)$$

In order to have a more common form he writes:

$$y = \frac{g}{2v_0^2}x^2 \quad (8)$$

Now the equation resembles the well known central position parabola equation $y = ax^2$. When coefficient a is for example 5, the parabola is very steep and vice versa, when a is small (for example 0,2) the curve is gently rising. We must take into account that now y-axe is oriented downward. The positive a gives us the picture equals to negative a using an up oriented y-axe coordinate system.

Does this mathematical observation fit the physics of the real world? In the case of higher velocity the slope is less steep. The upper points (Fig. 3, 4 and 5) correspond to higher velocity. The coefficient in the equation 8 is smaller in the case of higher velocity.

To the quantitative analysis

This is only qualitative analysis. Now the mathematics teacher is going to choose one point and measures its coordinates (establishing appropriate coordinate system and units). They are 12,0 cm and 4,3 cm (Fig. 5). But what is the scale of this picture? Mathematics teacher admits he entered the physics classroom with the help of cleaning lady a day before (Physics classroom is always locked!). There was nothing at the whiteboard so he just measured the height of the whiteboard. He calculated the ratio between the real objects and the picture: it is 1 : 14. The coordinates of encircled dot are calculated by multiplying the measured values by 14. Therefore they are: 1,68 m and 0,60 m.

Using the equation (8) we simply calculate the initial velocity. It is 4,8 m/s. The students look up in the physics notebook and find the same result.

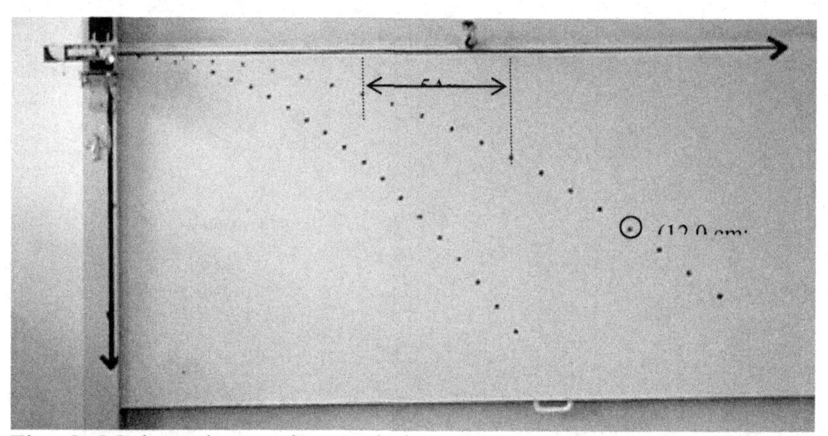

Fig. 5. Mathematics teacher encircled one dot and measured its coordinates. He also measured the displacement in horizontal direction between five consecutive points.

But mathematics teacher can find out also the time interval between two dots. Taking into account already calculated initial velocity and measuring the interval 5 x, he calculates the time interval: 0,0198 s. The inverse value corresponds to the frequency with which the pictures were taken. The frequency is almost 50 s^{-1} which corresponds to the frequency rate of an ordinary digital camcorder. Therefore, the time interval is about 0,020 s.

CONCLUSION

The present example of cooperating between physics and mathematics has been used and carefully examined. This approach motivated the students to put not only more attention during math lesson but also to see very productive and real life oriented mathematical structures – parabola in this case. Our aim is to improve the creative collaboration between physics and mathematics teaching. In our ScienceMath project we plan to produce more such interdisciplinary lessons as to better connect mathematics teaching to real life situations. We also intend to build up a system which could measure the successfulness of such a teaching approach and try it out on our teaching sequences.

References:

GOLEŽ, Tine (2005). Vodoravni met med fiziko in matematiko. *Fizika v šoli, 1, 13-20.*

9 781593 119843